Lecture Notes in Artificial Intelligence   8717

Subseries of Lecture Notes in Computer Science

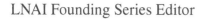
T0236161

Michael Mistry   Aleš Leonardis
Mark Witkowski   Chris Melhuish (Eds.)

Advances in Autonomous Robotics Systems

15th Annual Conference, TAROS 2014
Birmingham, UK, September 1-3, 2014
Proceedings

 Springer

Volume Editors

Michael Mistry
Aleš Leonardis
University of Birmingham
School of Computer Science
Edgbaston, Birmingham UK
E-mail: {m.n.mistry, a.leonardis} @bham.ac.uk

Mark Witkowski
Imperial College London
Circuits and Systems Group
Department of Electrical and Electronic Engineering
London, UK
E-mail: m.witkowski@imperial.ac.uk

Chris Melhuish
University of the West of England
Faculty of Computing
Engineering and Mathematical Sciences
Bristol, UK
E-mail: chris.melhuish@uwe.ac.uk

ISSN 0302-9743   e-ISSN 1611-3349
ISBN 978-3-319-10400-3   e-ISBN 978-3-319-10401-0
DOI 10.1007/978-3-319-10401-0
Springer Cham Heidelberg New York Dordrecht London

Library of Congress Control Number: 2014946279

LNCS Sublibrary: SL 7 – Artificial Intelligence

Typesetting: Camera-ready by author, data conversion by Scientific Publishing Services, Chennai, India

Printed on acid-free paper

Springer is part of Springer Science+Business Media (www.springer.com)

Preface

The 15th edition of the Towards Autonomous Robotic Systems (TAROS) conference was held during September 1–3, 2014, at the new Library of Birmingham, in the UK. The papers in these proceedings are those presented at TAROS 2014.

The TAROS series was initiated by Ulrich Nehmzow in Manchester in 1997 under the name TIMR (Towards Intelligent Mobile Robots). In 1999, Chris Melhuish and Ulrich formed the conference Steering Committee, which was joined by Mark Witkowski in 2003 when the conference adopted its current name. The Steering Committee has provided a continuity of vision and purpose to the conference over the years as it has been hosted by robotics centers throughout the UK. Under their stewardship, TAROS has become the UK's premier annual conference on autonomous robotics, while also attracting an increasing international audience. Sadly, Ulrich died in 2010, but his contribution is commemorated in the form of the TAROS prize named the "Ulrich Nehmzow Best Paper Award."

In 2014, our call for papers attracted 48 submissions (34 for full paper and 14 for extended abstracts) from institutions in 14 countries. Each paper was assigned to two Program Committee members and underwent a careful reviewing procedure, occasionally with support from additional experts. Based on the reviews and feedback from the committee, the organizers accepted 23 full papers and nine extended abstracts to be presented at the conference. The accepted submissions are published in these proceedings, with full papers allocated 12 pages and extend abstracts at two pages. The overall program covers various aspects of robotics, including navigation, planning, sensing and perception, flying and swarm robots, ethics, humanoid robotics, human–robot interaction, and social robotics.

We would like to thank all Program Committee members for their service to the robotics community and all authors for their contributions to TAROS. We also thank the University of Birmingham and its School of Computer Science, Claire Lindow for administrative support, the Institution of Engineering Technology (IET) for sponsoring an invited lecture, and Springer for sponsoring the best paper award. The local organizers additionally thank Mark Witkowski and Chris Melhuish for providing guidance and support.

We hope the reader finds this present volume of robotics papers insightful and engaging, and we look forward to seeing the continued progress and successes of the robotics community at TAROS conferences in the years to come.

July 2014

Michael Mistry
Aleš Leonardis
Mark Witkowski
Chris Melhuish

Organization

Conference Chairs

Michael Mistry   University of Birmingham, UK
Aleš Leonardis   University of Birmingham, UK

TAROS Steering Committee

Mark Witkowski   Imperial College London, UK
Chris Melhuish   Bristol Robotics Laboratory, UK

Program Committee (TAROS 2014)

Ajith Abraham   MIR Labs, USA
Lyuba Alboul   Sheffield Hallam University, UK
Joseph Ayers   Northeastern University, USA
Morteza Azad   University of Birmingham, UK
Tony Belpaeme   University of Plymouth, UK
Paul Bremner   Bristol Robotics Laboratory, University of the West of England, UK

Stephen Cameron   University of Oxford, UK
Lola Cañamero   University of Hertfordshire, UK
Anders Lyhne Christensen   Lisbon University Institute, Portugal
Geert De Cubber   Royal Military Academy of Belgium
Cédric Demonceaux   Université de Bourgogne, France
Tony Dodd   University of Sheffield, UK
Sanja Dogramadzi   University of West England, UK
Stéphane Doncieux   Université de Pierre Marie Curie, Paris, France
Marco Dorigo   Université Libre de Bruxelles, Belgium
Kerstin Eder   Bristol University, UK
Mathew H. Evans   University of Sheffield, UK
Michael Fisher   University of Liverpool, UK
Marco Gabiccini   University of Pisa, Italy
Antonios Gasteratos   Democritus University of Thrace, Greece
Swen Gaudl   University of Bath, UK
Ioannis Georgilas   Bristol Robotics Laboratory, University of the West of England, UK

Roderich Gross   University of Sheffield, UK
Dongbing Gu   University of Essex, UK

Hartmut Witte	Technische Universität Ilmenau FG Biomechatronik, Germany
Klaus-Peter Zauner	University of Southampton, UK
Michael Zillich	Technische Universität Wien, Austria
Sebastian Zurek	University of Birmingham, UK

Sponsoring Organizations

The University of Birmingham
Springer Publishing
The Institution of Engineering Technology (IET)

Table of Contents

Full Papers

Extended Abstracts

Modeling of a Large Structured Environment
With a Repetitive Canonical Geometric-Semantic Model*

Saeed Gholami Shahbandi and Björn Åstrand

Center for Applied Intelligent Systems Research (CAISR),
Intelligent Systems Lab, Halmstad University, Sweden

Abstract. AIMS project attempts to link the logistic requirements of an intelligent warehouse and state of the art core technologies of automation, by providing an awareness of the environment to the autonomous systems and vice versa. In this work we investigate a solution for modeling the infrastructure of a structured environment such as warehouses, by the means of a vision sensor. The model is based on the expected pattern of the infrastructure, generated from and matched to the map. Generation of the model is based on a set of tools such as closed-form Hough transform, DBSCAN clustering algorithm, Fourier transform and optimization techniques. The performance evaluation of the proposed method is accompanied with a real world experiment.

1 Introduction

Following the advances of the state of the art in autonomous vehicles, and growing research on the design and development of innovative solutions, *intelligent warehouses* emerge leveraging insights from several specialist domains. The Automatic Inventory and Mapping of Stock (AIMS) project targets the traditional warehouses where not necessarily infrastructures are designed or installed for automation. AIMS project intends to develop a process, through which an awareness of the surrounding environment is embedded in a "live" semantic map, for effective management of logistics and inventory (see fig. 1). Achieving this objective requires different sensors contributing to this semantic map in multiple layers. Forklift trucks enriched with such an awareness enable safe and efficient operations while sharing the workspace with humans. In such a shared workspace, compatibility between vehicles' knowledge, humans' cognition and Warehouse Management Systems (WMS) is important.

This paper focuses on mapping and modeling the infrastructure of the warehouse, as a foundation for addressing the location of both vehicles and storage of the warehouse. We present a method to extract structural pattern of the map that serves as the foundation of a multilayer geometric-semantic map to be used for logistic planning, Auto Guided Vehicle (AGV) path planning and WMS interaction. Semantic labels are subject to the context, in order to be functional

---

* This work as a part of AIMS project, is supported by the Swedish Knowledge Foundation and industry partners Kollmorgen, Optronic, and Toyota Material Handling Europe.

M. Mistry et al. (Eds.): TAROS 2014, LNAI 8717, pp. 1–12, 2014.

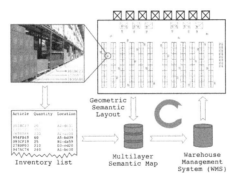

Fig. 1. The objective in AIMS project is to provide an awareness framework. The framework interacts with Warehouse Management System (WMS).

they shall be defined based on the environment and application. Accordingly we integrate multiple layers in this map, from geometric layout design of the environment to semantic annotation of the regions and infrastructures as a model. For this purpose a map of pillars of the pallet rack cells (see fig. 1) is generated, from a fish-eye camera based on a bearing-only mapping method. This map is used for inference and generation of a layout map of the warehouse, which integrates the desired conceptual meaning. Contribution of this work is a semantic-geometric model, based on an object map entailing infrastructure elements as landmarks. This is followed by an easement in inferences through a method for extracting and matching the model from and to the aforementioned map. Reliability and performance of the proposed model and accompanying method is demonstrated on a map acquired from a real world warehouse.

1.1 Related Works

In a time where robots are embedded in our daily life, robot's awareness of their surrounding is a crucial competence and semantic mapping is a particularly important aspect of it. That is because, while a geometrically consistent map is sufficient for navigation, it is not enough for task planning and reasoning. Many researchers have been contributing to this particular aspect, from different points of view.

Some tried to model the environment through the topology of open space in geometric map, like [6], where they employed a series of kernels for semantic labeling of regions. Some others like [5], [14] and [9] proposed spatial maps, enhanced conceptually by object recognition in regions of the map. [5] proposed a composition of two hierarchical maps, semantic and geometric anchored together. In [14] a framework of a multilayer conceptual map is developed, representing the spatial and functional properties of the environment. And [9] introduced a comprehensive framework of spatial knowledge representation for large scale semantic mapping.

While mentioned researchers aimed to derive semantic concept from the functionality of the objects into the map, some others such as [8], [12] and [7], introduced properties of the regions as semantic label. [8] annotates an occupancy map with the properties of the regions, either "building" or "nature", through data from range scanner and vision. In [12] properties of the environment such as terrain map and activity map, are embedded into a metric occupancy map. Concerning the global localization, [7] employed hybrid geometric-object map.

Mentioned works are proposed for cases where the global structure of the environment is not a concern, and semantic information is extracted locally. The conceptual meaning is the property or functionality of the content of those regions, and does not link to the structural model of the environment. Researchers have also taken into account the environment's structure. [13] attempted mastering the SLAM problem by a geometrical object-oriented map representation, through modeling the boundaries of obstacles with polygons, employing a Discrete Curve Evolution (DCE) technique. An interesting recent work [10], developed a method for detection, evaluation, incorporation and removal of structure constraint from latent structure of the environment into a graph-based SLAM. Two last examples take into account the structure of the environment, for improving the solution to SLAM problem and providing a more consistent map. However there is no conceptual meaning accompanying the extracted structure.

Spatial semantic of open space from occupancy map is an interesting aspect and we investigate it in another work. It does provide useful knowledge of the open space in a warehouse, such as connectivity, corridors, or crossing of the corridors. But it does not provide semantic labels for infrastructure, such as the entrance of a pallet rack cell (see fig. 2). Such an information is very useful when the articles are localized in the layout, for logistic and AGVs' task planning. The other approaches to semantic mapping, where the semantic labels are derived from objects in the region is not very beneficial to our work either. That is because the smallest entities of regions are pallet rack cells with same semantics. Stored articles in those pallet rack cells and their identities do not carry any conceptual meaning for their region. The objects that we are interested in are the infrastructure of the warehouse, such as pillars of the pallet rack cells which represent the structure of the environment (see fig. 2). Therefore a more suitable approach for us would be to create a map of the environment, using the infrastructure as landmarks, and then extracting those patterns that are meaningful for us.

1.2   Our Approach

This paper presents a canonical geometric model and describes how to match it into the latent structure of a map. Such a model enables the further processing of geometric-topological modeling of the environment, for the purpose of semantic annotation of structures and geometric layout extraction. Semantic concept is encoded into the model through the choice of landmarks in the map as shown in fig. 2. It is assumed that the environment is highly structured and a pattern is frequent enough, so that it is possible to effectively represent the whole structure by a set of this canonical model with different parameters. By choosing pillars as landmarks, opening of a pallet rack cell is implied by two neighboring landmarks, while the sequence of landmarks creates a layout map by the "boundaries" of the corridors

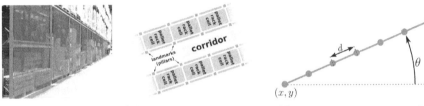

(a) warehouse scene   (b) top view of warehouse   (c) geometric semantic model

Fig. 2. Design of the model derived from the map's context, encoding semantic-geometric relation of landmarks

(a similar concept of representation as in [13]). This explains how the model in fig. 2 embeds a sufficient level of semantic and geometric knowledge while it is expressed only by a set of points uniformly distributed on a straight line. Different models are uniquely identified by the means of 5 parameters n, d, x, y, θ. The model and matching method is designed to represent independent lines of landmarks, therefore parameters of different models are independent. However the repetition of the model in the map of a warehouse, makes it possible to pose a global constraint on its parameters in order to achieve a set of models which are globally consistent.

In next section (2), the model and the matching method are explained in details. The process is demonstrated on a synthesized data in order to sketch the generality of the method where the models in one map are completely independent. Section 3 contains the method we adapt for mapping the environment from a fish-eye camera. Resulting map is modeled by the proposed method, while introducing global constraints for a real world map.

2   Model Generation and Fitting

Let's assume a $\mathcal{MAP}$ consisting of landmarks is given, where each landmark is described by its pose ($\mu = x, y$) and corresponding uncertainty modeled with a 2D normal distribution (with covariance Σ) as defined in equation 1.

$$\mathcal{MAP} = \{lm_i \mid lm_i := \mathcal{N}_i(X)\} \quad , \mathcal{N}_i(X) = \frac{\exp\left(-\frac{1}{2}(X-\mu_i)^T \Sigma_i^{-1}(X-\mu_i)\right)}{\sqrt{(2\pi)^2|\Sigma_i|}} \quad (1)$$

The expected pattern in this map was described and motivated in the introduction (section 1). This model as illustrated in fig. 2, represents a set of landmarks (pillars) aligned on one side of a corridor. We call these sets of landmarks $\mathcal{L}$, and we try to fit one model per set.

$$\mathcal{L}_i = \{\{lm_{ij}\}, \theta_i \mid lm_{ij} \in \mathcal{MAP}\} \quad (2)$$

In the definition of $\mathcal{L}_i$ (equation 2) lm_{ij} are landmarks aligned on the side a corridor, and θ_i is the angle of alignment. Model M shown in fig. 2c and expressed in equation 3 is designed to represent sets of landmarks $\mathcal{L}$, and each model is uniquely identified by 5 parameters (n, d, x, y, θ), where n is the number of landmarks in the corresponding $\mathcal{L}$, d is the distance between two consecutive landmarks, (x, y) are the coordination of the 1st landmark, and θ_i is the angle of alignment of the set.

$$M_i(n_i, d_i, x_i, y_i, \theta_i) := \{p_{ij}\} \quad , p_{ij} = \begin{bmatrix} x_i + n_j d_i \cos \theta_i \\ y_i + n_j d_i \sin \theta_i \end{bmatrix}, 0 \leq n_j < n_i \qquad (3)$$

First step in generating the models M is to segment the $\mathcal{MAP}$ into the sets of landmarks $\mathcal{L}_i$. For this purpose we have developed a closed-form of Hough transform in combination with a clustering algorithm. This technique not only locates the desired $\mathcal{L}$ by (θ, ρ), but also directly clusters the landmarks into different $\mathcal{L}$. After the segmentation of landmarks into sets, each set is projected into an axis passing through that set (see fig. 4b). This operation will map the 2D normal distributions of the landmarks into a 1D signal. An analysis of the resulting 1D signal in frequency domain will result in an estimation of the n and d of the model. Given the θ, n and d of the model by closed-form Hough transform and frequency analysis, x and y remain for extraction. An optimization would serve this purpose, where the first point of the set $\mathcal{L}$ serves as the initial guess of the optimization process.

2.1   Segmentation by Closed-Form Hough Transform

Transformation of a point from Cartesian space (x, y) into Hough space [3] (θ, ρ) is performed by equation 4.

$$\rho = x \cos \theta + y \sin \theta \qquad (4)$$

Conventional form of Hough transform is applied to discrete images. Hough space is also a discretized image where the value of each pixel (θ, ρ) is given by the summation of the value of all pixels (x, y) that satisfy the equation 4. The peaks in the Hough space represent lines in original image where points are aligned. But we are interested in more than that. We would like to know which particular points in Cartesian space contributed to an specific peak (θ, ρ) in Hough space. Therefore we introduce a closed-form solution of Hough transform to address that issue. This is realized by representing each point (x_i, y_i), with a corresponding sinusoid as expressed in equation 4. Next step is to intersect all resulting sinusoids and store the intersections with $(\theta_{ij}, \rho_{ij}, i, j)$. Where (θ_{ij}, ρ_{ij}) represent the location of the intersection in Hough space and i, j are the indices of intersecting sinusoids. Outcome of this step is a set of intersection points in Hough space as illustrated in fig. 3b.

Advantages of closed-form approach are, first, clustering the intersection points in Hough space directly corresponds to clustering aligned points in Cartesian space. Secondly, it prevents us from discretization of a continues Cartesian space.

From the Hough space, clustering the intersection points is straightforward. We employed the Density-based spatial clustering of applications with noise (DB-SCAN) [4] algorithm. This algorithm requires a value for the minimum number

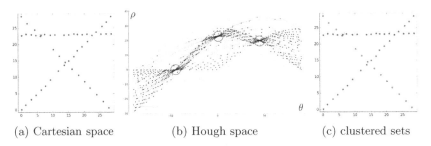

(a) Cartesian space   (b) Hough space   (c) clustered sets

Fig. 3. Closed-form Hough transform maps the points from Cartesian space into sinusoids in Hough space. A clustering is performed over intersections in Hough space (red circles) to cluster the aligned points in Cartesian space.

of samples per cluster, which is derived from the number of expected points in each alignment $\mathcal{L}$. Assuming a set of n point aligned in Cartesian space ($\mathcal{L}$), consequently there will be the same number of sinusoids in Hough space intersecting $\frac{n(n-1)}{2}$ times at the same location (θ, ρ). As a perfect alignment in noisy data is unlikely, minimum number of samples are set to half of that ($\frac{n(n-1)}{4}$) to guarantee a successful clustering.

The result of clustering in Hough space is mapped to Cartesian space, generating sets of points $\mathcal{L}_i$ as desired. Angle of alignment $\theta_i \in \left[\frac{-\pi}{2}, \frac{\pi}{2}\right]$, is therefore the first estimated parameter of the model.

2.2   Frequency and Length via Fourier Transform

In order to estimate two parameters of the model n, d, each set of landmarks $\mathcal{L}_i$ is projected to an axis passing through the center of mass, projecting the set of 2D normal distribution functions into a 1D signal. This projected signal is illustrated in fig. 4b and defined in equation 5.

$$signal_i = \sum_{lm_j \in \mathcal{L}_i} f_{lm_j}(\mu_j, \sigma_j)$$

$$f_{lm_j}(\mu_j, \sigma_j) = \frac{1}{\sigma_j\sqrt{2\pi}} e^{-\frac{(x-\mu_j)^2}{2\sigma_j^2}} \quad , \mu_j = \frac{\vec{v_1}.\vec{v_2}}{\|\vec{v_1}\|}, \sigma_j = \sigma_x \tag{5}$$

$$\begin{bmatrix} \sigma_x^2 & \rho\sigma_x\sigma_y \\ \rho\sigma_x\sigma_y & \sigma_y^2 \end{bmatrix} = R(\theta_j)\Sigma_j R^T(\theta_j)$$

In equation 5 vectors $\vec{v_1}, \vec{v_2}$ are those illustrated in fig. 4b. $\vec{v_1}$ is the projection axis itself and $\vec{v_2}$ is a vector from starting point of $\vec{v_1}$ to the point subjected to projection.

Assuming a uniform distribution of points in each set implies that, Fourier transform of the projection signal ($\mathcal{F}(f)$) has a dominant frequency. This frequency relates to the values n, d as expressed in equations 6.

$$n = f_0 \, , d = \frac{\|\vec{v_1}\|}{f_0}$$

$$f_0 = \arg\max \mathcal{F}(f) := \{f_0 \mid \forall f : \mathcal{F}(f) \leq \mathcal{F}(f_0)\} \tag{6}$$

2.3   Optimization and Model Fitting

Last parameters to uniquely define the model are (x, y). For that purpose, (x, y) are set to the pose (μ) of the first landmark in $\mathcal{L}$. Considering that the angle of alignment θ belongs to $\left[-\frac{\pi}{2}, \frac{\pi}{2}\right]$, first element of each $\mathcal{L}$ is the most left item, or the lowest in case of vertical lines. Then through an optimization process (x, y) are tuned. It should be noted that the first landmark in $\mathcal{L}$ is an initial guess for the optimization. This is based on the assumption that $\mathcal{L}$ does not consist of very far off outliers. Such outliers may bring the optimization into a local minima, causing a shift in model's position. Objective function of the optimization is given in equation 7. This function is a summation of all normal distribution functions of landmarks of the line ($\mathcal{N}_{ik} \in \mathcal{L}_i$) operating on all the points of the model ($p_{il} \in M_i$).

$$f_i(X) = \frac{1}{\underset{p_{il} \in M_i}{\sum_l} \underset{\mathcal{N}_{ik} \in \mathcal{L}_i}{\sum_k} \mathcal{N}_{ik}(p_{il})} \tag{7}$$

Since in the map of a real environment most of the models share 3 parameters (n, d, θ), for demonstrating the performance and generality of the method, it was applied to a synthesized data instead of a real map. Performance of the proposed method is evaluated over a map from real world in section 3. Unlike real environment the synthesized data in Figure 4a contains 3 $\mathcal{L}$ with all different parameters.

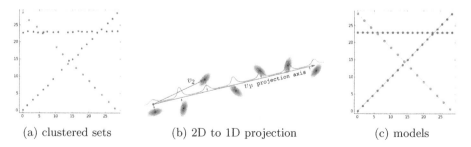

(a) clustered sets   (b) 2D to 1D projection   (c) models

Fig. 4. Modeling 3 segmented line sets ($\mathcal{L}$) from 4a to 4c. 4b shows 2D to 1D projection of a single $\mathcal{L}$ according to equation 5 for frequency analysis

Result of clustering is encoded in colors. While in a real world map a crossing between $\mathcal{L}$ is not expected, in this synthesized data the lines are crossing to demonstrate the performance of the closed-form Hough transform. In addition to a uniform noise added to the position of each landmark, 10% of landmarks are removed to evaluate the result of Fourier transform in estimating n and d. As it is observed in fig. 4c, the model would fill in the position of missing landmarks. Finally the values of the parameters are successfully computed by the method, in fig. 4 M_1 (20, 2.12, −0.1, 0.0, 0.78) colored red, M_2 (20, 1.50, 0.0, 23.0, 0.00) colored green and M_3 (15, 2.80, 0.0, 28.5, −0.78) colored cyan. It should be mentioned that in case of a real map not all minor assumptions are met, so global constraints are introduced over all parameters. However none of the global constraints are preset, but all are extracted from the map. This aspect is explored further in section 3.1.

3 Experimental Results and Discussion

This section describes the procedure of creating and modeling a map of a real warehouse. For this purpose we use a "Panasonic" 185° fish-eye lens mounted on "Prosilica GC2450" camera, installed on a AGV forklift truck in a warehouse. As the truck is provided with a localization system based on lasers and reflectors, we adopt an Extended Kalman Filter based bearing-only technique for mapping. Then the model developed in this work is employed to represent the map. To this end, a set of global estimations are extracted from the map and posed over the model's parameters as global constraints.

Pillar Detection. Considering the common color coding of the pillars in the warehouses, pillar detection starts with segmentation through color indexing [11], followed by calculation of the oriented gradient. The camera is pointing downward and all the pillars are parallel to the camera axis. Consequently pillars are pointing to the vanishing point of the camera in the images, hence gradient vectors of pillars' edges are perpendicular to lines passing through the vanishing point. Any other gradient vectors are considered non-relevant and filtered out. The gradient image is sampled over multiple concentric circles as illustrated in fig. 5b. Results of all sampling are accumulated in a signal as in fig. 5c, capturing the pattern of a pillar's appearance by two opposite peaks representing the edges of the pillar. Detecting the position of such a pattern returns the bearing to pillars. A continuous wavelet transform (CWT) based peak detection technique [2] is adopted for detection of pillars' pattern in the gradient signal. The method is based on matching a pattern encoded in a wavelet, by the help of CWT. The pattern representing two sides of the pillar in the gradient signal (see fig. 5c), could be modeled with a wavelet based on the 1st order derivative of a Gaussian function as illustrated in fig. 5d.

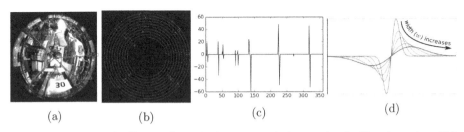

(a)   (b)   (c)   (d)

Fig. 5. Detection of pillars. 5a) original image and the result of pillar detection, 5b) circular sampling of gradient image. 5c) accumulation of circular sampling as a signal, illustrating the occurance of pillars. A wavelet in 5d, resembling the pattern of pillars in gradient signal, is used for pillar detection by CWT technique.

Bearing Only Mapping. Mapping is performed in an Extended Kalman Filter (EKF) fashion proposed in [1] for bearing-only SLAM. However, as the pose of the truck is provided through a laser-reflector based system, there is no need to localize the truck through EKF framework. Nevertheless, we include the state of the vehicle in the Kalman filter's state to ease the adaptation and system description. The truck's state is neither predicted nor updated in corresponding phases of the EKF, instead it comes directly from the localization system. This ensures that Kalman gain computation and prediction step are based on the "true" pose of the truck, and leaves the possibility to extend this implementation into SLAM if required.

A map generated through this framework, from the data logged in a real world warehouse is sketched in fig. 6. Colors in this map code the result of clustering of the line sets ($\mathcal{L}$), and blacks are outliers.

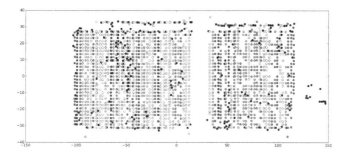

Fig. 6. Mapping 77 lines of pillars of a warehouse of size $250^m \times 72^m$

3.1 Global Constraints on the Models

Recalling the assumption of a structured environment, the set of models representing the boundaries of corridors are parallel, and have similar parameters n, d, θ. Even the starting points of the models are aligned on a straight line (like in fig. 6). We exploit this information to pose a set of global constraints, which results in a globally more consistent set of models of the environment.

Dominant Orientation refers to the fact that corridors in an environment have the same direction. This suggest that we could pose a global constraint on θ. This has a crucial importance for the clustering step as well, as the number of intersection points in Hough space (see fig. 3b) increases quadratically by the number of landmarks ($\frac{n(n-1)}{2}$). This means for a map consisting of more than 2000 landmarks like fig. 6, there will be more than 2 million intersection points in Hough space. Such a number of points can not be easily clustered. However, considering the assumption of dominant orientation, it is possible to reject a huge number of intersection points which are not within an acceptable range ($5°$) from dominant orientation. This filtering process is handled by discretization of the Hough space and finding the dominant orientations, where most of the peaks occur. After the filtering process, the clustering proceeds as explained earlier in closed-form.

Dominant Frequency. Similarly a constraint is posed over parameters n and d. This is also very helpful since in a real map, sometimes landmarks are not visible and hence not included in the map. If the percentage of missing landmarks become too big, it may yield a wrong estimation of the frequency. In order to estimate global constraint of these two parameters, first n and d are computed for all lines ($\mathcal{L}$) without constraint. Then the mode in the histogram of these parameters as shown in fig. 7 provides a global estimation of the parameters. After acquiring the global values from the map, models are generated again, this time with the global constraints.

Initialization Line. Assumption of a structured environment implies that the starting points of all $\mathcal{L}$ are located on the same line. This line is then used for initial guess in the optimization process. The initialization line is calculated by a linear regression among first landmarks in all $\mathcal{L}$, shown in fig. 7c. While each $\mathcal{L}$ comes from Hough space with its line equation, the exact position of the optimization's initial guess for each $\mathcal{L}$ is given by the intersection point of mentioned line and initialization line.

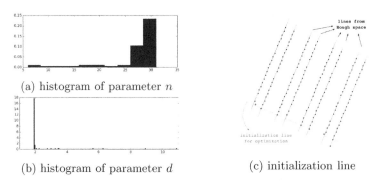

(a) histogram of parameter n

(b) histogram of parameter d

(c) initialization line

Fig. 7. Global constraints; Histograms of (n, d), mode in each shows the dominancy. Initialization line is a linear regression of first landmarks in all $\mathcal{L}$ (red points).

3.2   Discussion

Engaging all the global constraints, result of modeling the map given in fig. 6 is illustrated in fig. 8a. The final result of modeling is useful for geometric mapping of the boundaries of corridors, as well as semantic annotation of pallet rack cell on the side of corridors. A crop of the model augmented with truck's pose and reflectors is presented in fig. 8b. The reflectors are those pre-installed in the environment used for AGVs' localization. We use the position of reflectors as ground truth for accuracy estimation. A box plot in fig. 8c demonstrates the distances between 1143 reflectors and nearest point in the model. Taking into account the width of pillar's and reflector's mount, the distances are representing the errors of the model with an offset of about 1 decimeter.

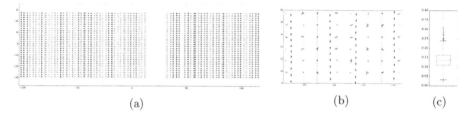

Fig. 8. Modeling a warehouse (see fig. 6) with geometric semantic models (all axes' unit in meter). 8b) shows a crop of model augmented with the truck's pose(black arrow) and reflectors (white triangles). 8c) shows distance between models and closest reflectors, with an offset about 1^{dm}.

It must be remembered, although the global constraints improve the global consistency of the models, yet it is only applicable if the infrastructures share similar geometry, that is to say the resulting models have the same parameters. In case of an environment that consists of multiple *regions* with different characteristics, those regions must be subjected to regional constraints.

4 Conclusion

Toward developing an awareness framework for effective management of logistics and inventory in a warehouse environment, a rich geometric-semantic map is set as a goal. In this work we address this objective on the level of infrastructure modeling. We suggest a choice of landmark for bearing-only mapping of a warehouse environment, leading to a map which integrates the infrastructure of the warehouse. This paper presents a canonical geometric-semantic model, along with a method for generating and matching these models into the map. The model is a set of points aligned on a straight line uniquely identified by 5 parameters. Geometrically it represents the boundaries of corridors, and semantically represents pallet rack cells. A series of tools such as closed-form Hough transform, DBSCAN clustering algorithm, Fourier transform and optimization techniques are employed to estimate the parameters of each model and match them with the map. The performance of the method is demonstrated over both synthesized data and real world map.

Proposed model and extracting method have interesting characteristics, such as modularity, generality, and representing both geometric and semantic knowledge. Most of the steps and parameters of the model are independent, therefore it is adjustable according to different scenarios. One example of such adjustment was given as global constraint in this work, where all of the models' parameters adopt to the global characteristic of the map for a better consistency. However the method has its own limits as discussed in 3.2, such as sensitivity to initial guess of the optimization, and consideration of global constraints for consistency.

We plan to develop this work further by fusing other sensory data into this map, such as introducing occupancy notation from range scanners. And also by developing a more dynamic model, allowing harmonics of the dominant frequency to participate, and handling those cases where landmarks are not uniformly distributed. Segmentation of map into regions based on dominant orientations would be helpful, if the environment consists of multiple orientations. Indeed we investigated this aspect in another work based on occupancy maps, and will introduce it to this method as soon as the occupancy notation is fused into this map.

References

1. Bailey, T.: Constrained initialisation for bearing-only slam. In: Proceedings of the IEEE International Conference on Robotics and Automation, ICRA 2003, vol. 2, pp. 1966–1971. IEEE (2003)
2. Du, P., Kibbe, W.A., Lin, S.M.: Improved peak detection in mass spectrum by incorporating continuous wavelet transform-based pattern matching. Bioinformatics 22(17), 2059–2065 (2006)
3. Duda, R.O., Hart, P.E.: Use of the hough transformation to detect lines and curves in pictures. Communications of the ACM 15(1), 11–15 (1972)
4. Ester, M., Kriegel, H.-P., Sander, J., Xu, X.: A density-based algorithm for discovering clusters in large spatial databases with noise, pp. 226–231. AAAI Press (1996)
5. Galindo, C., Saffiotti, A., Coradeschi, S., Buschka, P., Fernandez-Madrigal, J.A., González, J.: Multi-hierarchical semantic maps for mobile robotics. In: IEEE/RSJ International Conference on Intelligent Robots and Systems, IROS 2005, pp. 2278–2283. IEEE (2005)
6. Liu, Z., von Wichert, G.: Extracting semantic indoor maps from occupancy grids. Robotics and Autonomous Systems (2013)
7. Park, S., Kim, S., Park, M., Park, S.K.: Vision-based global localization for mobile robots with hybrid maps of objects and spatial layouts. Information Sciences 179(24), 4174–4198 (2009)
8. Persson, M., Duckett, T., Valgren, C., Lilienthal, A.: Probabilistic semantic mapping with a virtual sensor for building/nature detection. In: International Symposium on Computational Intelligence in Robotics and Automation, CIRA 2007, pp. 236–242. IEEE (2007)
9. Pronobis, A., Jensfelt, P.: Large-scale semantic mapping and reasoning with heterogeneous modalities. In: 2012 IEEE International Conference on Robotics and Automation (ICRA), pp. 3515–3522. IEEE (2012)
10. de la Puente, P., Rodriguez-Losada, D.: Feature based graph-slam in structured environments. Autonomous Robots, 1–18 (2014)
11. Swain, M.J., Ballard, D.H.: Color indexing. International Journal of Computer Vision 7(1), 11–32 (1991)
12. Wolf, D.F., Sukhatme, G.: Semantic mapping using mobile robots. IEEE Transactions on Robotics 24(2), 245–258 (2008)
13. Wolter, D., Latecki, L.J., Lakämper, R., Sun, X.: Shape-based robot mapping. In: Biundo, S., Frühwirth, T., Palm, G. (eds.) KI 2004. LNCS (LNAI), vol. 3238, pp. 439–452. Springer, Heidelberg (2004)
14. Zender, H., Martínez Mozos, O., Jensfelt, P., Kruijff, G.J., Burgard, W.: Conceptual spatial representations for indoor mobile robots. Robotics and Autonomous Systems 56(6), 493–502 (2008)

Monte Carlo Localization for Teach-and-Repeat Feature-Based Navigation

Matías Nitsche[1], Taihú Pire[1], Tomáš Krajník[2],
Miroslav Kulich[3], and Marta Mejail[1]

[1] Laboratory of Robotics and Embedded Systems, Computer Science Department,
Faculty of Exact and Natural Sciences, University of Buenos Aires, Argentina
[2] Lincoln Centre for Autonomous Systems, School of Computer Science, University
of Lincoln, UK
[3] Intelligent and Mobile Robotics Group, Department of Cybernetics, Faculty of
Electrical Engineering, Czech Technical University in Prague, Czech Republic
tkrajnik@lincoln.ac.uk, kulich@labe.felk.cvut.cz,
{mnitsche,tpire,marta}@dc.uba.ar

Abstract. This work presents a combination of a teach-and-replay visual navigation and Monte Carlo localization methods. It improves a reliable teach-and-replay navigation method by replacing its dependency on precise dead-reckoning by introducing Monte Carlo localization to determine robot position along the learned path. In consequence, the navigation method becomes robust to dead-reckoning errors, can be started from at any point in the map and can deal with the 'kidnapped robot' problem. Furthermore, the robot is localized with MCL only along the taught path, i.e. in one dimension, which does not require a high number of particles and significantly reduces the computational cost. Thus, the combination of MCL and teach-and-replay navigation mitigates the disadvantages of both methods. The method was tested using a P3-AT ground robot and a Parrot AR.Drone aerial robot over a long indoor corridor. Experiments show the validity of the approach and establish a solid base for continuing this work.

1 Introduction

The problem of autonomous navigation has been widely addressed by the mobile robotics research community. The problems of localization and mapping, which are closely related to the navigation task, are often addressed using a variety of approaches. The most popular is Simultaneous Localization and Mapping[1] (SLAM), with successful examples such as MonoSLAM[2] and more recently PTAM[3], which uses vision as the primary sensor. However, these approaches generally model both the environment and the pose of the robot with a full metric detail, using techniques such as Structure From Motion (SfM) or stereo-based reconstruction[4]. While successful, these methods are computationally demanding and have shortcomings which arise from the complexity of the mathematical models applied.

M. Mistry et al. (Eds.): TAROS 2014, LNAI 8717, pp. 13–24, 2014.

In terms of solving the navigation problem, it may not always be necessary to obtain a detailed metric description of the environment nor to obtain a full 6DoF pose. For example, teach-and-replay (T&R) techniques[5,6,7,8,9,10], where the robot is supposed to autonomously navigate a path that it learned during a tele-operated run, do not require to explicitly localize the robot.

While standard visual navigation approaches depend on a large number of correspondences to be established to estimate the robot position, feature-based T&R methods are robust to landmark deficiency, which makes them especially suitable for scenarios requiring long-term operation [11]. Moreover, the T&R methods are generally much simpler and computationally less-demanding. On the other hand, they are usually not able to solve the kidnapped-robot problem nor to localize a robot globally, requiring knowledge about the initial robot position. Also, most of them are susceptible to errors in dead-reckoning caused by imprecise sensors or wheel slippage.

To address these shortcomings, the use of particle-filters such as Monte Carlo Localization (MCL)[12] becomes attractive. MCL is able to efficiently solve the localization problem and addresses the global-localization and kidnapped-robot problems, and has been applied with success in vision-based applications[13,14] and robot navigation[6,10,15].

In this work, an existing visual T&R method[7,8] is improved by replacing localization by means of dead-reckoning with Monte Carlo Localization. The original method relied on the precision of it's dead reckoning system to determine the distance it has travelled from the path start. The travelled distance estimation determines the set of features used to calculate the robot lateral displacement from the path and thus influences the precision of navigation significantly. While the authors show [7] that estimating the distance with dead-reckoning is sufficient for robots with precise odometry and paths that do not contain long straight segments, these conditions might not be always met. By replacing this dead-reckoning with MCL, the localization error can be diminished, while not only tackling the kidnapped-robot problem but also allowing the navigation to start at an arbitrary point in the taught path.

2   Related Work

Teach-and-replay methods have been studied in several works. While some approaches are based on metric reconstruction techniques, such as stereo-based triangulation[9,16], several works employ so-called appearance-based or qualitative navigation[5,17,18].

This type of navigation uses simple control laws[5] to steer the robot according to landmarks remembered during a training phase. While some works approach the localization problem with image-matching techniques[19,20,21], others approaches use only local image features[18,17]. While the feature-based techniques are generally robust and reliable, image feature extraction and matching can be computationally demanding. Thus, strategies for efficient retrieval of landmarks are explored, such as with the LandmarkTree-Map[18].

However, feature- or dead-reckoning- based localization is error-prone and several visual T&R navigation methods propose to improve it by using advanced techniques such as Monte Carlo Localization (MCL). One of these works corresponds to [15] where MCL is applied together with an image-retrieval system to find plausible localization hypotheses. Another work[6] applies MCL together with FFT and cross-correlation methods in order to localize the robot using an omni-directional camera and odometry. This system was able to robustly repeat a 18 km long trajectory. Finally, a recent work[10] also employs MCL successfully with appearance-based navigation.

Based on similar ideas, the method proposed in this work applies MCL to a visual T&R navigation method[7] that proved its reliability in difficult scenarios including life-long navigation in changing environments[11]. Despite of simplicity of the motion and sensor models used in this work's MCL, the MCL proven to be sufficient for correct localization of the robot. In particular, and as in [10,6], the localization only deals with finding the distance along each segment of a topological map and does not require determining orientation. However, in contrast to the aforementioned approaches, global-localization and the kidnapped-robot problem are considered in our work.

3 Teach-and-Replay Method

In this section, the original teach-and-replay algorithm, which is improved in the present work, is presented. During the *teaching* or mapping step, a feature-based map is created by means of tele-operation. The *replay* or navigation phase uses this map to repeat the learned path as closely as possible. For further details about the method please refer to works[7,8].

3.1 Mapping

The mapping phase is carried out by manually steering the robot in a turn-move manner. The resulting map thus consists of a series of linear segments, each of a certain length and orientation relative to the previous one. To create a map of a segment, salient image-features (STAR/BRIEF) from the robot's onboard camera images are extracted and tracked as the robot moves. Once a tracked feature is no longer visible, it is stored as a landmark in the current segment's map. Each landmark description in the map consists of its positions in an image and the robot's distance relative to the segment start, both for when the landmark is first and last seen. Finally, the segment map contains the segment's length and orientation estimated by dead-reckoning.

In the listing 1, the segment mapping algorithm is presented in pseudo-code. Each landmark has an associated descriptor l_{desc}, pixel position l_{pos_0}, l_{pos_1} and robot relative distance l_{d_0}, l_{d_1}, for when the landmark was first and last seen, respectively.

Algorithm 1. Mapping Phase

Input: F: current image features, d: robot distance relative to segment start, T:
      landmarks currently tracked
Output: L: landmarks learned current segment
foreach $l \in T$ **do**
    $f \leftarrow$ find_match(l,F)
    if *no match* **then**
        $T \leftarrow T - \{\, l \,\}$ `/* stop tracking */`
        $L \leftarrow L \cup \{\, l \,\}$ `/* add to segment */`
    else
        $F \leftarrow F - \{\, f \,\}$
        $t \leftarrow (l_{desc}, l_{pos_0}, f_{pos}, l_{d_0}, d)$ `/* update image coordinates & robot position */`

foreach $f \in F$ **do**
    $T \leftarrow T \cup \{\, (f_{desc}, f_{pos_0}, f_{pos_0}, d, d) \,\}$ `/* start tracking new landmark */`

3.2 Navigation

During the navigation phase, the robot attempts to repeat the originally learned path. To do so, starting from the initially mapped position, the robot moves forward at constant speed while estimating its traveled distance using dead-reckoning. When this distance is equal to the current segment's length, the robot stops and turns in the direction of the following segment and re-initiates the forward movement. To correct for lateral deviation during forward motion along a segment, the robot is steered so that that its current view matches the view perceived during the training phase. To do so, it continuously compares the features extracted from the current frame to the landmarks expected to be visible at the actual distance from the segment's start.

The listing 2 presents the algorithm used to traverse or 'replay' one segment. Initially, the list T of expected landmarks at a distance d from segment start is created. Each landmark $l \in T$ is matched to features F of the current image. When a match is found, the difference between the feature's pixel position f_{pos} is compared to an estimate of the pixel-position of l at distance d (obtained by linear interpolation). Each of these differences is added to a histogram H. The most-voted bin, corresponding to the mode of H, is considered to be proportional to the robot's lateral deviation and is therefore used to set the angular velocity ω of the robot. Thus, the robot is steered in a way that causes the mode of H to be close to 0. Note that while for ground robots the horizontal pixel-position difference is used, it is possible to apply it also with vertical differences, which allows to deploy the algorithm for aerial robots[8] as well.

While only lateral deviations are corrected using visual information, it can be mathematically proven[7] that the error accumulated (due to odometric errors) in the direction of the previously traversed segment is gradually reduced if the currently traversed segment is not collinear with the previous one. In practice, this implies that if the map contains segments of different orientations, the robot position error during the autonomous navigation is bound.

Algorithm 2. Navigation Phase

Input: L: landmarks for present segment, s_{length}: segment length, d: current
  robot distance from segment start
Output: ω: angular speed of robot
while $d < s_{length}$ **do**
    $H \leftarrow \emptyset$ `/* pixel-position differences */`
    $T \leftarrow \emptyset$ `/* tracked landmarks */`
    foreach $l \in L$ **do**
        if $l_{d_0} < d < l_{d_1}$ **then**
            $T \leftarrow \cup \{ l \}$ `/* get expected landmarks according to d */`

    while T *not empty* **do**
        $f \leftarrow$ find_match(l, F)
        if *matched* **then**
            `/* compare feature position to estimated current landmark position by`
            `interpolation                                                        */`
            $h \leftarrow f_{pos} - \left((l_{pos_1} - l_{pos_0}) \frac{d - l_{d_0}}{l_{d_1} - l_{d_0}} + l_{pos_0} \right)$
            $H \leftarrow \{ h \}$
        $T \leftarrow T - \{ l \}$
    $\omega \leftarrow \alpha$ mode(H)

However, this restricts the method to maps with short segments whenever significant dead-reckoning errors are present (eg. wheel slippage). Otherwise, it is necessary to turn frequently to reduce the position error [22]. Moreover, the method assumes that the autonomous navigation is initiated from a known location. This work shows that the aforementioned issues are tackled by applying MCL.

4   Monte Carlo Visual Localization

In the present work the position estimate d of the robot relative to the segment start is not obtained purely using dead-reckoning. Rather than that it is obtained by applying feature-based MCL. With MCL, if sufficient landmark matches can be established, the error of the position estimation will remain bounded even if a robot with poor-precision odometry will have to traverse a long segment. Moreover, the MCL is able to solve the global localization problem which allows to start the autonomous navigation from arbitrary locations along the taught path.

The robot localization problem consists in finding the robot state $\mathbf{x}_k$ at time k given the last measurement z^k (Markov assumption) and a prior state $\mathbf{x}_{k-1}$. MCL, as any bayesian-filter based approach, estimates the probability density function (PDF) $p(\mathbf{x}_k|z^k)$ recursively, i.e. the PDF calculation uses information about the PDF at time $k - 1$. Each iteration consists of a *prediction* phase, where a motion model is applied (based on a control input u_{k-1}) to obtain a predicted PDF $p(\mathbf{x}_k|z^{k-1})$, and an *update* phase where a measurement z^k is applied to obtain an estimate of $p(\mathbf{x}_k|z^k)$. Under the MCL approach, these PDFs are represented by means of samples or *particles* that are drawn from the estimated probability distribution that represents the robot's position hypothesis.

In the present work, since the goal of the navigation method is not to obtain an absolute 3D pose of the robot, the state vector is closely related to the type of map used. In other words, since the environment is described by using a series of segments which contain a list of tracked landmarks along the way, the robot state $\mathbf{x}$ is simply defined in terms of a given segment s_i and a robot distance from the segment's start d:

$$\mathbf{x} = [s_i \; d].$$

The orientation and lateral displacement of the robot is not modeled in the state since these are corrected using the visual input. This simplification leads to a low dimensional search-space that does not require to use a large number of particles.

The MCL algorithm starts by generating m particles p_k uniformly distributed over the known map. As soon as the robot starts to autonomously navigate, the prediction and update phases are continuously applied to each particle.

4.1   Prediction Step

The prediction step involves applying a motion-model to the current localization estimate represented by the current value of particles p_k, by sampling from $p(\mathbf{x}_k|\mathbf{x}_{k-1}, u_k)$, where u_k is the last motion command. The motion model in this work is defined as:

$$f([s_i \; d]) = \begin{cases} [s_i, d + \Delta d + \mu] & (d + \Delta d) < \mathrm{length}(s) \\ [s_{i+1}, (d + \Delta d + \mu) - \mathrm{length}(s)] & \mathrm{else} \end{cases}$$

where Δd is the distance traveled since the last prediction step and is a random variable with Normal distribution, i.e. $\mu \sim \mathcal{N}(0, \sigma)$. The value of σ was chosen as $10\,\mathrm{cm}$. This noise term is necessary to avoid premature convergence (i.e. a single high-probability particle chosen by MCL) and to maintain the diversity of localization hypotheses. Finally, this term can also account for the imprecision of the motion model which does not explicitly account for the robot heading.

The strategy to move particles to the following segment, which is also used in other works[10], may not necessarily be realistic since it does not consider the fact that the robot requires rotating to face the following segment. However, not propagating the particles to the following segment means that the entire particle set needs to be reinitialized every time new segment traversal is started, which is impractical.

4.2   Update Step

In order to estimate $p(\mathbf{x}_k|z^k)$, MCL applies a sensor model $p(z_k|\mathbf{x}_k)$ to measure the relative importance (weight) of a given particle p_i. Particles can then be re-sampled considering these weights, thus obtaining a new estimate of $\mathbf{x}_k$ given the last measurement z^k. In this work, z_k corresponds to the features F extracted at time k. The weight w_i of a particle p_i is computed in this work as:

$$w_i = \frac{\#\text{matches}(F, T)}{|T|}$$

where T is the set of landmarks expected to be visible according to p_i. In other words, the weight of a particle is defined as the ratio of visible and expected landmarks. The weights w_i are then normalized in order to represent a valid probability $p(z_k|p_i)$. It should be noted that it is possible for a particle to have $w_i = 0$ (ie. no features matched). Since every particle requires to have a non-zero probability of being chosen (i.e. no particles should be lost), particles with $w_i = 0$ are replaced by particles generated randomly over the entire map.

While this sensor model is simple, the experiments proved that it was sufficient to correctly match a particle to a given location.

4.3   Re-sampling

To complete one iteration of MCL, a new set of particles is generated from the current one by choosing each particle p_i according to its weight w_i. A naive approach for this would be to simply choose each particle independently. However, a better option is to use the low-variance re-sampling algorithm [1]. In this case, a single random number is used as a starting point to obtain M particles honoring their relative weights. Furthermore, this algorithm is of linear complexity.

The low-variance sampler is used not only to generate particles during the update step, but also to generate uniformly distributed particles in the complete map (during initialization of all particles and re-initialization of zero-weight particles). In this case, the set of weights used consists of the relative length of each segment in the map.

4.4   Single Hypothesis Generation

While MCL produces a set of localization hypotheses, the navigation needs to decide which hypothesis is the correct one. Furthermore, to asses the quality of this type of localization compared to other methods (such as a simple dead-reckoning approach) it is necessary to obtain a single hypothesis from the complete set.

To this end, the proposed method to obtain a single hypothesis is to compute the mean of all particle positions over the complete map. This is the usual choice for uni-modal distributions but is not suitable in general for an approach such as MCL which may present multiple modalities. However, when MCL converges to a single hypothesis, the mean position of all particles is a good estimate for the true position of the robot. In these cases, all particles are found to be clustered around a specific region and the population standard deviation is similar to the σ value used as the motion model noise. Therefore, it is possible to check if the standard deviation of all particles is less than $k\sigma$ and if so, the mean position will be a good estimate of the true robot position. In other cases, if the standard deviation is higher, the mean may not be a good position estimate. This can happen when the robot is "lost" and when the underlying distribution is multi-modal (i.e. when the environment is self-similar and the current sensor readings do not uniquely determine the robot position).

Finally, since an iteration of MCL can be computationally expensive and single-hypothesis estimates can only be trusted whenever the std. dev. is small enough, dead-reckoning is used to update the last good single-hypothesis whenever MCL is busy computing or it has failed to find a good position estimate.

5   Experiments

The proposed system was implemented in C/C++ within the ROS (Robot Operating System) framework as a set of separate ROS modules. In order to achieve high performance of the system, each module was implemented as a ROS *nodelet* (a thread).

Tests were performed over a long indoor corridor using a P3-AT ground and a Parrot AR.Drone aerial robot. The P3-AT's training run was $\sim$ 100 m long while the AR-Drone map consisted of a $\sim$ 70 m long flight. The P3-AT used wheel encoders for dead-reckoning and a PointGrey 0.3MP FireflyMV color camera configured to provide 640×480 pixel images at 30 Hz. The AR.Drone's dead reckoning was based on its bottom camera and inertial measurement unit and images were captured by its forward camera that provides 320×240 pixel images at 17 Hz. A Laptop with a quad-core Corei5 processor running at 2.3 GHz and 4 GB of RAM was used to control both of the robots.

Both robots were manually guided along the corridor three times while recording images and dead-reckoning data. For each robot, one of the three datasets was used for an off-line training run and the remaining ones were used for the replay phase evaluation (we refer to the remaining two datasets as 'replay 1' and 'replay 2'). Since this work focuses on the use of MCL for improving the original teach-and-replay method, only the localization itself is analyzed. Nevertheless, preliminary on-line experiments with autonomous navigation were performed using MCL which yielded promising results.[1]

In figure 1 the position-estimates of the robot obtained by dead-reckoning and MCL are compared. The standard deviation of the particle population in relation to the threshold of 5σ is also presented, see 1.

When analyzing figures 1(a),1(b), it becomes evident that the odometry of the P3AT is very precise and MCL does not provide a significant improvement. The standard deviations (see figures 1(a),1(b)) for these two cases show an initial delay caused by initialization of the MCL. Once the MCL particles converge, the position estimate becomes consistent and remains stable during the whole route.

On the other hand, in the AR.Drone localization results (see figures 1(c),1(d)) there's a noticeable difference between the MCL and dead-reckoning estimates due to the imprecision of the AR.Drone's dead-reckoning. During the 'replay 2',

---

[1] Note to reviewers: experiments are in progress, more exhaustive results will be ready for the camera-ready version.

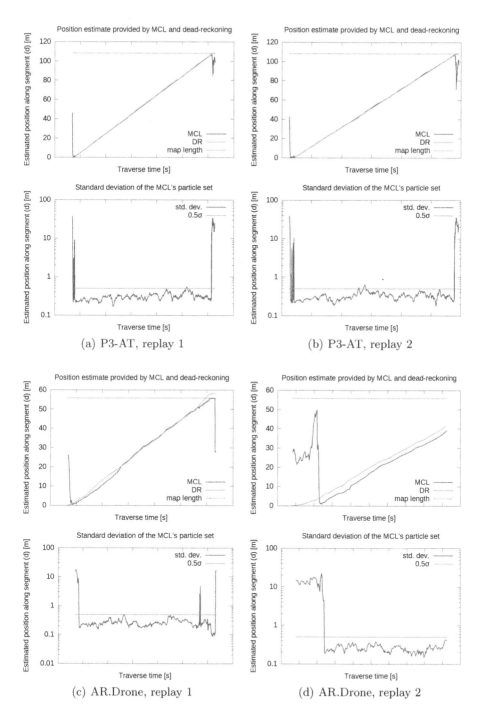

Fig. 1. Comparison of position estimates obtained from dead-reckoning (DR) and MCL, with corresponding standard deviation of the MCL's particle set

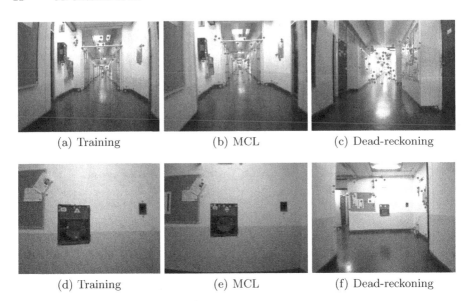

(a) Training   (b) MCL   (c) Dead-reckoning

(d) Training   (e) MCL   (f) Dead-reckoning

Fig. 2. Parrot AR.Drone view during replay and training, for matching distance estimates obtained with MCL and dead-reckoning: **(a)-(c)**: replay 2, $d \approx 0$; **(d)-(f)**: replay 1, $d \approx$ length(s)

the AR.Drone was placed a few meters before the initial mapping point, which is reflected by MCL, but not by the dead-reckoning, see Figure 1(d).

When analyzing the AR.Drone's 'replay 1' (see Figure 1(c)) it can be seen that after the initial convergence of MCL, the position estimates differ only slightly. However, there is a particular difference at the segment end where the drone actually overshot the ending point. In this case, MCL correctly estimates that the robot has reached the end, while with dead-reckoning the estimate goes beyond the segment length. Eventually, the MCL estimate diverges but this happens since the robot landed very close to a wall and suddenly no more recognizable feature were visible.

A similar situation occurred on 'replay 2' (1(d)) where the initial displacement of the AR.Drone is corrected by MCL and its difference to the odometry remains prominent. In this case, the MCL did not offer a suitable position hypothesis until the drone reached the segment start (notice the high standard deviation of the particles on 1(d)). One can see that while the MCL position estimate converges to 0 as soon as the robot passes the segment start, the position estimate provided by the dead-reckoning retains the offset throughout the entire segment.

Since no ground-truth is available to verify the correctness of the position estimates, we present the images taken when the MCL or dead-reckoning reported the drone to be at the segment start or end, see figures 2(a)-2(c) and 2(d)-2(f). It can be seen that in the case of MCL position estimation, the images captured by the robot are more similar to the images seen during the training phase. In the case of dead-reckoning, the views differ from the ones captured during the training phase which indicates a significant localization error.

6   Conclusions

This article has presented a combination of a teach-and-replay method and Monte Carlo Localization. The Monte Carlo Localization is used to replace the dead-reckoning-based position estimation that prohibited to use the teach-and-replay method by robots without a precise odometric system or in environments with long straight paths. Moreover, the ability of the MCL to deal with the kidnapped robot problem allows to initiate the 'replay' (i.e. autonomous navigation) from any location along the taught path. The experiments have shown that MCL-based position estimate remains consistent even in situations when the robot is displaced from the path start or its odometric system is imprecise.

To further elaborate on MCL's precision, we plan to obtain ground-truth position data by using an external localization system introduced in [23], In the future, we plan to improve the method by more robust calculation of the MCL's position hypothesis, more elaborate sensor model and improved feature extraction (eg. fixed number of features per frame). We also consider approaches such as Mixture Monte Carlo [24], that may provide further advantages.

Finally, the use of MCL opens the possibility of performing loop-detection, which will allow to create real maps rather than just remember visual routes. Using loop-detection, the robot would automatically recognize situations when the currently learned segment intersects the already taught path. Thus, the user will not be required to explicitly indicate the structure of the topological map that is learned during the tele-operated run.

Acknowledgements. The work is supported by the EU project 600623 'STRANDS' and the Czech-Argentine bilateral cooperation project ARC/13/13 (MINCyT-MEYS).

References

1. Thrun, S., Burgard, W., Fox, D.: Probabilistic robotics. MIT Press (2005)
2. Davison, A.J., Reid, I.D., Molton, N.D., Stasse, O.: MonoSLAM: real-time single camera SLAM. IEEE Transactions on Pattern Analysis and Machine Intelligence 29, 1052–1067 (2007)
3. Klein, G., Murray, D.: Parallel Tracking and Mapping on a camera phone. In: 2009 8th IEEE International Symposium on Mixed and Augmented Reality (2009)
4. Hartley, R., Zisserman, A.: Multiple view geometry in computer vision (2003)
5. Chen, Z., Birchfield, S.: Qualitative vision-based mobile robot navigation. In: Proceedings International Conference on Robotics and Automation, ICRA, pp. 2686–2692. IEEE (2006)
6. Zhang, A.M., Kleeman, L.: Robust appearance based visual route following for navigation in large-scale outdoor environments. The International Journal of Robotics Research 28(3), 331–356 (2009)
7. Krajník, T., Faigl, J., Vonásek, V., Kosnar, K., Kulich, M., Preucil, L.: Simple yet stable bearing-only navigation. Journal of Field Robotics 27(5), 511–533 (2010)
8. Krajnik, T., Nitsche, M., Pedre, S., Preucil, L.: A simple visual navigation system for an UAV. In: Systems, Signals and Devices, pp. 1–6 (2012)

9. Furgale, P., Barfoot, T.: Visual teach and repeat for longrange rover autonomy. Journal of Field Robotics (2006), 1–27 (2010)
10. Siagian, C., Chang, C.K., Itti, L.: Autonomous Mobile Robot Localization and Navigation Using a Hierarchical Map Representation Primarily Guided by Vision. Journal of Field Robotics 31(3), 408–440 (2014)
11. Krajník, T., Pedre, S., Přeučil, L.: Monocular Navigation System for Long-Term Autonomy. In: Proceedings of the International Conference on Advanced Robotics, Montevideo. IEEE (2013)
12. Dellaert, F., Fox, D., Burgard, W., Thrun, S.: Monte Carlo localization for mobile robots. In: IEEE International Conference on Robotics and Automation, vol. 2 (1999)
13. Niu, M., Mao, X., Liang, J., Niu, B.: Object tracking based on extended surf and particle filter. In: Huang, D.-S., Jo, K.-H., Zhou, Y.-Q., Han, K. (eds.) ICIC 2013. LNCS, vol. 7996, pp. 649–657. Springer, Heidelberg (2013)
14. Li, R.: An Object Tracking Algorithm Based on Global SURF Feature. Journal of Information and Computational Science 10(7), 2159–2167 (2013)
15. Wolf, J., Burgard, W., Burkhardt, H.: Robust vision-based localization by combining an image-retrieval system with monte carlo localization. Transactions on Robotics 21(2), 208–216 (2005)
16. Ostafew, C.J., Schoellig, A.P., Barfoot, T.D.: Visual teach and repeat, repeat, repeat: Iterative Learning Control to improve mobile robot path tracking in challenging outdoor environments. In: 2013 IEEE/RSJ International Conference on Intelligent Robots and Systems, pp. 176–181 (November 2013)
17. Ok, K., Ta, D., Dellaert, F.: Vistas and wall-floor intersection features-enabling autonomous flight in man-made environments. In: Workshop on Visual Control of Mobile Robots (ViCoMoR): IEEE/RSJ International Conference on Intelligent Robots and Systems (2012)
18. Augustine, M., Ortmeier, F., Mair, E., Burschka, D., Stelzer, A., Suppa, M.: Landmark-Tree map: A biologically inspired topological map for long-distance robot navigation. In: 2012 IEEE International Conference on Robotics and Biomimetics (ROBIO), pp. 128–135 (2012)
19. Ni, K., Kannan, A., Criminisi, A., Winn, J.: Epitomic location recognition. IEEE Transactions on Pattern Analysis and Machine Intelligence 31(12), 2158–2167 (2009)
20. Cadena, C., McDonald, J., Leonard, J., Neira, J.: Place recognition using near and far visual information. In: Proceedings of the 18th IFAC World Congress (2011)
21. Matsumoto, Y., Inaba, M., Inoue, H.: Visual navigation using view-sequenced route representation. In: Proceedings of the 1996 IEEE International Conference on Robotics and Automation, vol. 1, pp. 83–88 (April 1996)
22. Faigl, J., Krajník, T., Vonásek, V., Preucil, L.: Surveillance planning with localization uncertainty for uavs. In: 3rd Israeli Conference on Robotics, Ariel (2010)
23. Krajník, T., Nitsche, M., Faigl, J., Vank, P., Saska, M., Peuil, L., Duckett, T., Mejail, M.: A practical multirobot localization system. Journal of Intelligent and Robotic Systems, 1–24 (2014)
24. Thrun, S., Fox, D., Burgard, W., Dellaert, F.: Robust Monte Carlo localization for mobile robots. Artificial Intelligence 128, 99–141 (2001)

An Improved Cellular Automata-Based Model for Robot Path-Planning

Giordano B. S. Ferreira[1], Patricia A. Vargas[2], and Gina M. B. Oliveira[1]

[1] Bio-Inspired Computation Lab, Universidade Federal de Uberlandia, Brazil
gina@facom.ufu.br
[2] Robotics Lab, Heriot-Watt University, Edinburgh, UK
P.A.Vargas@hw.ac.uk

Abstract. Cellular automata (CA) are able to represent high complex phenomena and can be naturally simulated by digital processors due to its intrinsic discrete nature. CA have been recently considered for path planning in autonomous robotics. In this work we started by adapting a model proposed by Ioannidis *et al.* to deal with scenarios with a single robot, turning it in a more decentralized approach. However, by simulating this model we noticed a problem that prevents the robot to continue on its path and avoid obstacles. A new version of the model was then proposed to solve it. This new model uses CA transition rules with Moore neighborhood and four possible states per cell. Simulations and experiments involving real e-puck robots were performed to evaluate the model. The results show a real improvement in the robot performance.

Keywords: Autonomous Robotics, Cellular Automata, Path Planning.

1 Introduction

Path planning is one of the most-studied tasks in autonomous robots [1]. Traditional approaches to path planning are: route maps [2], cell decomposition [3] and potential field [4]. Since they are totally discrete models, cellular automata (CA) [5], [6] have also been recently considered for path planning [7], [8], [9], [12], [13], [14], [15], [16], [17], [18]. The decentralized CA architecture permits the development of high-distributed solutions for path-planning. In this work, we investigate the model proposed by Ioannidis and collaborators [14]. The authors use a team of robots to navigate between two points on a 2D environment. The team formation is controlled by a master-slave architecture. This characteristic turns the model into a centralized control system which contradicts the idea of totally distributed CA-based architectures.

We start by presenting an adaptation to the model in [14] to apply it using scenarios with a single robot. The purpose of such simplification was to better investigate the resultant robot's behavior by using only CA rules and no centralized control architecture. The desirable pattern is to try to keep a single robot navigating in the axis formed by the initial position and the goal whenever it is possible. We could observe the behavior of the CA rules without the influence of the team pattern control strategy. However, when we simulated this model using the Webots platform [10]

M. Mistry et al. (Eds.): TAROS 2014, LNAI 8717, pp. 25–36, 2014.
© Springer International Publishing Switzerland 2014

we noticed a problem that prevented the robot to continue on its path and avoid obstacles. We called it as the *corner deadlock*. It happens when the robot faces an obstacle and tries to overcome it by applying the CA deviation rules. The robot then gets stuck by the current combination of rules and the imprecise sensor readings, thus rotating in the same axis and not moving forward. In the present work, a new local decision making model is proposed for path-planning starting from Ioannidis *et al.* original model. We show that the identified problem can be solved by including one new state per cell to be used by deviation rules in corner situations and modifying such rules to deal with this new state. These modifications help the robot to overcome the obstacles and return to its original axis of movement after several steps. This better performance was observed in simulations and real experiments using real e-puck robots [11].

2   Cellular Automata and Robot Path-Planning Approaches

Cellular automata are discrete dynamical systems composed by simple components with local interactions. It has a regular lattice of N cells, each one with an identical pattern of local connections to other cells. CA transition rule determines the state of the cell i at time $t + 1$ depending on the states of its neighborhood at time t including cell i [5]. Formally, let Z^d be a d-dimensional lattice and Σ a finite set of states. Lattice configuration is a map $c: Z^d \rightarrow \Sigma$, which specifies the states of all lattice cells. The state of cell i in a time t is denoted by s_i^t, where $s_i^t \in \Sigma$. The state s_i^t of cell i, together with the states of all cells connected with i is called neighborhood of cell i and denoted by η_i^t. The transition rule of cellular automata is defined by a function $\Phi: \Sigma^n \rightarrow \Sigma$ where n is the neighborhood size, that is, $\Phi(\eta_i)$ gives the next state s_i^{t+1} for cell i, as a function of η^t [5]. A one dimensional CA defines its neighborhood using a radius specification. Elementary CA has 1D lattice, 2 states and radius-1 neighborhood. Transition rule defines the output bit that is the new state of the central cell for each possible neighborhood. In elementary CA, the transition rule is defined by 8 output bits. Figure 1 shows a transition rule defined by the output bits for each one of the eight possible 3-cells binary neighborhood and the lattice configuration at the initial time and after the synchronous application of the rule over all cells.

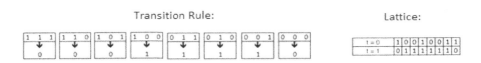

Fig. 1. Example of elementary CA: 1D-lattice, binary states and radius 1 neighborhood

Considering 2D lattices, two types of neighborhood are commonly used: von Neumann formed by the central cell and its four neighbors in the orthogonal directions, and Moore, which also includes the diagonal cells on the neighborhood, totalizing 9 cells (including the central cell itself). Considering binary CA, the transition rule for 2D lattices is formed by 512 bits using Moore neighborhood. Thus, 2D rules are usually defined in a totalistic description [5], instead of a list of output bits.

Cellular automata can be exactly simulated by digital processors due to its intrinsic discrete nature. Besides, they are able to represent high complex phenomena [6]. Those characteristics lead CA to be considered for path-planning and several models can be found in the literature. Using their similarities, we classified them into six distinct approaches: *Power Diffusion* [16]; *Goal Attraction* [17]; *Voronoi's Diagram* [18]; *Distance Diffusion* [7], [8]; *Local decision making* [12], [13], [14], [15]; and *Message Sending* [9]. The new model proposed here belongs to the local decision making approach, which is characterized by the usage of robot's sensors at each step to evaluate the neighborhood and the application of a CA rule to decide the next movement. There is no previous knowledge of the environment.

3   Adaptation of the Previous Model for a Single Robot

The main objective of the model proposed by Ioannidis *et al.* [14] is to navigate a team of robots in a straight line from their initial positions to their final goals and deviate from eventual obstacles during navigation. The model also tries to keep the team in a desirable pattern while navigating. The environment is a two-dimensional space formed by square cells with side *l*. The robot can be larger than *l* but for the algorithm it occupies a single cell (where its center is localized). The neighborhood at each time step was elaborated using the robot's infrared proximity sensors. Two navigation patterns were evaluated in [14]: the triangular and the linear formations. The authors performed experiments both in simulation and with real e-pucks to validate their results. Two different types of 2D CA rules are employed in [14]: the first type is used to deviate from obstacles and the second to perform the formation control. The application of such rules is performed through a master-slave control architecture in [14], which takes in account the relative position of other robots as well and the desirable pattern. The server robot receives all robots' positions and decides what will be the rule to apply to each member of the team. Therefore, this process is performed by a centralized control system which contradicts the idea of totally distributed CA-based architectures. In this work we present the adaptation performed to the model in [14] to apply it using scenarios with a single robot. The purpose of such simplification is to better investigate the resultant robot's behavior by using only CA rules and no central-ised control architecture.

The desirable navigation pattern for this single robot is to cross the environment in a straight line whenever is possible, deviating from eventual obstacles. The path-planning algorithm is divided into three stages: (i) construction of the neighborhood (ii) decision about which rule will be applied and (iii) application of the rule resulting in the change of state of the neighborhood central cell and the effective robot action. The three possible cell states are *Free* (*F*), *Obstacle* (*O*) and *Robot* (*R*). The Moore neighborhood is constructed using robot's sensors readings. *E-puck* robots [11] with 8 infrared sensors were employed both in simulated and real experiments. The neighborhood is a vector $a = [a_0, a_1, a_2, a_3, a_4, a_5, a_6, a_7, a_8]$ which represents in a linear way the spatial neighborhood in Figure 2.

The current robot's rotation angle (Θ_R) is also stored and it can take 5 possible values in relation to its crossing axis. For example, if the robot's goal is to navigate in an axis from the south to the north, the possible values and interpretations for Θ_R are: $0°$ (the robot points to a_1), $45°$ (it points to a_2), $90°$ (it points to a_3), $-45°$ (it points to a_8), $-90°$ (it points to a_7). Two kinds of discrete robot's movements are possible at each time step: (i) it moves to an adjacent cell keeping its current orientation or (ii) it rotates keeping its current orientation. In a movement action, the distance to be crossed is defined by the width cell (l): l for cardeal neighbor cells and $\sqrt{2}\, l$ for diagonal ones. Cellular automata transition rules define the next robot movement at each time step and it updates cell states. Each transition rule is a function of the current neighborhood and the rotation angle (Θ_R). There are 2 types of rules: those used to avoid collisions and those applied to formation control. Each robot identifies its neighborhood and which kind of rule will be applied. If there is an obstacle, a deviation rule will be applied. Otherwise, a formation control rule will be applied.

a_8	a_1	a_2
a_7	a_0	a_3
a_6	a_5	a_4

Fig. 2. Spatial neighborhood

There are two types of deviation rules: (i) those that makes the robot to move 1 cell; (ii) those that rotates the robot. The rules able to move the robot must be applied in pairs: the first rule is executed for the neighborhood for which the central cell is in the state *Robot*, indicating that the robot is within this cell (the next state will be *Free*) and the second one is executed for the neighborhood for which the central cell is in the state *Free* (this state will be changed to *Robot*). Figures 3.a and 3.b exemplifies two scenarios of movement that must be applied in pairs. The first one is the scenario involving the neighborhood, in which the robot is in the central cell. The second one, is the scenario involving the neighborhood with the central cell free, which will receive the robot in the next time step. Robot's position and orientation are represented by an arrow, obstacles by shadow cells and free space by white cells. Suppose that the robot's crossing axis is defined from the south to the north. Figure 3.a indicates that if there are two obstacles in cells a_1 and a_2 and the current Θ_R is $-90°$, the robot must move to the cell a_7, keeping its orientation. Figure 3.b indicates that if there is a free cell at the central position, the robot is within cell a_3, there is an obstacle in cell a_2 and the current Θ_R is $-90°$, the robot must move to the central cell, keeping its orientation.

Besides the change of the central cell state in both rules, the trigger of those rules also activates the movement dynamics in a way that the robot will cross a distance equivalent to one cell in the desirable direction. The transition rules that determine a robot rotation must be applied in two kinds of scenarios: (i) when the robot's orientation is in the same of its crossing axis ($\Theta_R = 0°$) and it identifies an obstacle in its front that forbids its progression as the scenario in Figure 3.c; (ii) when the robot's orientation is different from its crossing axis, indicating that it did a previous rotation to deviate from some obstacles and it does not identify any obstacle in its neighborhood that forbids to return to its crossing axis' orientation as the scenario in Figure 3.d. Figure 3.c shows a scenario where the robot was moving to the north and its sensors identifies obstacles in 3 cells (a_1, a_7 and a_8). In such case, the robot must rotate to $90°$ aiming to avoid the obstacle. Figure 3.d shows a scenario where the robot was moving to the west indicating that it identified an obstacle in some previous steps and it was forced to rotate. However, at the current time step, the only identified obstacle is on cell a_2, which enables the robot to rotate and to return to its original orientation (to the north).

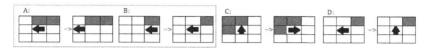

Fig. 3. Scenarios relative to deviation obstacle rules supposing that the robot's crossing axis is to the north

In [14], the authors presented the deviation rule set on a linear representation, where each rule represents a single scenario. Figure 4 show examples of these linear rules relative to the scenarios of Figure 3. However, if the rule set is written is such a way, it will be necessary to do an exhaustive list of all possible configurations of the current state. Considering only the rules in which the central cell is in state *Robot*, there are 5×2^8 (1280) different possible rules. Besides, as presented in next section, in the new model proposed here, the number of states per cell was increased from 3 to 4, leading to a larger number of scenarios to be listed. Therefore, we adopted a totalistic representation [5] of the rule set, which generalizes several scenarios in a unique description. Figure 5 shows the transition rule implemented in the adapted model.

Current State										Next State	
a0	a1	a2	a3	a4	a5	a6	a7	a8	Θ_t	a0	Θ_{t+1}
r	o	o	f	f	f	f	f	f	90°	f	90°
f	f	o	r	f	f	f	f	f	90°	r	90°
r	o	f	f	f	f	f	o	o	0°	r	-90°
r	f	o	f	f	f	f	f	f	90°	r	0°

Fig. 4. Examples of linear transition rules corresponding to the scenarios in Figure 3

The algorithm in [14] was elaborated for a team of robots. This characteristic was modified for the navigation of a single robot. In that way, it was possible to observe the robot behavior without the influence of the centralized master-slave control used in [14] to implement the team cooperation. Besides, the robot behavior can be assumed to be exclusively due to the CA rules. Therefore, the formation control used in

[14] needed to be modified because it was based on multiple robots and a constant check whether it was necessary to change their crossing axis. This change of goal is determined by the master robot which is responsible to define the rule to be applied to each robot team at each moment. As there is only a unique robot in the scenario, there is no change of axis and the robot tries to keep its own crossing axis. Whenever it is necessary to escape from its crossing axis to avoid obstacles, the formation control rules make the robot returns to its previous path after deviating from the obstacles.

State	Angle	Neighborhood	New State	New Angle
FREE	0°	a5 = ROBOT & a3 & a7 = FREE	ROBOT	0°
		Otherwise	FREE	0°
	90°	a3=ROBOT & a5 = FREE & a2 = OBSTACLE	ROBOT	-90°
		Otherwise	FREE	-90°
	-90°	a7=ROBOT & a5 = FREE & a8 = OBSTACLE	ROBOT	90°
		Otherwise	FREE	90°
OBSTACLE	ANY	ANY	OBSTACLE	ANY
ROBOT	0°	a1 & a2 & a8 = FREE	FREE	0°
		a1 = OBSTACLE & a2 & a3 = FREE	ROBOT	-90°
		a1 & (a2 OR a3) = OBSTACLE & a7 & a8 = FREE	ROBOT	90°
		a1 & a7 = OBSTACLE & a3 = FREE	ROBOT	-90°
		a1 & a3 = OBSTACLE & a7= FREE	ROBOT	90°
		a8 = OBSTACLE & a3 = FREE	ROBOT	-90°
		a2 = OBSTACLE & a7 = FREE	ROBOT	90°
	90°	a1 = OBSTACLE & a7 & a6 = FREE	FREE	90°
		a1 = FREE	ROBOT	0°
	-90°	a1 = OBSTACLE & a3 & a4 = FREE	FREE	-90°
		a1 = FREE	ROBOT	0°

Fig. 5. Adapted model: deviation transition rules

The formation control rules are presented in Figure 6. Since it tries to keep the robot on its axis, it is necessary to store its original horizontal position called hor_r. The column which is the current robot's horizontal position is called x_r. The formation control rules try to keep x_r equal to hor_r and they are applied only if there is no obstacle and the robot is on its original orientation. If the robot position indicates it is on its axis, the robot will move forward. On the other hand, if the robot is to the outside of its crossing axis, the formation control rule will make the robot moves towards its axis through diagonal steps.

The adapted model was initially implemented in the Webots simulator [10] using e-puck robots [11]. Figure 7 shows a simulation experiment performed using 1 robot and 1 obstacle: these images were generated by Webots except for the blue arrows included to show the robot orientation in each time step illustrated. The robot's initial behavior was as expected. However, a problem always happened when the robot faced an obstacle and tried to overcome it by applying the CA deviation rules. The robot got stuck by the combination of rules and imprecise sensor readings, thus rotating in the same axis and not moving forward. This particular problem was named *corner deadlock*. A sequence of movements illustrating the problem in simulation can be seen in Figure 7 (a-f). The same problem could be observed in the real robot experiment. The sequence of movements can be seen in Figure 8 (a-g). After we identified this problem by simulation and confirmed it using real e-pucks, we returned carefully to Ioannidis' works to check if there was some comments related to this behavior and how it was fixed, since they also used e-pucks robots in their experiments. However, we couldn't find any mention to such problem and the robots seemed to successfully overcome the corners in the reported experiments. One point that caught our attention in [14] was that in all reported experiments whenever a robot faces a corner it moves closer to a robot neighbor. In such case, the formation control changed the robot's referential axis and instead of trying to overcome the obstacle the robot started to walk in diagonal steps to the changed axis. However, there are other possible scenarios not shown in [14] where the features that caused the *corner deadlock* could also happen. For example, if the robot deviate from an obstacle to the left and it is the team member in the left side of the formation, the robot would not have any other member to change their referential axis and it would have to overcome the obstacle. In this case, we are confident that a standard e-puck will present the same difficult with corners and the looping of rotations should be expected. Another evidence of this difficulty is that on a posterior paper [15], the authors reported that they used non-standard e-puck robots with 36 sensors. Probably in such configuration the robot would identify the corner with precision. However, using the standard e-puck robot with 8 sensors it is critical and should be deal with. Besides, even using other robot's architectures, it is common to have some "blind regions", specially in the later half of their bodies. Therefore, an improvement in the previous model is necessary to deal with the *corner deadlock*.

Situation	a0	a1	a2	a3	a4	a5	a6	a7	a8	Θt	a0	Θt+1
			Neighborhood at time t								Central cell (t+1)	
$hor_r - x_r = 0$	r	f	f	f	f	f	f	f	f	0°	f	0°
$hor_r - x_r = 0$	f	f	f	f	f	r	f	f	f	0°	r	0°
$hor_r - x_r > 0$	r	f	f	f	f	f	f	f	f	0°	f	0°
$hor_r - x_r > 0$	f	f	f	f	f	f	r	f	f	0°	r	0°
$hor_r - x_r < 0$	r	f	f	f	f	f	f	f	f	0°	f	0°
$hor_r - x_r < 0$	f	f	f	f	r	f	f	f	f	0°	r	0°

Fig. 6. Formation control rules

(a)   (b)   (c)   (d)   (e)   (f)

Fig. 7. Simulation with the adapted model

4   A New Path-Planning Model

Due to the *corner deadlock* observed in the experiments performed with the adapted model, we carried out some modifications to propose a new path-planning algorithm. This new model preserves the major characteristics of the adapted model: (a) it is based on local transition rules of CA to decide the next robot step at each time step; (b) there are two types of CA rules being that the first type is used when the robot must avoid an obstacle in its neighborhood and the second type is used to keep the robot navigating on its crossing axis. The major modification was to include a new state to the model. This new state identifies that the robot has come from a previous rotation to deviate from an obstacle when it rotates to its original orientation again, as in the scenario of Figure 8.e. In the new model, when the robot is on a changed orientation and did not identify any obstacle as in Figure 8.d, the robot returns to its original orientation like in Figure 8.e, bust changing to the state *Rotated_robot*, indicating that it comes from a previous rotation due to some obstacle identification in an early step. In such state, if the robot does not identify an obstacle in its front, the state changes again to *Robot* and it will start to move by formation control rules. On the other hand, if the robot identifies an obstacle in its front, the robot will rotate again to avoid it but keeping the state *Rotated_robot*. Once in its new orientation, the robot identifies it as a "possible corner scenario" and it will make a new step to the left to deviate from the obstacle, even if its side sensors indicate that the obstacle was overcome. Therefore, the new model includes a state *Rotated_robot* and the consequent modifications in the deviation rules to deal with the new state. Figure 9 shows the new set of deviation rules used. The formation control rules were not modified and they were the same presented in Figure 6 for the adapted model. Comparing figures 5 and 9, one can observe the modifications performed in the deviation rules to deal with the new state *Rotated_robot*.

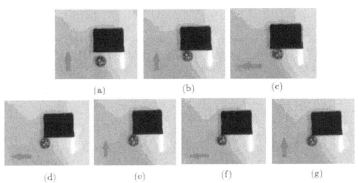

Fig. 8. Real experiment with the adapted model

The new model was first simulated using Webots. Figure 10 shows a simulation using the same initial scenario of Figure 7. The scenarios in which the robot is at state *Rotated_robot* are marked with "*". It identifies an obstacle, makes a rotation (Figure 10.b) and moves up to the corner (Figure 10.c) where the sensors do not identify the obstacle. Subsequently, the robot rotates to the original orientation (Figure 10.d).

However, in this new version, the state is also changed to *Rotated_robot*, indicating that it came from a previous rotation to avoid some obstacle. Therefore, when the obstacle is identified, it turns to the left but keeping the current state (Figure 10.e). In this situation, the deviation rules make the robot to walk one more step to the left, even with no sensors' reading (Figure 10.f) and its state is changed to *Robot* again. This process will be repeated again (Figure 10.g and 10.h) and the robot will do another step to the left (Figure 10.i) and returned to the original orientation (Figure 10.j).

The process is composed by the following steps - (i) rotation to $0°$ changing to state *Rotated_Robot*, (ii) rotation to $90°$ keeping the state *Rotated_Robot*, (iii) move one cell to the left changing to state *Robot* - will be repeated until the robot does not identify any obstacle in its front when it is in the original orientation. When it happens, the robot will start to move forward and changes its state to *Robot* again. Therefore, it is possible to notice that the *corner deadlock* was solved with the inclusion of the new state *Rotated_robot* and its related rules presented in Figure 9.

State	Angle	Neighborhood	New State	New Angle
FREE	0°	(a5 = ROBOT OR ROTATED_ROBOT) & a3 & a7 = FREE	ROBOT	0°
		Otherwise	FREE	0°
	-90°	a3=ROBOT & a5 = FREE & a2 = OBSTACLE	ROBOT	-90°
		a3 = ROTATED_ROBOT	ROBOT	-90°
		Otherwise	FREE	-90°
	90°	a7=ROBOT & a5 = FREE & a8 = OBSTACLE	ROBOT	90°
		a7 = ROTATED_ROBOT	ROBOT	90°
		Otherwise	FREE	90°
OBSTACLE	ANY	ANY	OBSTACLE	ANY
ROBOT	0°	a1 & a2 & a8 = FREE	FREE	0°
		a1 = OBSTACLE & a2 & a3 = FREE	ROBOT	-90°
		a1 & (a2 OR a3) = OBSTACLE & a7 & a8 = FREE	ROBOT	90°
		a1 & a7 = OBSTACLE & a3 = FREE	ROBOT	-90°
		a1 & a3 = OBSTACLE & a7 = FREE	ROBOT	90°
		a8 = OBSTACLE & a3 = FREE	ROBOT	-90°
		a2 = OBSTACLE & a7 = FREE	ROBOT	90°
	-90°	a1 = OBSTACLE & a7 & a6 = FREE	FREE	-90°
		a1 = FREE	ROTATED_ROBOT	0°
	90°	a1 = OBSTACLE & a3 & a4 = FREE	FREE	90°
		a1 = FREE	ROTATED_ROBOT	0°
ROTATED_ROBOT	-90°	a7 = FREE	FREE	-90°
		Otherwise (There is no path)	ROTATED_ROBOT	-90°
	90°	a3 = FREE	LIVRE	90°
		Otherwise (There is no path)	ROTATED_ROBOT	90°

Fig. 9. Deviation rules for the new model

After simulation, we embedded the algorithm code related to the new model on a real e-puck robot. Figure 11 shows some images of the real experiment. Figure 11.a to 11.d shows the navigation until the robot achieves the obstacle's corner. Figure 11.e

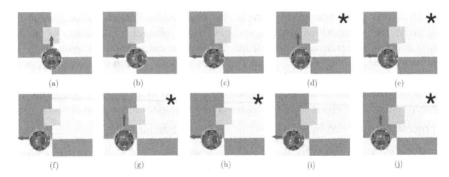

Fig. 10. Simulation with the first version of the new model

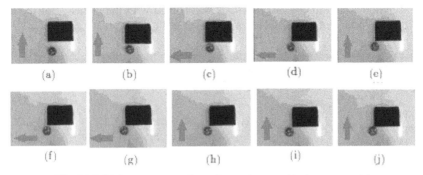

Fig. 11. Initial movements of a real experiment with the new model

shows the moment that the e-puck returns to the original orientation and changes its state to *Rotated_robot*. As an obstacle is identified, it rotates keeping its current state (Figure 11.f). The e-puck moves 1 step to the left (Figure 11.g) and returns to its original orientation (Figure 11.h). As the obstacle is not identified, the robot changes to state *Robot* and moves 2 steps forward, using diagonal movements (Figure 11.i and 11.j). After this, the robot will identify the obstacle in its front again and the sequence of rotations and state changes will be used to move to the left until the robot could totally deviate the obstacle's corner and start to navigate toward the original path.

Therefore, simulations and real experiments have shown that the inclusion of the state *Rotated_robot* solved the *corner deadlock*. The e-puck was able to deviate the obstacle and returned to its crossing axis at the end of the experiment. Figure 11 only highlights the initial movements of the robot to escape from its obstacle corner. Figure 12 shows pictures from a similar experiment using the new path-planning model proposed here from a different perspective: the crossing axis was marked on the table where the e-puck was navigating only to facilitate the understanding, but the robot does not use it to make any decision. The robot's decision making is done exclusively using the infrared sensors, its CA rules (Figures 6 and 9) and in the robot's current neighborhood. It was possible to confirm that the new model was able to overcome the *corner deadlock* problem.

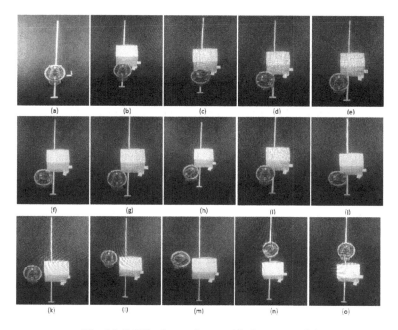

Fig. 12. Full Real experiment with the new model

5   Conclusion and Future Work

We proposed an improved cellular automata-based model for the path-planning of mobile robots. We started from the analysis of the model from Ioannidis et al. [14]. Our adapted model was just based on decentralized CA rules instead of using a centralized architecture controlling these rules as in the original model. We also used only one robot instead of a team of robots. During the first simulations a major problem was spotted and we called it *corner deadlock*. Although this problem might be very simple to solve using other local path-planning algorithms [19], we wanted to further investigate a solution using the CA rules in a decentralized way. The decentralized CA architecture permits the development of high-distributed solutions for path-planning. The main motivation behind the use of CA for path planning relies on the application of CA in more complex scenarios involving multiple robots.

We are currently investigating a better set of states and CA rules so that the robot improves its performance even further. As future work we plan to concentrate on applying the new improved and refined CA model to a swarm of robots that could efficiently perform path-planning while keeping the team formation on a more realistic dynamical environment.

Acknowledgments. GBSF is grateful to CNPq scholarship. GMBO is grateful to Fapemig, CNPq and CAPES for financial support. The authors would like to thank Dr Micael Couceiro for his valuable advice.

References

1. Arkin, R.C.: Behavior-based robotics. MIT Press (1998)
2. Zhang, Y., Fattahi, N., Li, W.: Probabilistic roadmap with self-learning for path planning of a mobile robot in a dynamic and unstructured environment. In: Mechatronics and Automation (ICMA), pp. 1074–1079 (2013)
3. Ramer, C., Reitelshofer, S., Franke, J.: A robot motion planner for 6-DOF industrial robots based on the cell decomposition of the workspace. In: 2013 44th International Symposium on Robotics (ISR), pp. 1–4 (2013)
4. Jianjun, Y., Hongwei, D., Guanwei, W., Lu, Z.: Research about local path planning of moving robot based on improved artificial potential field. In: Control and Decision Conference (CCDC), pp. 2861–2865 (2013)
5. Mitchell, M.: Computation in cellular automata: A selected review. Non-standard Computation, 385–390 (1996)
6. Oliveira, G., Martins, L.G.A., de Carvalho, L.B., Fynn, E.: Some investigations about synchronization and density classification tasks in one-dimensional and two-dimensional cellular automata rule spaces. Electronic Notes in Theoretical Computer Science 252, 121–142
7. Behring, C., Bracho, M., Castro, M., Moreno, J.A.: An Algorithm for Robot Path Planning with Cellular Automata. In: Proc. of the 4th Int. Conf. on Cellular Automata for Research and Industry, pp. 11–19 (2000)
8. Soofiyani, F.R., Rahmani, A.M., Mohsenzadeh, M.: A Straight Moving Path Planner for Mobile Robots in Static Environments Using Cellular Automata. In: Int. Conf. on Computational Intelligence, Communication Systems and Networks, pp. 67–71 (2010)
9. Rosenberg, A.: Cellular ANTomata. Paral. and Distributed Processing and Applications, 78–90 (2007)
10. Cyberbotics. Webots 7: robot simulator (2013),
    http://www.cyberbotics.com/overview
11. E-puck Education Robot, http://www.e-puck.org
12. Akbarimajd, A., Lucas, C.: A New Architecture to Execute CAs-Based Path-Planning Algorithm in Mobile Robots. In: IEEE Int. Conf. on Mechatronics, pp. 478–482 (2006)
13. Akbarimajd, A., Hassanzadeh, A.: A novel cellular automata based real time path planning method for mobile robots. Int. Journal of Engineering Research and Applications (2011)
14. Ioannidis, K., Sirakoulis, G.C., Andreadis, I.: A Cellular Automaton Collision- Free Path Planner Suitable for Cooperative Robots. In: Panhellenic Conf. on Informatics (2008)
15. Ioannidis, K., Sirakoulis, G., Andreadis, I.: Cellular ants: A method to create collision free trajectories for a cooperative robot team. Robotics and Autonomous Systems (2011)
16. Shu, C., Buxton, H.: Parallel path planning on the distributed array processor. Parallel Computing 21(11), 1749–1767 (1995)
17. Marchese, F.: A reactive planner for mobile robots with generic shapes and kinematics on variable terrains. In: Proc. Advanced Robotics, ICAR 2005, pp. 23–30 (2005)
18. Tzionas, P., Thanailakis, A., Tsalides, P.: Collision-free path planning for a diamond-shaped robot using two-dimensional CA. IEEE Trans. on Rob. and Automation (1997)
19. Vargas, P.A., Di Paolo, E.A., Harvey, I., Husbands, P.: The Horizons of Evolutionary Robotics. MIT Press (2014)

Bioinspired Mechanisms and Sensorimotor Schemes for Flying: A Preliminary Study for a Robotic Bat

Carmine Tommaso Recchiuto[1], Rezia Molfino[2], Anders Hedenströem[3],
Herbert Peremans[4], Vittorio Cipolla[5], Aldo Frediani[5,6], Emanuele Rizzo[5], and
Giovanni Gerardo Muscolo[1,2,*]

[1] Humanot S.r.l., Prato, Italy
info@humanot.it
[2] Scuola Politecnica, University of Genova, PMAR Lab, Genova, Italy
{molfino,muscolo}@dimec.unige.it
[3] Lunds Universiteit, Department of Theoretical Ecology, Lund, Sweden
Anders.Hedenstrom@biol.lu.se
[4] Universiteit Antwerpen, Active Perception lab, Antwerp, Belgium
herbert.peremans@uantwerpen.be
[5] SkyBox Engineering S.r.l., Pisa, Italy
{v.cipolla,info,e.rizzo}@skyboxeng.com
[6] Università di Pisa, Department of Aerospace Engineer, Pisa, Italy
a.frediani@dia.unipi.it

Abstract. The authors present a critical review on flying motion and echolocation in robotics, directly derived from research on biology and flight mechanics of bats, aimed at designing and prototyping a new generation of robotic technologies.

To achieve this goal, the paper starts with an in-depth review on real bats, studying and proposing models derived from the analysis of the aerodynamics of bat wings combined together with the investigation of the bat echolocation system.

These models will form the basis for the implementation of the conceptual design of the proposed robotic platform, consisting of a new complete bat body, light and smart body structures, provided with a set of sensors and actuated joints, displaying echolocation, sensing and coordination capabilities.

Keywords: Aerial robots, bio-inspiration, bio-robotics, echolocation, robotic bat, smart materials, flapping wings, micro aerial vehicles, perception.

1 Introduction

As more and more widely recognized today, both science and engineering can benefit from joint research in biomimetics: on one hand, biology supplies the knowledge and models on the biological side to provide inspiration for developing biomimetic systems;

---

* Corresponding author.

M. Mistry et al. (Eds.): TAROS 2014, LNAI 8717, pp. 37–47, 2014.
© Springer International Publishing Switzerland 2014

on the other hand, the development of biomimetic systems provides insight about performance and constraints into the animals they are inspired from [1], and can even represent a powerful tool for experimental investigation on the living organism itself [2]. Animals can represent a source of inspiration for robotics and flying animals can be used as starting point for the development of novel unmanned micro-aerial vehicles.

Among flying animals, bats have impressive motor capabilities and orientation ability. The coordination between their flying motion and their sonar system make them great aerial hunters, able to move in unknown and complex environments. From a strictly engineering point of view, bats present interesting characteristics, such as:

- sustained flight
- sensitive wings
- echolocation in complex natural environments
- flexible finger bones.

Because of these features, the bat's structure is an inspiration in robotics, both for the development of bio-inspired artificial sonar systems [3] and for the development of lightweight flying artefacts [4].

In this work we propose a novel approach for biological and robotics research, aimed at investigating and understanding the principles that give rise to the bat sensory-motor capabilities and at incorporating them in new design approaches and technologies for building physically embodied, autonomous, lightweight flying artefacts.

The main idea is that a better understanding of these kinematic, perception and behavioural models of bats and their foundations is a key for a successful transfer to the design and development of new robotic platforms able to reproduce some of these abilities. Thus, one of the principal aims of this work is an in-depth investigation on bats kinematics, dynamic control and echolocation systems. Succeeding in these aims will open up a variety of novel opportunities for research and novel applications in robotics. On the other hand the future robotic platforms based on this work will serve for the validation of novel hypotheses and models describing abilities of bats and, thus, giving animal researchers the possibility to implement and validate behavioural models and hypotheses in a completely new way.

The paper is structured in the following way: we will start presenting the biological point of view and the possible novel models that could be developed; then the focus will move on the description of the planned work for the realization of the robotic prototype, with details on the bioinspired flight, the artificial sonar system, the bioinspired mechanisms and the sensorimotor schemes.

2   From Biology to Bio-Robotics: A Critical Review

2.1   Biological Models

The overall objective of this critical review is to open new ways for basic bat biology research, and develop novel kinematics and behavioral models that lead to new strategies to implement these models in a new robotic platform inspired by bats.

During flapping flight the wings of these animals generate an aerodynamic footprint as a time-varying vortex wake in which the rate of momentum change represents the aerodynamic force [5,6]. Each wing generates its own vortex loop and at moderate and high flight speeds, the circulation on the outer wing and the arm wing differs in sign during the upstroke, resulting in forward lift (thrust) on the hand wing. At the transition between up- and downstroke there is the shedding of a small vortex loop from outer wing that generates an upwash (i.e. negative lift), which is due to the wing motion during the stroke reversal. Small bats also generate leading edge vortices (LEV) during the downstroke that enhance the lift in addition to the quasi-steady lift [7]. These observations open for new strategies for the study of equivalent natural and engineered flying devices [5].

Novel models related to aerodynamic forces, kinematics and echolocation models could be developed based on wind tunnel studies of real bats. To achieve this aim, the Glossophaga soricina (Pallas's Long-tongued Bat), a small-sized bat commonly used in research can be used as real-bat model. This species is a South and Central American bat; it is a relatively small bat species at 10-14 g, with a short visible tail, only about a quarter of an inch (0.6 centimeters), while the head and the body length is about 4.5 to 5.9 cm long.

These models could be translated in design guidelines for the robotic bat. The analysis of the kinematics motion patterns can be synthesized with echolocation patterns in order to develop new sensorimotor control schemes. The use of fluid dynamics methods, such as Particle Image Velocimetry, high speed cameras for time resolved 3D-photogrammetry in a wind tunnel, may allow to investigate the wings' motions during the whole flapping cycle. Important features characteristic of bat flight are changing camber, angle of attack (AoA) and orientation of the leading edge flap (part of the wing membrane controlled by the "thumb" and index finger, Propatagium and Dactylopatagium brevis). The change of kinematic parameters during flight can be connected to time-resolved data of the vortex wake, in order to understand the relative contribution of kinematic parameters to the aerodynamic force. From another point of view, this paper proposes a novel way to investigate biological models. A great difficulty of doing measurements with live animals is that a parametric study is very hard or even impossible to achieve, and animal safety considerations sometimes give restrictions. The solution proposed in this paper is to use a robotic artefact that will not have these limitations and thereby could give new information about bat flight and provide the possibility to explore aerodynamic regions also outside the natural flight envelope of bats. The robotic platform that will be developed can represent a substitute for the realization of behavioural tests in which it is not ethically correct to use real bats, without compromising animal safety.

2.2   Bioinspired Flight

The last decade has seen many exciting developments in the area of Micro unmanned Aerial Vehicles (MAVs) that are between 0.1-0.5 meters in length and 0.1-0.5 kilograms in mass.

The greater part of research to develop practical MAVs can be generally categorized into three fundamental approaches. The first and most widely used is the airplane-like configuration with rigid wing and propeller-driven thrust. A second approach is using rotary wings to design helicopter-like MAVs while other researchers used the bio-inspired option of flapping wings. Among the most successful examples of rigid wing MAVs is the Black Widow [8] a 15 cm flying wing having every component custom built, including a sophisticated gyro-assisted control system.

Unfortunately, scaling down of large aircraft is accompanied by a multitude of complications due to a significant change in flow regime at small size fliers. The size of biological flyers as well as of MAVs positions them in a range of low Reynolds numbers (Re < 10,000), which induces complex flow phenomena due to viscous effects. While some researchers have endeavoured to design and test rigid and rotary wings MAVs, others have searched for a way to improve the aerodynamics of low Reynolds numbers aircraft looking at the biological fliers for inspiration, building "flapping wing" systems. A flapping wing is a wing that accomplishes, at the same time, the role of producing lift and the role of producing thrust. In other words, flight of these flying mammals is possible thanks to this integration of lift and thrust in the wing. From a merely aerodynamic point of view, the flow around a bat's wing is highly unsteady, characterized by a low Reynolds number (typical Reynolds numbers are of the order of 10^3-10^4). In this flow regime the classical potential theory of aerodynamics can be applied only with large approximation. Proposed models in literature dealing with unsteady potential theory applied to bat's flight, have been derived from industrial models already employed for sail modelling; therefore they are unable to model flow generated by the movement of the bat's wing.

One example of bio-inspired MAVs with flapping flight is the Microbat developed by Sirirak [9] (Fig. 1).

Fig. 1. Battery-powered prototype of Microbat [9]

The particular flight path and manoeuvring capabilities of bats suggest that bats are able to control wing stall in a very efficient way; in this respect, the model of the morphing wing finds a real application. All flight parameters as speed, attitude, number of flapping per second, section's wing twists, spatial position of every point of the wings, position of center of mass and of aerodynamic center shall be available at each instant in order to provide a feedback to the control system.

The shape of the wing during flapping stroke is imperative to generate lift and thrust. The wings of bats are shaped in an unusual way with respect to other flying animals. From a structural point of view, the particularities of the bat's wings are related to the skin membrane and the arm's skeleton that presents many degrees of freedom. Waldman et al. [10] developed a computational analysis for a bat-like membrane airfoil behaviour assuming inviscid potential flow and compared it with in vivo measurements of membrane wings of bats during flight. This study shows that bat wings undergo a change in their area of about 100% during flapping stroke in order to maintain a high level of performance for a variety of flight modes and flow conditions ranging from soaring to hovering flight. The wings will be designed to operate efficiently throughout a wide velocity range from hovering (0 m/s) up to a flight speed of at least 8 m/s. Along this velocity range, the wingbeat frequency ranges from about 13 Hz (hovering) to about 9 Hz (fast forward flight). Each wing will be also composed by the humerus and radius bones, digits connected to the wrist joint and a membrane modeled by a thin latex or silicon sheet, with the stiffeners constructed using carbon reinforced epoxy. Since bone stiffness varies much throughout a bat wing, an exact match in stiffness is almost impossible, and therefore the wings will be designed too stiff rather than too compliant. The dimensions of the model wing will be based on the wing geometry at mid downstroke, estimated from kinematics measurements of real bats.

Anders Hedenström and his team already designed a bat-inspired wing, RoBat [11] used to mimic flight kinematics of Leptonycteris yerbabuenae (Fig. 2) and the wings that will be used for the current application will rely heavily on this.

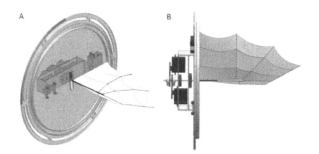

Fig. 2. The RoBat wing, drive system and mounting plate [11]

3   Conceptual Design of a Bio-Robotic Bat

3.1   Artificial Sonar System

The sonar system of the future robotic bat could be designed to provide the robot with the sensory capabilities to match the platform's ability to fly. The most important fundamental limitation of sonar is the low measurement rate due to the slow speed of sound [12]. Hence, to arrive at real-time control of complex behavior, e.g. flight, maximum information needs to be extracted from the few measurements that can be collected within the relevant response time. In most current state-of-the-art robotic sonar systems multiple objects are localised in 3D by combining echo travel times from multiple receivers and/or multiple emissions [13, 14, 15]. A recent bio-inspired robotic sonar system [3] has demonstrated 3D localisation of multiple reflecting objects based on a single binaural sonar measurement. This is made possible by also decoding the spatial cues introduced through the binaural filtering of the echoes caused by the sonar's head and outer ear shape.

All current robotic sonar systems, whether bio-inspired or not, extract spatial cues from individual pulse-echo pairs. However, prey classification experiments [16] show that modulation patterns, i.e. acoustic flow cues, extracted from echo trains reflected off fluttering insects contain target identity information. In addition to object class information, acoustic flow cues [17] also contain information about the geometric relationship between objects, the surface geometry of those objects and the motion of the objects relative to the sensor. Psychophysical experiments show that bats are indeed sensitive to such modulation patterns, e.g. intensity, time-of-flight, spectral content modulations [18]. Furthermore, the timescale of these modulation patterns (100-1000ms) fits better than the timescale of individual echoes (1-10ms) with the timescale the bat's motor system operates on. Therefore we conjecture that specific features of the extracted modulation patterns can be used quite naturally to provide motion control feedback.

Hence, we propose to investigate both how this spatial information encoded by egomotion induced modulation patterns can be extracted from echo trains as well as show how it can be used most efficiently in flight control. In the end, this will allow acoustic flow analyses of the echo trains collected during 2D and 3D movements to replace individual echo feature analyses that are the current state of the art in robot sonar. The investigation of the acoustic flow cues that result during such motions is entirely novel and has the potential to generate important new fundamental insights into biosonar.

Fig.3. (left) The sonar-forming antennae "robat", based on 3-D scans of bat ears, developed by the University of Antwerpen from the original (right) CIRCE robotic bat head

3.2   Actuators

Actuators represent the real bottleneck in many robotic applications, including the biomimetic field. Currently available actuators are mainly electromagnetic and their performance is far from the one achieved, for example, by natural muscles. Nevertheless, in the last few years new and promising technologies are emerging thus offering new possibilities to fill the gap between natural muscles and artificial artefacts. A thorough study could be performed on smart materials and in particular on shape memory alloys, a lightweight alternative to conventional actuators, that responds to the heat from an electric current. The heat actuates micro-scale wires, making them contract like 'metal muscles'. During the contraction, the powerful muscle wires also change their electric resistance, which can be easily measured, thus providing simultaneous action and sensory input. This dual functionality will help cut down on the robotic-bat's weight, and can allow the robot to respond quickly to changing conditions as perfectly as a real bat. Shape Memory Alloys (SMA) [19] are increasingly used in biomimetic robot as bioinspired actuators. There are several kind of alloys that can be used with different performances. The response time depends on the time needed to pass from martensitic to austenitic phase and consequently it depends to the current that passes through the wires, to the dimensions and to the thermal coefficient. A trade off between current and velocity has to be found, but an acceptable frequency of contraction/relaxation can be reached with a well designed geometry. SMA presents high force to mass ratio, light weight, compactness and for these reasons represents a very interesting technology in the prosthetic field [20]. Nevertheless, SMA shows several limitations, such as: difficulty in controlling the length of the fibers as they undergo the phase transition first of all due to their hysteresis; dependence of the bandwidth on heating and cooling rates; limited life cycle. Despite that, there are several examples (especially in the biomimetic field) that demonstrate the real possibility to use this technology for actuation. Literature analysis and results on first studies demonstrated that SMA wires can reach potentially high performances very close to the requirements; moreover, they are easy to manufacture, manageable and compliant. Even if nominally the performances of SMAs are not the best of the class, it is possible to improve them playing on chemical properties and geometrical parameters. One of the drawback of this technology is that the maximum force is proportional to the diameter of the SMA. In this application, high forces are needed, but the contraction/ stress cycle time highly increases with the wire's diameter. Therefore, the novel actuator design will combine many single SMA wires in a single actuator, with the advantage of the scalability of pull force and length [21]. First results have been obtained [22, 23] and they highlight the potentiality of this technology (Fig. 4). In order to reproduce the wing-morphing motion, the goal is to control the muscles for achieving morphing-trajectories based on the biological study of bat flight. As far as mechatronic design, much of the work could be devoted to speed-up the operation frequency of the SMA-based muscles (going further 2.5-3Hz) for a feasible flapping-morphing synchronization during flight. Some of the challenges involved in this process are related to the modeling and mechatronics design of bat-wing motion merged with SMA actuation, and SMA hysteresis control that ensure the morphing frequency desired.

To address these challenges, we propose an antagonistic pair of SMA arrays acting as the triceps and biceps muscles along the humerus bone. The work could be also oriented towards the synchronization of flapping and morphing control based on aerodynamics analysis using the wind tunnel testbed.

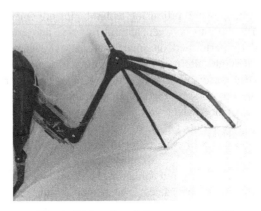

Fig. 4. Right-wing skeleton protytpe [22]

3.3   Sensors

Regarding the sensory system, the technological challenge is to create a system deeply interacting with the environment that shows performance comparable to those of real bats in terms of efficiency, accuracy, and small size. A huge number of physical sensors could interact with the environment and could both allow the integration of innovative sensors in the artificial platform and also achieve, through appropriate data processing and fusion algorithms, full flexibility and functionality while in a complex context. High resolution and high-speed sensors such as miniaturized accelerometers and inertial sensors, load cells, pressure and acoustic transducers, as well as highly accurate strain gauges could be used at this aim. Advances in Micro-Electro-Mechanical Systems (MEMS) technology have offered the opportunity to develop both single highly reliable sensors and sensor arrays with small dimensions based mainly on capacitive and piezoresistive principles. Several of them have implemented normal and shear force sensing even in small arrays on a flexible substrate for non-planar surfaces. These technologies lacked, however, mechanical compliance and conformance. One of the approaches for the implementation of tactile skins consists in combining single silicon microsensor elements with elastomeric materials, in which the key to functional tactile devices depends on the type of packaging design [24, 25]. Moreover, wing-surface airflow sensors and the appropriate feedback for the actuating mechanism could be thoroughly investigated.

3.4   Sensorimotor Schemes

The principle of sensory-motor coordination was inspired by John Dewey, who as early as 1896, had pointed out that sensation cannot be separated from action, and that perception must be considered as a whole, rather than input-processing-output. Recent work in artificial intelligence and cognitive science has demonstrated that through coordinated interaction with the environment, an agent can structure its own sensory input [26]. In this way, correlated sensory stimulation can be generated in different sensory channels -- an important prerequisite for perceptual learning and concept development.

Because physical interaction of an agent with its environment leads to structured patterns of sensory stimulation, neural processing is significantly facilitated, demonstrating the tight coupling of embodiment and cognition. Categorization -- the ability to make distinctions -- is the foundation of any form of cognition. Edelman [27] clearly mapped out how sensory-motor coordinated behaviour leads to categorization. Lakoff and Johnson [28] make a strong case that all kinds of cognition, even very abstract ones, are tightly coupled to embodiment. These psychological and philosophical arguments have been further elaborated in a robotics context using ideas from non-linear dynamics and information theory, who showed more generally how cognition can emerge from the dynamic interaction of an agent with its environment. Using case studies from locomotion and sensory-motor tasks, they also discuss how something like a body schema can serve as an intermediary between embodiment and cognition [29].

In the field of humanoid robotics, this paradigm has triggered a number of interesting research opportunities. A natural global dynamical structure emerges from natural physical interactions among multiple sensory-motor flows between the agent and its environment. This structure was thus exploited to allow for "body-schema" acquisition and self-exploration of sensory-motor learning on a humanoid robot. The formation of a body schema through sensory-motor interaction between the robot and environment was also utilized to develop learning of body versus non-body discrimination in [30] and learning about objects through visual processing and active grasping [31]. Active perception was explored in [32] where object manipulation using a humanoid's arm actively generates the sensory-motor experience needed to simplify the object segmentation and learning tasks of the robot's visual system. Sensory-motor coordination strategies were also utilized to develop reaching and grasping capabilities [33].

The real-time perception of the robot environment with an adequate balance between movement and sensory capabilities is crucial to ensure a smooth interaction [34]. At this aim, the paper suggests a strict and massive integration between the sensors network, the sonar apparatus and the actuators control system. The final behavioral control strategy must utilize all sensors feedback and the information about the environment from the sonar system to assure the stable and coordinated motion of the robot. First of all, an analysis of basic kinematics and kinetic trajectories of the bat's body and wings should be performed. Based on the experimental data, novel forward and inverse kinematics and dynamics models of the biomechanics of the bat can be developed and they should mimic the principal bat motions: wing flapping, wing pitching, leading edge flapping. Proprioceptive feedback control loops should be investigated as the basis for motion control and stability: the measure of the SMA's electrical resistance and of the wing-surface airflow sensors must be taken into account as a parameter to adjust wings motion. This may constitute the inner layer of the overall control structure, based on a hierarchical control, while the echolocation system may constitute the outer layer of the autonomous control system.

4   Conclusion

This paper presents a novel approach for the investigation of particular aspects of bat biology and morphology. The approach consists in the implementation of a complex environment for the measurements of biological parameters of animals and the design and realization of bat-inspired robotic prototypes, endowed with an artificial echolocation system, flapping wings and an autonomous control.

We expect that a bat-robot so structured can have a really wide impact: indeed, it can open new ways in robotics, improving the performance of robots in terms of locomotion and adaptability, with the investigations on novel bioinspired control strategies and artificial echolocation system and in a long-term vision, robotic devices can be seen as substitutes for real animals in behavioural experiments.

References

1. Webb, B., Consi, T.: Biorobotics: Methods and Applications. MIT Press (2001)
2. Ijspeert, A., Crespi, A., Ryczko, D., Cabelgruen, J.M.: From swimming to walking with a salamander robot driven by a spinal cord model. Science 315, 1416–1420 (2007)
3. Schillebeeckx, F., de Mey, F., Vanderelst, D., Peremans, H.: Biomimetic sonar: Binaural 3D localization using artificial bat pinnae. The International Journal of Robotics Research 30(8), 975–987 (2011)
4. Chung, S.J., Dorothy, M.: Neurobiologically Inspired Control of Engineered Flapping Flight. Journal of Guidance, Control, and Dynamics 33(2), 440–453 (2010), doi:10.2514/1.45311
5. Hedenström, A., Johansson, L.C., Wolf, M., von Busse, R., Winter, Y., Spedding, G.R.: Bat flight generates complex aerodynamic tracks. Science 316, 894–897 (2007)
6. Hubel, T.Y., Riskin, D.K., Swartz, S.M., Breuer, K.S.: Wake structure and wing kinematics: the flight of the lesser dog-faced fruit bat, Cynopterus brachyotis. J. Exp. Biol. 213, 3427–3440 (2010)
7. Muijres, F.T., et al.: Leading-edge vortex improves lift in slow-flying bats. Science 319(5867), 1250–1253 (2008)
8. Grasmeyer, J.M., Keennon, M.T.: Development of the black widow micro air vehicle. Progress in Astronautics and Aeronautics 195, 519–535 (2001)
9. Pornsin-Sirirak, T.N., et al.: MEMS wing technology for a battery-powered ornithopter. In: The Thirteenth Annual International Conference on Micro Electro Mechanical Systems, MEMS 2000. IEEE (2000)
10. Waldman, R.M., et al.: Aerodynamic behavior of compliant membranes as related to bat flight. In: 38th AIAA Fluid Dynamics Conference and Exhibit Seattle, Washington, USA (2008)
11. Koekkoek, G., et al.: Stroke plane angle controls leading edge vortex in a bat-inspired flapper. Comptes Rendus Mecanique (2012)
12. Peremans, H., de Mey, F., Schillebeeckx, F.: Man-made versus biological in-air sonar systems. In: Barth, F., et al. (eds.) Frontiers in Sensing: From Biology to Engineering. Springer, Berlin (2012) ISBN 978-3-211-99748-2

13. Chong, K., Kleeman, L.: Feature-based mapping in real, large scale environments using an ultrasonic array. The International Journal of Robotics Research 18(1), 3 (1999)
14. Jimenez, J., Mazo, M., Urena, J.: Using PCA in time-of-flight vectors for reflector recognition and 3-D localization. IEEE Transactions on Robotics 21(5), 909–924 (2005)
15. Krammer, P., Schweinzer, H.: Localization of object edges in arbitrary spatial positions based on ultrasonic data. IEEE Sensors Journal 6(1), 203–210 (2006)
16. Fontaine, B., Peremans, H.: Compressive sensing: A strategy for fluttering target discrimination employed by bats emitting broadband calls. The Journal of the Acoustical Society of America 129(2), 1100–1110 (2011)
17. Jenison, R.: On acoustic information for motion. Ecological Psychology 9(2), 131–151 (1997)
18. Genzel, D., Wiegrebe, L.: Time-variant spectral peak and notch detection in echolocation-call sequences in bats. J. Exp. Biol. 211, 9–14 (2008)
19. Funakubo, H.: Shape memory alloys. Gordon and Breach Science Publishers (1987)
20. Bundhoo, V., Park, E.J.: Design of an artificial muscle actuated finger toward biomimetic prosthetic hands. In: Proceedings International Conference on Advanced Robotics, pp. 368–375 (2005)
21. Kratz, R., Stelzer, M., Friedmann, M., von Stryk, O.: Control approach for a novel high power-to-weight ratio SMA muscle scalable in force and length. In: IEEE/ASME Intl. Conf. on Advanced Intelligent Mechatronics (AIM), September 4-7 (2007) (to appear)
22. Colorado, J., Barrientos, A., Rossi, C.: Biomechanics of morphing wings in a Bat-robot actuated by SMA muscles. In: Proceedings of the International Workshop on bio-inspired robots, Nantes, France, April 6-8 (2011)
23. Colorado, J., Barrientos, A., Rossi, C., Breuer, K.: Biomechanics of smart wings in a bat robot: Morphing wings using SMA actuators. Bioinspiration & Biomimetics (2012)
24. Muhammad, H.B., Oddo, C.M., Beccai, L., Recchiuto, C., Anthony, C.J., Adams, M.J., Carrozza, M.C., Hukins, D.W.L., Ward, M.C.L.: Development of a bioinspired MEMS based capacitive tactile sensor for a robotic finger. Sensors and Actuators A: Physical
25. Oddo, C.M., Beccai, L., Muscolo, G.G., Carrozza, M.C.: A Biomimetic MEMS-based Tactile Sensor Array with Fingerprints integrated in a Robotic Fingertip for Artificial Roughness Encoding. In: Proc. IEEE 2009 International Conference on Robotics and Biomimetics (Robio 2009), Guilin, Guangxi, China, December 19-23 (2009)
26. Pfeifer, R., Bongard, J.: How the body shapes the way we think: a new view of intelligence. MIT Press (2007) ISBN 0-262-16239-3
27. Edelman, G.: Neural Darwinism. Basic Books, New York (1987)
28. Lakoff, G., Johnson, M.: Philosophy in the Flesh. Basic Books, New York (1999)
29. Pfeifer, R., Iida, F., Gomez, G.: Morphological computation for adaptive behaviour and cognition. International Congress Series 1291, 22–29 (2006)
30. Yoshikawa, Y., Tsuji, Y., Hosoda, K., Asada, M.: Is it my body? Body extraction from un-interpreted sensory data based on the invariance of multiple sensory attributes. In: Proceedings of the IEEE/RSJ International Conference on Intelligent Robots and Systems, pp. 2325–2330. IEEE Press, Piscataway (2004)
31. Natale, L., Orabona, F., Berton, F., Metta, G., Sandini, G.: From sensorimotor development to object perception. In: International Conference of Humanoids Robotics (2005)
32. Metta, G., Fitzpatrick, P.: Early integration of vision and manipulation. Adaptive Behavior 11(2), 109–128 (2003)
33. Lopes, M., Santos-Victor, J.: Learning sensory-motor maps for redundant robots. In: Proc. IEEE/RSJ International Conference on Intelligent Robots and Systems, IROS (2006)
34. Muscolo, G.G., Hashimoto, K., Takanishi, A., Dario, P.: A Comparison between Two Force-Position Controllers with Gravity Compensation Simulated on a Humanoid Arm. Journal of Robotics 2013, Article ID 256364, 14 pages (2013)

Evolutionary Coordination System for Fixed-Wing Communications Unmanned Aerial Vehicles

Alexandros Giagkos[1], Elio Tuci[1], Myra S. Wilson[1], and
Philip B. Charlesworth[2]

[1] Department of Computer Science, Llandinam Building,
Aberystwyth University, Aberystwyth, Ceredigion, SY23 3DB, UK
{alg25,elt7,mxw}@aber.ac.uk
[2] Technology Coordination Centre 4, Airbus Group Innovations
Celtic Springs, Newport, NP10 8FZ, UK
phil.charlesworth@eads.com

Abstract. A system to coordinate the movement of a group of unmanned aerial vehicles that provide a network backbone over mobile ground-based vehicles with communication needs is presented. Using evolutionary algorithms, the system evolves flying manoeuvres that position the aerial vehicles by fulfilling two key requirements; i) they maximise net coverage and ii) they minimise the power consumption. Experimental results show that the proposed coordination system is able to offer a desirable level of adaptability with respect to the objectives set, providing useful feedback for future research directions.

Keywords: Evolutionary algorithms, unmanned aerial vehicles, coordination strategies.

1 Introduction

This paper presents an initial investigation into the coordination of multiple unmanned aerial vehicles (UAVs) that provide network coverage for multiple independent ground-based vehicles, when imperfect connectivity is experienced. Imperfect communication can be due to the mobility of ground-based vehicles thus leading to long distances between them and the aerial vehicles, limited radio frequency power, and other communication failures. Under these conditions, defining flying strategies able to react to topological changes and ensure relaying of data between ground-based vehicles is a complex problem. Fig. 1 is an illustration of a scenario of a group of 3 aerial vehicles that provide network coverage to a number of ground-based vehicles.

Rapid, unexpected changes to the topology require a coordination system with a ligh level of adaptability. It has to be able to generate flying manoeuvres and formations according to the movement patterns of the ground-based vehicles and their communication needs. Power consumption plays a key role in the success of such a demanding mission. The aim of this research work is to design a decision unit that generates flying manoeuvres that offer network coverage to support as

M. Mistry et al. (Eds.): TAROS 2014, LNAI 8717, pp. 48–59, 2014.

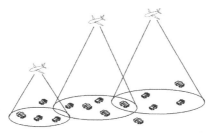

Fig. 1. Communication links are provided to ground-based vehicles. The overlap coverage is found at intersections.

many ground-based vehicles as possible, while the power management follows a reasonable trend. Two objectives are identified. Firstly, to maximise the net coverage by decreasing the overlaps between two or more aerial vehicles. Secondly, as the power consumption is related to the distance between the antennae of the transmitter and the receiver, it is important to control the vertical flying of the aerial vehicles such that the slant distances between them and the ground-based vehicles are minimised. The proposed system employs evolutionary algorithms (EAs) [2] as the adaptable decision unit. The results of a first series of simulation experiments are presented in this paper, illustrating the effectiveness of EAs in solving the problem by relocating the groups of aerial vehicles. The outcome of the study indicates that EAs are an efficient way of coordination that fulfils the two research objectives.

The rest of the paper is structured as follows. Examples of related work are presented in section 1.1, followed by a brief discussion of the aerial vehicle models in terms of performing feasible manoeuvres and communication link budget in section 2 and 3, respectively. Experimental results and their discussion is given in section 4, and in section 5 the paper concludes by addressing the future research directions.

1.1 Background

Researchers have explored the possibility of using evolutionary computation to solve path planning and coordination problems for single or groups of aerial vehicles. In [12], a flyable path plan for multiple UAV systems using genetic algorithms (GAs) is presented, suitable for controlling a specific mission area with vehicles flying at a constant altitude via a number of control points. The Bézier curves technique is used to smooth flying trajectories, resulting in a flight that allows each UAV to move very close to the control points.

[1] considers the coordination of multiple permanently flying (in small circles) UAVs in order to support an ad hoc communication network over multiple ground-based users. The authors propose an EA-based solution that allows each UAV agent to adjust its output power levels and antenna orientation, such that the download effectiveness to the end users will be maximised. The ground area covered by each UAV is determined by the gain of the vehicle's antenna.

Low boresight gain allows a wider area to be covered but with a lower signal, whereas a high gain antenna transmits with higher signal power in the centre of line-of-sight target, but covers a smaller area.

In [8], the authors employ a GA-based approach for a UAV path planning problem within dynamic environments. The authors define a good solution as the path that optimises three components (distance, obstacles, and path length). The genetic representation consists of a series of manoeuvres that are planned according to a maximum turn rate as well as an acceleration/deceleration maximum value, corresponding to the UAV flight.

The work described in [11] proposes the use of B-Spline curves [7] as the way to represent the trajectory of a UAV flight. Generally, the continuous curve of a B-Spline is defined by control points, which delimit the smoothly joined B-Spline curve's segments. The authors argue that unlike Bézier curves, B-Spline curves are more suitable to represent a feasible UAV route, as an update in one of the control points changes only its adjacent segments due to its local propagation.

In [3], the authors investigate a search method for multi-UAV missions related to surveillance and searching unknown areas, using EAs. Their work allows several UAVs to dynamically fly throughout a search space and autonomously navigate by avoiding unforeseen obstacles, without a priori knowledge of the environment. Although the proposal assumes central administration and user control from take-off time to the end of the mission, the authors employ an EA-based approach to generate appropriate coordinates for UAV relocation.

After studying the evolutionary path planner for single UAVs in realistic environments, [5] propose a solution to the coordination of multiple UAV flying simultaneously, while minimising the costs of global cooperative objectives. As long as the UAVs are able to exchange some information during their evaluation step, the proposed system is able to provide off-line as well as on-line solutions, global and local respectively.

2   The UAV Kinematics and Communication Models

The main methodological aspects with respect to the kinematics of the aerial vehicles and the link budget that characterises the communication links, are briefly described in this section. A more detailed description of the methods used for this study can be found in [9].

In the simulation model, an aerial vehicle is treated as a point object in the three-dimensional space with an associated direction vector. At each time step, the position of an aerial vehicle is defined by a latitude, longitude, altitude and heading $(\phi_c, \lambda_c, h_c, \theta_c)$ in the geographic coordination system. A fixed-wing aerial vehicle flies according to a 6DOF model with several restrictions, ranging from weight and drag forces to atmospheric phenomena, that affect its motion. However, as this work focuses on the adaptive coordination with respect to the communication network, a simplified kinematics model based on simple turns is considered for the restrictions of both horizontal and vertical motions [4]. For security and safety reasons, the model is designed to allow flight within a predefined corridor, such that the model denies altitude additions or subtractions in cases where the maximum or minimum permitted altitude is reached. Notice that no collision avoidance mechanism is currently employed.

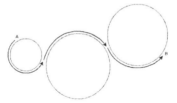

Fig. 2. A manoeuvre of 3 segments of different durations and bank angles, from a starting point A and the finishing point B. Direction of flying is dictated by the bank angle.

As the flying vehicle is fixed-wing, it may either perform a turn circle manoeuvre with a tight bank angle in order to keep its current position, or implement a manoeuvre generated by the EA decision unit. Taking inspiration from the Dubins curves and paths [6], when implemented, a manoeuvre will generate a trajectory consisting of 3 segments, as depicted in Fig. 2. Each segment can be a straight line, turn left or turn right curve, depending on the given bank angle.

The EA is free to select the duration for any of the segments as long as the overall duration remains equal to the time of one revolution circle manoeuvre. This strategy ensures synchronisation between all aerial vehicles within the group. With a bank angle of 75 degrees and a constant speed of 110 knots, The time for one revolution is approximately 6 seconds. The aerial vehicles perform 2 turn circle manoeuvres before they are allowed to implement the latest generated solution from the EA. This time window ensures that the artificial evolution will have reached a result, while at the same time the aerial vehicles will fly in a controlled and synchronised way, keeping their previous formation. Furthermore, the time window ensures that the aerial vehicles will have enough time to exchange fresh GPS data and ultimately communicate the resulting solution on time.

Networking is achieved by maintaining communication links between the aerial backbone and as many ground-based vehicles as possible. The communication links are treated independently and a transmission is considered successful when the transmitter is able to feed its antenna with enough power, such that it satisfies the desirable quality requirements. It is assumed that aerial vehicles are equipped with two radio antennae. One isotropic able to transmit in all directions, and a horn-shaped one able to directionally cover an area on the ground. It is also assumed that all vehicles are equipped with a GPS and can broadcast information about their current position at a reasonable interval (default 3 seconds). In this section, focus is primarily given to the communication between aerial vehicles and ground-based vehicles using the former horn-shaped antennae, as it dictates the effectiveness of the communication coverage of the mission and the power consumption of a flying mission.

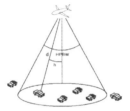

Fig. 3. Slant distance d and angle h of a communication link

In order for a ground-based vehicle to be covered, it needs to lie within the footprint of at least one aerial vehicle. As shown in Fig. 3, a footprint is determined by the altitude of the aerial vehicle as well as its antenna's half-power beamwidth (HPBW) angle. The higher the aerial vehicle flies, the wider its footprint is on the ground, the greater the area covered. The slant angle h of the ground-based vehicle with respect to the aerial vehicle is calculated by applying spherical trigonometry on the available GPS data that each network participant broadcasts. The following piecewise function is then used to decide whether a ground-based vehicle lies within the footprint (coverage profile p).

$$L(p) = \begin{cases} 1, & h < \frac{HPBW}{2} \\ 0, & h \geq \frac{HPBW}{2} \end{cases} \tag{1}$$

Similarly, the slant distance d can be calculated. The greater the distance between the transmitter and the receiver, the higher the signal power required to support the communication.

3 The Evolutionary Algorithm and Fitness Function

A centralized, on-line, EA-based approach is considered for the coordination of the group of aerial vehicles. The decision making for the next set of manoeuvres for the group is made by a single aerial vehicle, nominated as master. Taking advantage of the underlying network, it is assumed that every 3 seconds the master aerial vehicle is able to receive messages carrying the last known positions and direction vectors of the group as well as the ground-based vehicles. Data updates may be received from relaying aerial vehicles and directly from the ground-based vehicles within the master's footprint, and are tagged such that the master aerial vehicle and in turn the EA decision unit is fed with up-to-date knowledge of the topology. Once the EA has evolved a new set of manoeuvres, the master aerial vehicle is responsible for broadcasting the solutions to the whole group, using the network. As this work mainly focuses on providing network coverage to ground customers, it is assumed that there is no packet loss and that a dynamic routing protocol allows flawless data relaying within the topology. The process flow of receiving, generating and distributing solutions amongst the group members is depicted in Fig. 4. Notice that the EA runs independently from the controller of the master aerial vehicle (threaded), which in practice allows the master aerial vehicle to complete its turn circle manoeuvre, while waiting for a solution.

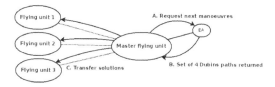

Fig. 4. At the completion of a turn circle manoeuvre, the master aerial vehicle queries the EA for the next set of manoeuvres for the whole group. The solution is then communicated to the rest of the group using the network. If the EA is not ready to generate a solution due to lack of up-to-date information, the returned solution is a set of turn circle manoeuvres, forcing the group to maintain its current position.

As described in section 2, a flying manoeuvre is described by a Dubins path of 3 segments. Each segment comprises a bank angle and the duration for which the segment's manoeuvre is to be performed. Furthermore, a Dubins path may request a change to the vertical plane, thus require an alteration to the current aerial vehicle altitude. The information is stored to the chromosome's genes, as shown below.

$$\boxed{\beta_1}\boxed{\delta t_1}\boxed{\beta_2}\boxed{\delta t_2}\boxed{\beta_3}\boxed{\delta t_3}\boxed{b}\boxed{\delta h}$$

The first six genes describe the horizontal motion and the duration of each of the 3 segments of the Dubins path and are stored as floating point values. The seventh gene b, as well as the last Δh, control the vertical behaviour of the aerial vehicle. When the former is set to 0, the aerial vehicle flies at the same altitude (level flight). If it is set to 1, then the vertical motion is considered and the aerial vehicle is expected to change its altitude by Δh within the duration of the Dubins path, $\sum_{i=1}^{3}(\delta t_i)$.

An evolutionary algorithm using linear ranking is employed to set the parameters of the paths [10]. We consider populations composed of $M = 100$ teams, each consisting of $N = 4$ individuals (the number of aerial vehicles in the flying group). At generation 0 each of the M teams is formed by generating N random chromosomes. For each new generation following the first one, the chromosomes of the best team ("the elite") are retained unchanged and copied to the new population. Each of the chromosomes of the other teams is formed by first selecting two old teams using roulette wheel selection. Then, two chromosomes, each randomly selected among the members of the selected teams are recombined with a probability of 0.3 to reproduce one new chromosome. The resulting new chromosome is mutated with a probability of 0.05. This process is repeated to form $M - 1$ new teams of N chromosomes each.

The fitness function f is designed to optimise two objectives. Firstly, it maximises the net coverage by minimising overlap (i.e., the footprints' intersections).

This is in favour of supporting as many ground-based vehicles as possible using the available number of aerial vehicles. Secondly, it minimises the average altitude of the group. Reducing altitude also reduces the slant distances (marked as d in Fig. 3) between the supporting aerial vehicle and the supported ground-based vehicles which in turn lowers the power consumption. The fitness f is used to compute the performance of a group, thus the fitness score of a set of flying manoeuvres, and is expressed as:

$$f = \frac{C_{net} - C_{overlap}}{G} \times \left(1 - norm\left(\frac{\sum_{i=0}^{U}(h_i)}{U}\right)\right) \qquad (2)$$

where C_{net} and $C_{overlap}$ are the net and overlap coverage scores respectively, U is the number of aerial vehicles or genomes per group, G the number of ground-based vehicles, and h_i the resulting altitude of the ith aerial vehicle. Finally, $norm$ returns the normalised mean of the altitudes within the permitted flying corridor, and is a value between 0 and 1.

In order to measure power consumption, an abstract network traffic model is implemented. As previously stated, it is assumed that the communication link between an aerial vehicle and a supported ground-based vehicle is successful when the former is able to feed its antenna with enough power to satisfy the desired link quality. That is, at each time step the transmission of 3 UDP datagrams from each aerial vehicle down to any ground-based vehicle it currently covers is simulated, using a downline data rate of 2Mbit/s and frequency of 5GHz to finally ensure E_b/N_0 of 10db. Ultimately, the number the available communication links that can be accessed depends on the position of the aerial backbone and the number of ground-based vehicles being currently covered. In this way, the power consumption per time step is measured.

4   Experiments and Results

The experiments presented in this paper are simulation-based and target aerial missions where both aerial vehicles and ground-based vehicles are placed randomly around a centre point of pre-defined latitude and longitude. All ground-based vehicles move according to a biased Random Way Point model (see [9] for details). Biases in the movements are introduced according to the following strategy. Each time a ground-based vehicle has to generate a new random bearing, there is a 75% chance of moving towards a bounded range of angles. Hence, although individual ground-based vehicles travel randomly in different directions, as a cluster they move in a biased fashion towards a random direction, which is selected in every interval. For the experiments presented in this paper, this interval is set to the duration of the simulation divided by 3. All aerial vehicles start with an initial available power of 250 Watts and fly within the flying corridor of altitudes between 1500 and 22000 ft. An angle of 75° is defined as a turn circle manoeuvre bank angle. Simulations last for 1800 seconds,

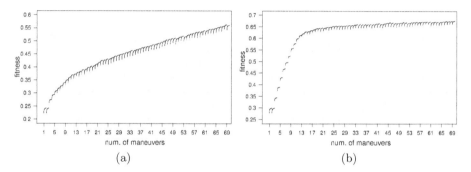

(a)   (b)

Fig. 5. Fitness of the best solutions after 200 generations for 69 generated consequence sets of flying manoeuvres are shown, for (a) *Exp A* and (b) *Exp B*. Each segment represents the averaged fitness scores (over 20 simulation experiments) and its growth from the first to the last generation.

enough time for the EA to produce a number of manoeuvres that lead to formations of interesting behaviours.

Two sets of simulation experiments targeted to slightly different scenarios designed to illustrate the adaptability of the system are presented. In *Exp A*, all aerial vehicles start by flying at low altitude and support a small number of ground-based vehicles. In this scenario, the aerial vehicles are expected to increase their altitude in order to subsequently increase the net coverage. In *Exp B*, all aerial vehicles start by flying at high altitude and support a large number of ground-based vehicles with increased overlaps between the aerial vehicles' footprints. The system is expected to push the aerial vehicles to decrease their altitude, saving power and reducing the number of ground-based vehicles that lie within the overlapping footprint areas.

Table 1 summarises the configuration parameters of the experiments in terms of initial positions and manoeuvres. For each scenario, 20 differently seeded simulations experiments were conducted. Fig. 5 summarises the dynamics observed in both scenarios from an evolutionary algorithm point of view. In particular, the figures depict the fitness trend for each set of flying manoeuvres over 200 generations. Each segment represents the average fitness of the best solution (i.e., the set of manoeuvres), averaged over 20 simulation runs, for *Exp A* (Fig. 5a), and *Exp B* (Fig. 5b) respectively. For each scenario, the aerial vehicles are asked to execute a sequence of 69 flying manoeuvres, generated by the EA decision

Table 1. Initial configuration of the two experiments

Parameter	Exp A		Exp B	
	Aerial	Ground	Aerial	Ground
No. of units:	4	300	4	300
Radius:	6 km	6 km	3 km	3km
Altitude:	2000-2300 ft	0 ft	6000-6500 ft	0 ft
Heading:	80°	0-360°	80°	0-360°
Speed:	110 kts	5-20 mph	110 kts	5-20 mph

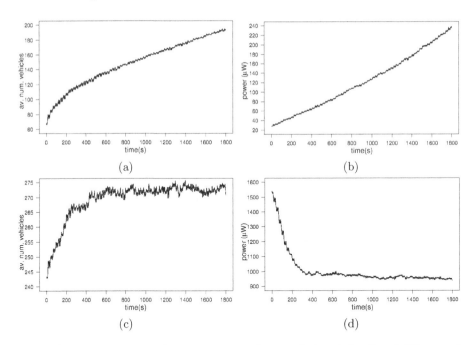

Fig. 6. Average coverage and power consumption are depicted, for *Exp A* (Fig. a and b) and *Exp B* (Fig. c and d) respectively. Coverage is expressed in terms of number of ground-based vehicles that are covered by at least one aerial vehicle.

unit. The starting point of each segment refers to the average fitness of the best solution at generation 1. The end point of each segment refers to the average fitness of the best solution at generation 200.

Looking at the segments in both figures, it can be clearly shown that at each new solution, the EA tends to progressively find a better set of manoeuvres. This indicates that the decision unit is able to efficiently relocate the aerial vehicles in order to find good positions with respect to the current status of the system (i.e., the distribution of ground-based vehicles, the current coverage and the current relative positions of the aerial vehicles). Notice that the initial score of each segment is always lower than the final point of the previous. This is reasoned due to the fact that by the time the aerial vehicles reached the desired position, the ground-based vehicles have already changed theirs, unexpectedly altering the dynamics of the system. This leads to the conclusion that the previously best solution yields less fitness than that predicted by the EA. However, it is shown that the EA tends to increase the fitness score for the sequence of solutions and thus is shown to progressively return better results during the cruise flight of the aerial vehicles. This important observation is made when looking at the general trends in both figures, as they reason about the strategies that aerial vehicles exploit to perform the task. In *Exp A*, the best solution at the end of the 200 generations for each consequence solution (i.e., set of manoeuvres) is progressively better than the previous best (Fig. 5a).

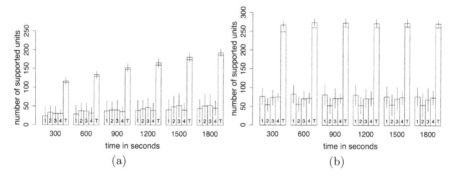

Fig. 7. Individual and total coverage as a function of time for *Exp A* (a) and *Exp B* (b). The horizontal segment line in the total bars shows the overlapping number of ground-based vehicles. Error bars show the standard deviation from the means over 20 runs.

In Fig. 6a and 6b, where the average coverage and average power consumption are depicted, it is seen that the increase in coverage indicates that the aerial vehicles gradually build better formations and keep them to fulfil the two objectives. Initially, when starting from relatively low altitudes (*Exp A*) the group of aerial vehicles tends to fly higher in order to support more ground-based vehicles (Fig. 6a). As expected, better coverage comes with a small increase in power consumption (Fig. 6b). After starting from high altitudes (*Exp B*) the EA returns solutions that tend to fly lower (Fig. 6c) with a consequent decrease in the power required to provide the communication links (Fig. 6d). The behaviour of minimising the altitude seems to last for the first 20 solutions (Fig. 5b), after which the group of aerial vehicles has reached optimal flying altitudes and formation that offer the possibility of a good coverage to power consumption trade-off. Furthermore, the solutions after the first 20 tend to optimise only the relative positions of aerial vehicles, in order to track the movement of the ground-based vehicles and minimise overlap. This seems to have a minor effect on the power consumption which tends to be constant after the first 400s of simulation time (Fig. 6d).

Network coverage management is clearly shown in Fig. 7a and 7b, for *Exp A* and *Exp B* respectively. The results complement the previous observations as they depict an increment to the group's net coverage (marked as "T"). It is seen that the latter is rather insensitive to the rapid changes of the flying formation. Particularly in *Exp B* the overlap coverage is found to decrease even though the aerial vehicles are asked to decrease their altitude.

Finally, Fig. 8 shows the trajectories of aerial vehicles and ground-based vehicles in terms of their average positions. The centroids of the ground-based vehicles cluster as well as the one of the group of aerial vehicles are shown to follow a similar trend and move closely, highlighting the success of the algorithm in adapting according to the movement and needs of the ground-based vehicles. This figure refers to a single experimental run in *Exp A*. The ground-based vehicles change their overall direction twice during the simulation, as defined by their biased random waypoint mobility model described in previous sections.

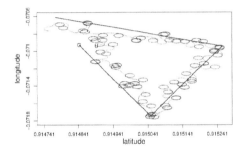

Fig. 8. Continuous line represents the trajectory of the centroid of the ground-based vehicles (as estimated by the whole cluster), whereas the dotted line refers to the movement of the centroid of the airborne group. Letters "G" and "U" indicate the starting points of the two centroids, ground-based vehicles and aerial vehicles respectively.

5   Conclusion

This paper proposes a coordination system for a group of aerial vehicles designed to form a communication network backbone that connects ground-based vehicles in an ad hoc, highly dynamic fashion. An EA is used as the decision unit responsible for generating relocation information for the group. The EA's fitness function aims to optimise two objectives. Namely, to maximise the net coverage by reducing the overlap between the footprints of each aerial vehicle, whilst minimising the average power consumption to support the communication network. Altitude is strongly related to the power required to provide a link of a satisfactory quality. First developments and experimental results are reported, illustrating promising adaptive behaviour.

There are two future directions that are considered for this research work. Due to the strong dependence on the underlying network, a fully decentralised approach is required. Although several communication and decision protocols may be added to the system to reduce the risk of losing the master aerial vehicle (use of cluster heads, token, etc.), the system should be able to minimise inter-aerial vehicle control overhead communication and thus maximise utilisation of the links for data payload transmissions.

In addition, the use of multi-objective evolutionary algorithm (MOEA) in the system is undoubtedly expected to increase the efficiency of the group evaluation within the EA, by allowing multiple objectives to be clearly defined and optimised simultaneously based on the desired flying strategies subject to mission-related, group-based, and individual-based constraints. The system will then be thoroughly examined under challenging experiments, and it will be compared with existing non-GA-based approach such as [4].

Acknowledgements. This work is funded by EADS Foundation Wales.

References

1. Agogino, A., HolmesParker, C., Tumer, K.: Evolving large scale uav communication system. In: Proceedings of the Fourteenth International Conference on Genetic and Evolutionary Computation Conference, GECCO 2012, pp. 1023–1030. ACM, New York (2012)
2. Bäck, T.: Evolutionary Algorithms in Theory and Practice: Evolution Strategies, Evolutionary Programming, Genetic Algorithms. Oxford University Press, Oxford (1996)
3. Carruthers, B., McGookin, E.W., Murray-Smith, D.J.: Adaptive evolutionary search algorithm with obstacle avoidance for multiple uavs. In: Zítek, P. (ed.) Proc. 16th IFAC World Congress (2005)
4. Charlesworth, P.B.: Simulating missions of a uav with a communications payload. In: 2013 UKSim 15th International Conference on Computer Modelling and Simulation (UKSim), pp. 650–655 (April 2013)
5. de la Cruz, J.M., Besada-Portas, E., Torre-Cubillo, L., Andres-Toro, B., Lopez-Orozco, J.A.: Evolutionary path planner for uavs in realistic environments. In: Proceedings of the 10th Annual Conference on Genetic and Evolutionary Computation, GECCO 2008, pp. 1477–1484. ACM (2008)
6. Dubins, L.E.: On plane curves with curvature. Pacific Journal of Mathematics 11(2), 471–481 (1961)
7. Farin, G.: Curves and Surfaces for CAGD: A Practical Guide, 5th edn. Morgan Kaufmann Publishers Inc., San Francisco (2002)
8. Gao, X.-G., Fu, X.-W., Chen, D.-Q.: A genetic-algorithm-based approach to uav path planning problem. In: Proceedings of the 5th WSEAS International Conference on Simulation, Modelling and Optimization, SMO 2005, pp. 523–527. World Scientific and Engineering Academy and Society (WSEAS), Stevens Point (2005)
9. Giagkos, A., Tuci, E., Wilson, M.S., Charlesworth, P.B.: Evolutionary coordination system for fixed-wing communications unmanned aerial vehicles: supplementary online materials (April 2014),
   http://www.aber.ac.uk/en/cs/research/ir/projects/nevocab
10. Goldberg, D.E.: Genetic Algorithms in Search, Optimization and Machine Learning. Addison-Wesley, Reading (1989)
11. Hasircioglu, I., Topcuoglu, H.R., Ermis, M.: 3d path planning for the navigation of unmanned aerial vehicles by using evolutionary algorithms. In: Proceedings of the 10th Annual Conference on Genetic and Evolutionary Computation, GECCO 2008, pp. 1499–1506. ACM, New York (2008)
12. Sahingoz, O.K.: Flyable path planning for a multi-uav system with genetic algorithms and bézier curves. In: 2013 International Conference on Unmanned Aircraft Systems (ICUAS), pp. 41–48 (2013)

Multi-agent Environment Exploration with AR.Drones

Richard Williams, Boris Konev, and Frans Coenen

Department of Computer Science, University of Liverpool, Liverpool L69 7ZF
{R.M.Williams1,Konev,Coenen}@liv.ac.uk

Abstract. This paper describes work on a framework for multi-agent research using low cost Micro Aerial Vehicles (MAV's). In the past this type of research has required significant investment for both the vehicles themselves and the infrastructure necessary to safely conduct experiments. We present an alternative solution using a robust, low cost, off the shelf platform. We demonstrate the capabilities of our system via two typical multi-robot tasks: obstacle avoidance and exploration. Developing multi-agent applications safely and quickly can be difficult using hardware alone, to address this we also present a multi-quadcopter simulation based around the Gazebo 3D simulator.

1 Introduction

Aerial robotics is a very exciting research field with many applications including exploration, aerial transportation, construction and surveillance. Multi-rotors, particularly quadcopters, have become popular due to their stability and maneuverability, making it easier to navigate in complex environments. Their omnidirectional flying capabilities allow for simplified approaches to coordinated pathfinding, obstacle avoidance, and other group movements, which makes multi-rotors an ideal platform for multi-robot and multi-agent research. One major difficultly in multi-rotor research is the significant investment of both time and capital required to setup a safe, reliable framework with which to conduct research. This discourages researchers from conducting any practical experimentation [9] which leads to a knowledge gap.

This work attempts to bridge that gap and encourage more practical research by providing a framework based on a low cost platform that does not require dedicated infrastructure or expensive platforms to conduct safe, effective experiments. For this work the Parrot AR.Drone[1] quad-rotor, a relatively low cost commercial toy developed for augmented reality games, was chosen as the experimental platform. We develop a two-tiered software architecture for facilitating multi-agent research with AR.Drone. At the lower level of our architecture we provide the basic robotics tasks of agent localisation, position control and obstacle avoidance. To do so, we extend the PTAM [2] key-frame-based monocular SLAM system to provide localisation and mapping functions for multiple agents.

---

[1] http://ardrone2.parrot.com/

M. Mistry et al. (Eds.): TAROS 2014, LNAI 8717, pp. 60–71, 2014.

An Extended Kalman Filter (EKF) is used for state estimation and a PID controller provides position control and path following. At the higher level we implement collision avoidance based on the Optimal Reciproal Collision Obstacle (ORCA) approach [18] and a multi-agent approach to autonomous flying where each quadcopter agent communicates with it's peers to achieve goals such as exploring an environment. The development and practical evaluation of meaningful higher level multi-agent procedures requires extended experimentation, which is hindered by the short flying time of the AR.Drone. To address this issue, we also develop a simulation environment based on the Gazebo [11] 3D simulator. We conduct both simulated and real world experiments to demonstrate the use of our framework in a multi-agent exploration scenario and validate the veracity of the simulation.

The remainder of the paper is organised as follows. In Section 2 we discuss other approaches to using AR.Drone in single- or multi-robot research. In Section 3 we present our framework in detail. Section 4 introduces the multi-agent exploration scenario and the results and conclusion are discussed in Sections 5 and 6 respectively.

2   Related Work

While the low cost and highly robust construction make AR.Drone an ideal platform for academic research, it is primarily a toy meant to be under human control at all times, whose built-in functionality lacks the precision required for the applications mentioned above. For example, while the AR.Drone features optical flow based horizontal velocity estimation, which can be integrated for position estimation (dead reckoning), it is subject to drift. This is unsurprising as the inclusion of an optical flow sensor on the AR.Drone is mainly for position hold rather than position estimation. Therefore a more robust drift-free position estimation solution is required.

We are aware of three distinct solutions, which have been successfully tested on the AR.Drone: using external high precision positioning devices, using fiducial markers and using visual SLAM. Motion capture systems, such as the Vicon system[2], make use of a system of high-speed infrared cameras and retro-reflective markers to estimate the full 3D pose of arbitrary bodies with sub-millimetre precision. While a motion capture system would provide the highest accuracy it comes with a significant price tag. Additionally many of these systems are limited in the number of targets they can track (usually 4–6).

Fiducial markers, or *tags*, can either be placed on the robots themselves and tracked using a fixed system of cameras [21] or the markers may be placed at known locations in the environment and tracked by on-board cameras (e.g. [16]). This system has the advantage of a high level of accuracy at very low cost. However it does require significant setup and does limit the range of the robots so as to be certain of their position the drones must always be able to see at least one marker.

---

[2] http://vicon.com/

A third option is a feature-based localisation system, most notable is the monocular SLAM system PTAM (Parallel Tracking and Mapping). This system makes use of FAST [14] visual features and as such does not require fiducial markers to be placed in the environment at know locations but simply requires an environment with sufficient texture such as an office or lab space. PTAM has been successfully employed for quadcopter localisation in in a single agent configuration in a number of projects [4, 5].

The main drawback for multi-robot needs is the lack of multi-camera support in the original PTAM. Castle et al. [2], however, successfully demonstrated that in an augmented reality setting, given the decoupled nature of the tracking and map making processes which is a hallmark of the PTAM approach, it is possible to implement multiple camera tracking and mapping (using a single shared map) with relatively small effort. Castle et al. applied the problem of tracking a single user in an augmented reality setting using two cameras. To the best of out knowledge no one has previously applied the approach to multiple independent cameras or multi-robot localisation and mapping.

3   Framework Architecture

Before giving detail on the low- and high-level architecture of our framework, we briefly mention the two modifications of the original AR.Drone system that were necessary for our framework to work. The drone features two cameras, the front facing camera is capable of streaming images at 640×360 and is fitted with a 92° wide angle lens. The downward facing camera is primarily used for the on-board optical flow and captures images at 160×120 which is upscaled to 640×360 for streaming. Our multi-robot scenario requires omni-directional flying, and we found it difficult to keep control of the drone and avoid obstacles based on either of the video streams: the resolution of the downward facing camera was too low for reliable localisation, and the drone was flying into walls and obstacles while moving in the direction not covered by the front-facing camera. Therefore, we re-arranged the front-facing camera to look down. This is a very straightforward modification due to the modular structure of the AR.Drone. As a result, the drone can reliably avoid obstacles as long as they come upwards from the floor.

Another technical difficulty that we met was communication. The drone communicates via 802.11n WiFi which is used to stream control and sensor data as well as a single compressed camera feed. The high bandwidth of the information stream leads to communication latencies, especially when multiple drones are deployed at the same time. Additionally, we found that there is a lot of interference on the 2.4GHz frequency range used by the drone due to various WiFi networks in the proximity of our lab. We tried different options and it turned out that enabling the WPA2 security protocol on the drone and the access point dramatically reduce the latency. Implementing this modification involves installing a cross-compiled version of the open-source application wpa_supplicant and reconfiguring the network settings of the AR.Drone.

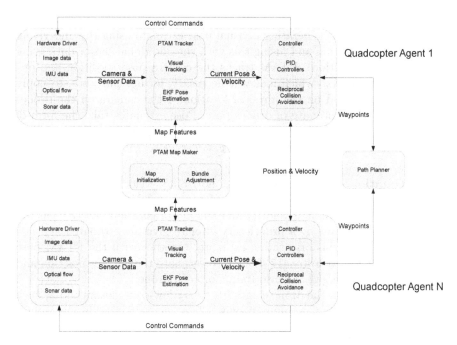

Fig. 1. The System Architecture

Our framework has been implemented in C++ and integrated into the Robot Operating System (ROS) [13], each main component within our system has been implemented as a separate ROS node meaning that in theory each component could be run on a separate processor/computer adding to the scalability of our framework.

3.1   Low-Level Architecture

In this section we describe the main components of our system. A general overview of the key components of our system is shown in (Fig. 1). In what follows we describe in detail the monocular visual SLAM system for mapping and camera pose estimation, the Extended Kalman Filter (EKF) for camera pose and inertial sensor data fusion and the PID based way-point controller.

Localisation and Mapping. The framework makes use of the Monocular keyframe-based SLAM package, PTAM, inspired by the work described in [10]. The original system introduced in [10] allowed only a single map to be built using images from a single camera; it was later extended [2] to allow multiple maps to be built using images from multiple cameras. We make use of this additional functionality within our framework to allow multiple quadrotors to localise using, and contribute to, the same map. To this extent, we have extended the system to create a PTAM tracking thread for each quadcopter. Each tracking thread is responsible for generating a visual pose estimate using the underlying PTAM tracking algorithm and fusing this with the additional sensor data using the EKF (see next section). PTAM tracks the camera pose (position and orientation) within the map which makes it necessary, for our purposes, to apply a transformation (based on the physical location of the camera on the drone) to obtain the PTAM estimate of drones pose. Pose estimates take the form of position (x, y, z) and orientation (roll, pitch and yaw angles) i.e. PTAM pose for drone i at time t is given by $\boldsymbol{ptam}_t^i = (x, y, z, \Phi, \Theta, \Psi)$.

State Estimation. In addition to the horizontal velocity estimation the AR.Drone also features an IMU (Inertial Measurement Unit) which measures the body acceleration, angular velocity and angular orientation. These measurements are fused on-board by a proprietary filter to give the estimated attitude of the drone. We use this orientation estimate, together with the PTAM pose in an Extended Kalman Filter [17] (EKF) to estimate the current position and orientation w.r.t to the global world frame for each drone.

The sensor data from the drone comes from very low-cost sensors and as such is subject to a significant amount of noise. The accuracy of the pose generated by PTAM can also vary depending on many factors such as the quality of the map, ambient lighting conditions and the motion of the drone (fast movements cause significant problems for rolling shutter cameras, like the ones on the AR.Drone). Filtering is a common approach to extract useful information from noisy/unreliable data, our framework makes use of an EKF to fuse the noisy drone sensor data $(\boldsymbol{imu}_t^i)$ with uncertain PTAM poses $(\boldsymbol{ptam}_t^i)$ to produce a more precise estimate of the drones pose given by $\boldsymbol{ekf}_t^i$.

The AR.Drone communicates via a WiFi link and there is no synchronisation between video frames and IMU data, thus it is necessary to explicitly handle the synchronisation of the data. Engel et al [5] do this by keeping a history of observations between the time the last visual SLAM estimate was available up to the current moment. When a new estimate is required the filter is rolled forward up-to the required time, integrating all available measurements from the observation buffer. This approach compensates for the delay/missed observations that occur from the unreliability of the WiFi link to the drone. Our framework extends this approach to a multi-robot scenario.

Position Control and Path Following. We make use of traditional PID controllers for x,y,z position and Yaw control. Input consists of the desired x,y,z theta (or a sequence of positions for path following), an independent PID controller for each degree of freedom calculates the control command necessary to achieve the goal position. The gains for each PID controller were calibrated experimentally and the values tested on multiple quadcopters.

Simulation. For simulations we chose the Gazebo multi-robot simulator as it provides capabilities to model complex 3d environments, reliably model multiple

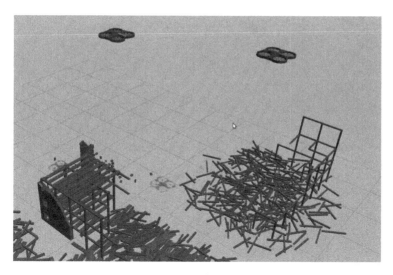

Fig. 2. The simulation environment

flying vehicles and generate realistic sensor data including camera images. Gazebo also integrates cleanly into the distributed infrastructure in ROS which means we are able to test our framework on both the simulated and real robots without altering the framework.

Meyer et al. [12] introduced a number of UAV-specific sensor plug-ins for Gazebo such as barometers, GPS receivers and sonar rangers. In addition to this they have created a comprehensive quadrotor simulation that includes accurate flight dynamics. For this work we focused on accurate sensor modeling rather than flight dynamics and as such our simulator uses a simplified flight dynamics model that does not accurately model the aerodynamics and propulsion behavior of the AR.Drone. We make use of Meyer et al's simulator plug-ins to replicate the sensor suite on the AR.Drone.

3.2   Higher-Level Architecture

We have implemented as easy to use interface to control the quadcopters based on ROS messages and actions. The action paradigm in ROS allows the execution of an action, e.g. move to a waypoint or follow a path, and provides facilities to monitor execution of the action, determine the success/failure and preempt a running action. High level agent code can be written in any of the supported ROS programming languages including Python, C++, Java and Lisp using the same interface.

Obstacle Avoidance. In our initial experiments, for simplicity, we were fixing the specific altitude at which each drone could operate thus avoiding colliding with one another. This choice, however, limits drastically the number of drones we could safely fly (especially indoors); therefore, a more robust solution was

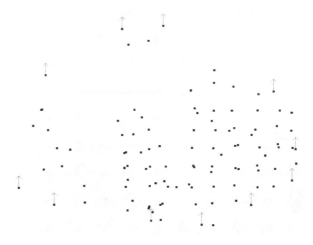

Fig. 3. An example of frontier point extraction showing the downsampled point-cloud and the frontier points (red arrows)

required. We assume that drones are the only dynamic obstacles in the environment. The velocity obstacle [7] (VO) is a representation of the set of all unsafe velocities i.e. velocities that will eventually result in a collision. Van den Berg et al. reasoned that in a multi-agent environment the other agents are not just dynamic obstacles but reasoning agents who themselves take steps to avoid obstacles. This leads to the notion of the reciprocal velocity obstacle [1] (RVO) where each agent takes half the responsibility for avoiding the collision under the assumption that all other agents reciprocate by taking complementary avoiding action. To enable efficient calculation of safe velocities van den Berg et al. introduced the Optimal Reciprocal Collision Avoidance (ORCA) [18] approach which uses an efficient linear program to calculate the half-planes of collision free velocities for each other agent. The intersection of all half-planes is the set of collision free velocities.

We use the ORCA approach with a simple agent communication model, each agent transmits it's current position and velocity to all agents. At each time step an agent will read in all available messages and calculate a safe velocity based on it's current goal and all the other agent positions and velocities. In ORCA each agent is represented by a sphere of fixed radius, however this does not take position uncertainty into account. We apply a similar approach to [8], where the radius of the robots is varied according the uncertainty within the particle filter used for localisation. In our framework we take into account not only the localisation uncertainty but, as our flying robots also lack the comparatively accurate velocity estimates from wheel odometry, we also account for the uncertainty in velocity estimation. The main limitation of this approach is the reliance on communication between the agents, an alternative would be to use relative sensing similar to Conroy et al's [3] work. However a drawback to that approach is the limited field of view of the AR.Drones cameras mean only head on collisions would be avoided, this is exacerbated by the fact the we relocated the front facing camera to look down.

4   Cooperative Exploration with Auction Mechanisms

Our first goal with the framework was to develop a way to deploy a team of AR.Drones to autonomously explore an environment. More formally given a map m, consisting of k feature-points, how can we extract a set of *interest points* $i \subseteq k$ so that by assigning an AR.Drone to explore these interest points extends the existing map m. This presents two challenges:

1. How to extract the set of interest points and
2. How to assign these tasks to the team of AR.Drones in an efficient manner

To address these two challenges we have developed an extension of our framework that uses a frontier-based approach for interest point extraction and a Sequential Single Item (SSI) auction approach for task assignment.

4.1   Interest Point Extraction

One of the most widely used exploration algorithms is the Frontier based algorithm introduced in [19] and extended to multiple robots in [20]. We apply a similar idea in our exploration strategy within our system; however, instead of an occupancy grid map model we have a sparse pointcloud of natural features.

Frontier point extraction from this sparse pointcloud is achieved by first downsampling the pointcloud, to reduce the granularity. This has the added benefit of improving the efficiency of the subsequent steps as the feature-map grows.

In order to build a reliable map each drone must be certain of its position when adding new features, this means that each drone must keep a portion of the existing feature map visible at all times in order to maintain good visual tracking. Therefore the interest points must be sufficiently close to previously mapped areas while still being sufficiently far away so new features can be discovered. To achieve this we first project the set of feature points onto the ground plane and compute the outliers/boundary points, these points represent the boundary between know/mapped areas and unoccupied areas. Due to the down-sampling step these points are sufficiently close to already mapped areas while still being close enough to maintain stable tracking. Figure 3 shows an example of this frontier point extraction process.

4.2   Auction Mechanism

Assigning frontier points to drones is done by making use of a simple Sequential Single Item (SSI) auction mechanism, which we adapted as follows:

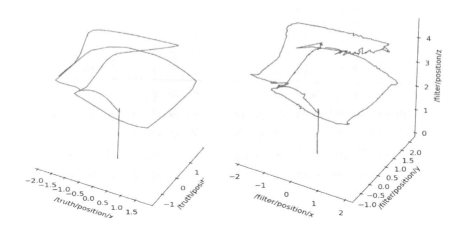

Fig. 4. Plot of the ground truth position (left) and the estimated position (right) Pose Estimation RMSE: 2.65cm

- On initiation the first set of frontier points is extracted from the current map.
- Points are auctioned to the available drones. All points are auctioned off and all drones bid on every point. Bids consists of path costs (linear distance between all points on the path), where the path includes all the previous points a particular drone has won. Therefore a drone bids on the cost of adding the latest point to its existing goals, rather than on the properties of these points.
- Once all points have been won the drones visit each of the frontier points adding a new features to the map as they are discovered.
- After successfully completing their missions another round commences and another frontier point extraction is carried out on the new expanded map.
- This process continues until no new points are added to the map or the drones run low on battery.

5   Evaluation

Sufficient work has already gone into verifying the performance of PTAM-based AR.Drone localisation w.r.t the ground truth [5,6,15]. Our system does not perform significantly differently from any of these system and as such our testing extended to simply verifying our system performs in line with the results others have reported. To that end we conducted several manual flight tests in simulation and compared the estimated path to simulated ground truth. Figure 4 shows the results of one such test. As expected the performance is simulation

Table 1. A table showing the results from 3 explorations experiments using one, two and three drones, respectively. The table shows the time taken and the number of: Frontier Points (FP), Map Points (MP) Deleted Map Points (DP).

	#AR.Drones =1				#AR.Drones =2				#AR.Drones =3		
Time	FP	MP	DP	Time	FP	MP	DP	Time	FP	MP	DP
900	86	21814	7567	437.76	74	21055	5820	310.84	79	18010	5017
900	192	21069	6582	467.47	82	20687	5852	353.23	79	19048	5949
900	161	22193	6813	434.01	69	17938	4746	316.33	74	17666	4559
900	146	19370	5992	403.89	74	17080	5063	487.04*	81	18298	5242
900*	125	12987	3767	508.34	86	19417	6167	332.15	74	16319	5200

is marginally better than other reported results this can be accounted for by the lack of distortion and lighting effects in the camera images and the lack of physical disturbances such as air wind, air flow disturbances from other drones.

Collision Avoidance. We conducted collision avoidance experiments both in simulation and in the real world. The simulated experiments featured 2-4 AR.Drones and 75 experiments were conducted. The drones were instructed to follow intersecting paths, and the collision avoidance mechanism was tasked with resolving the situation. Of the 75 simulated experiments conducted 73 collisions were avoided, the 2 collisions that did occur were due to complete PTAM tracking loss. The limited flying time and available resources made running real experiments more difficult so we were only able to complete 12 experiments with the 2 AR.Drones we have available. Out of the 12 experiments 3 resulted in collisions, with 9 collisions successfully avoided. Interestingly the 3 cases of collision were a result of WiFi communications latency i.e the robots took avoiding measures but did so too late to avoid the collision.

Exploration. We conducted three sets of exploratory experiments with teams of 1, 2 and 3 drones. The experiments were conducted in the simulated environment shown in Figure 2, the area consists of 20 x 17 metre area with the take-off/landing location roughly in the centre. The environment model consisted of piles of planks and damaged structures to simulate the exploration of a disaster site. Additionally the use of large numbers of similar models was designed to be particularly challenging for a Visual-SLAM based localisation framework. The results from the exploration experiments are shown in Table 1, a timeout of 15 minutes was set for all experiments. The table shows the number of froniter points visited during each experiment; the map points points give an indication of map size and quality and the deleted points are map features found to be unreliable for tracking. It is interesting to note that while 1 AR.Drone is not able to completely map the environment within the time limit the number of map points found is larger particularly when comparing the results from the single AR.Drone experiments to those of the 3 AR.Drone experiments. This is accounted for by the fact that a single AR.Drone exploring the frontiers of a map often has to cross from one side of the map to the other and is able to find additional points within the existing map. Another interesting case is the two simulated runs marked with a '*'. These cases highlight the utility of using multiple agents for exploration tasks, there the single robot became lost and the time taken to recover meant the AR.Drone mapped significantly less (1200 points) of the environment. In the 3 robot case and this resulted in some additional time but they were still able to produce a complete map of similar size to the other 3 AR.Drone experiments. More details of these experiments including videos can be found at the following web-page: `http://cgi.csc.liv.ac.uk/~rmw/drone.html`

6   Conclusion

In this paper we have presented a framework for multi-agent research using a low-cost quadcopter platform. We demonstrate the use of the framework on two typical multi-robot tasks: collision avoidance and environment exploration. Future work will involve improving the robustness of the collision avoidances to tracking loss of a single/multiple drones by defining un-safe zones to avoid when a drone has become lost. As mentioned we would also like to investigate more complex interest point extraction and auctioning mechanisms to improve the consistency and reliability of our exploration system. The main limitation of the framework is the reliance of wireless communication for state estimation and control, this limits the range of the AR.Drones and introduces problems of latency and data loss. The release of many small ARM-based development boards such as the Raspberry Pi[3] offer low cost and low power-consumption solutions to adding on-board processing to the AR.Drone. This would allow us to move time critical tasks such as tracking, control and collision avoidance on-board the AR.Drone.

References

1. Van den Berg, J., Lin, M., Manocha, D.: Reciprocal velocity obstacles for real-time multi-agent navigation. In: IEEE International Conference on Robotics and Automation, ICRA 2008, pp. 1928–1935. IEEE (2008)
2. Castle, R., Klein, G., Murray, D.W.: Video-rate localization in multiple maps for wearable augmented reality. In: 12th IEEE International Symposium on Wearable Computers, ISWC 2008, pp. 15–22. IEEE (2008)
3. Conroy, P., Bareiss, D., Beall, M., van den Berg, J.: 3-d reciprocal collision avoidance on physical quadrotor helicopters with on-board sensing for relative positioning. In: IEEE/RSJ International Conference on Intelligent Robots and Systems (2013) (submitted)

---

[3] `http://www.raspberrypi.org/`

4. Dijkshoorn, N.: Simultaneous localization and mapping with the ar. drone. Ph.D. thesis, Masters thesis, Universiteit van Amsterdam (2012)
5. Engel, J., Sturm, J., Cremers, D.: Camera-based navigation of a low-cost quadrocopter. IMU 320, 240 (2012)
6. Engel, J., Sturm, J., Cremers, D.: Scale-aware navigation of a low-cost quadrocopter with a monocular camera. Robotics and Autonomous Systems (2014)
7. Fiorini, P., Shiller, Z.: Motion planning in dynamic environments using velocity obstacles. The International Journal of Robotics Research 17(7), 760–772 (1998)
8. Hennes, D., Claes, D., Meeussen, W., Tuyls, K.: Multi-robot collision avoidance with localization uncertainty. In: Proceedings of the 11th International Conference on Autonomous Agents and Multiagent Systems, vol. 1, pp. 147–154. International Foundation for Autonomous Agents and Multiagent Systems (2012)
9. Kendoul, F.: Survey of advances in guidance, navigation, and control of unmanned rotorcraft systems. J. Field Robot. 29(2), 315–378 (2012)
10. Klein, G., Murray, D.: Parallel tracking and mapping for small ar workspaces. In: 6th IEEE and ACM International Symposium on Mixed and Augmented Reality, ISMAR 2007, pp. 225–234. IEEE (2007)
11. Koenig, N., Howard, A.: Design and use paradigms for gazebo, an open-source multi-robot simulator. In: Proceedings of the IEEE/RSJ International Conference on Intelligent Robots and Systems, IROS 2004, vol. 3, pp. 2149–2154. IEEE (2004)
12. Meyer, J., Sendobry, A., Kohlbrecher, S., Klingauf, U., von Stryk, O.: Comprehensive simulation of quadrotor uavs using ros and gazebo. In: Noda, I., Ando, N., Brugali, D., Kuffner, J.J. (eds.) SIMPAR 2012. LNCS, vol. 7628, pp. 400–411. Springer, Heidelberg (2012)
13. Quigley, M., Conley, K., Gerkey, B., Faust, J., Foote, T., Leibs, J., Wheeler, R., Ng, A.Y.: Ros: An open-source robot operating system. In: ICRA Workshop on Open Source Software, vol. 3 (2009)
14. Rosten, E., Drummond, T.W.: Machine learning for high-speed corner detection. In: Leonardis, A., Bischof, H., Pinz, A. (eds.) ECCV 2006, Part I. LNCS, vol. 3951, pp. 430–443. Springer, Heidelberg (2006)
15. Sa, I., He, H., Huynh, V., Corke, P.: Monocular vision based autonomous navigation for a cost-effective mav in gps-denied environments. In: 2013 IEEE/ASME International Conference on Advanced Intelligent Mechatronics (AIM), pp. 1355–1360. IEEE (2013)
16. Sanchez-Lopez, J.L., Pestana, J., de la Puente, P., Campoy, P.: Visual quadrotor swarm for imav 2013 indoor competition. In: ROBOT 2013: First Iberian Robotics Conference, pp. 55–63. Springer International Publishing (2014)
17. Thrun, S., Burgard, W., Fox, D., et al.: Probabilistic robotics, vol. 1, pp. 48–54. MIT Press, Cambridge (2005)
18. van den Berg, J., Guy, S.J., Lin, M., Manocha, D.: Reciprocal n-body collision avoidance. In: Pradalier, C., Siegwart, R., Hirzinger, G. (eds.) Robotics Research. STAR, vol. 70, pp. 3–19. Springer, Heidelberg (2011)
19. Yamauchi, B.: A frontier-based approach for autonomous exploration. In: Proceedings of the 1997 IEEE International Symposium on Computational Intelligence in Robotics and Automation, CIRA 1997, pp. 146–151. IEEE (1997)
20. Yamauchi, B.: Frontier-based exploration using multiple robots. In: Proceedings of the Second International Conference on Autonomous Agents, pp. 47–53. ACM (1998)
21. Zickler, S., Laue, T., Birbach, O., Wongphati, M., Veloso, M.: SSL-vision: The shared vision system for the robocup small size league. In: Baltes, J., Lagoudakis, M.G., Naruse, T., Ghidary, S.S. (eds.) RoboCup 2009. LNCS, vol. 5949, pp. 425–436. Springer, Heidelberg (2010)

H_∞ Path Tracking Control for Quadrotors Based on Quaternion Representation

Wesam Jasim and Dongbing Gu

School of Computer Science and Electronic Engineering, University of Essex
Wivenhoe Park, Colchester, UK
{wmjasi,dgu}@essex.ac.uk
http://www.essex.ac.uk/csee/

Abstract. In this work, the path tracking problem of quadrotors is investigated. A quadrotor is represented by unit quaternion and modeled with added disturbance. Given full and accurate location information, a nonlinear H_∞ control law is proposed and its stability is analyzed by using Lyapunov stability theorem. The added disturbance includes parameter changes and force disturbance. The simulation result demonstrates the closed loop path tracking system is stable with and without the added disturbance.

Keywords: Path tracking, H_∞ control, Quaternion, Quadrotor UAV.

1 Introduction

The path tracking problem of a quadrotor is to develop a state feedback controller to control the quadrotor to track a given trajectory. It includes two components: position estimation and tracking control. In this paper, we focus on the tracking control problem where accurate position states are available. The tracking control is also built on the top of attitude stabilization of the quadrotor where the quadrotor's attitude is stabilized by an attitude state feedback controller.

In the past several years many researchers have used the quaternion representation to describe the dynamic model of quadrotors. Several control methods have been used to control both the attitude stabilization and translational motion such as path tracking of quadrotors. The path tracking problem for quadrotors based on quaternion representation using filtered backstepping was addressed in [1], in which a second order quaternion filter was used to calculate the quaternion rate and the angular rate. A backstepping control using a decoupling quaternion parametrization was proposed by [2]. The controller depends on the decoupling of two rotations of quaternion. Global tracking control using one-step ahead backstepping controller depends on standard backstepping and Lyapunov's theorem was implemented and tested in simulation for a quadrotor by [3]. The dynamic model of the quadrotor was represented by a combination of quaternions and Euler angles.

Nonlinear robust tracking control based on quaternions in particular Euclidean set was presented to follow the position and attitude commands. The controller

M. Mistry et al. (Eds.): TAROS 2014, LNAI 8717, pp. 72–84, 2014.

was simulated to reject considerable disturbances when it follows a complicated trajectory [4]. Parra-Vega V. et al. [5] proposed two novel model-free second order sliding mode controllers for attitude stabilization and path tracking. The work composed of two theorems: the first one was to ensure semi-global exponential and robust path tracking of a closed-loop system with zero yaw angle, while the second one was deduced for final stability. A sliding mode surface controller based-on quaternion representation for time-parametrization control of a quadrotor was presented by [6]. The controller was obtained with the exponential and terminal stabilization. A PD gathering with second order sliding mode controllers was implemented to solve the problems of path tracking control and the attitude stabilization, respectively, in [7]. A path tracking controller using a so-called quasi-static feedback linearization for the quadrotor dynamics based on quaternion representation was presented in [8] to reduce the translational dynamics order. It has two phases: the first relating to the altitude and consequently to the thrust and the second to split up the altitude from the other two directions.

Path tracking of quadrotors are often affected by external disturbances, such as payload changes, mass changes, wind disturbance, inaccurate model parameters, etc. Therefore, a quadrotor controller must be robust enough in order to reject the effect of disturbances and cover the changes caused by the uncertainties. Robust state feedback controllers are very demanding in the path tracking control problem. H_∞ control approach is able to attenuate the disturbance energy by measuring a ratio between the energy of cost vector and the energy of disturbance signal vector [9]. It has been applied in [10] for the quadrotor control problem and it has been applied in [11] with input coupling based on Euler angles representation. In this paper, we will use H_∞ control approach to design a state feedback controller for the path tracking control problem of quadrotors. To the best of our knowledge, this is the first use of H_∞ control approach for the path tracking problem of quadrotors based on quaternion representation. Quaternion representation does have advantages over other representations such as avoiding the gimbal lock and being a 4 numbers representation which is easier to obtain an affine system, incontrast to the 9 numbers of a cosine matrix.

In the following, Section 2 presents the basics of unit quaternion and the quadrotor dynamic system. Section 3 provides a review on H_∞ optimal control approach. The main result of this paper is given in Section 4. Section 5 provides simulation results. Our conclusion and future work are given in Section 6.

2 Dynamic System

2.1 Quaternion

The attitude or orientation of a quadrotor is described by a rotation with an angle α above the axis $\mathbf{k} \in \mathbb{R}^3$. The corresponding rotation matrix is R, which falls in the special orthogonal group of degree three $SO(3) = \{R \in \mathbb{R}^{3\times3} | R^T R = RR^T = I, \det(R) = 1\}$ where I is a 3×3 unit matrix. It is also represented by a unit quaternion $Q = [q_0, q_1, q_2, q_3]^T = [q_0, \mathbf{q}^T]^T$ where $\mathbf{q} = [q_1, q_2, q_3]^T$ is the vector part and q_0 is the scalar part of the quaternion, and $q_0^2 + \mathbf{q}^T \mathbf{q} = 1$. Given angle α and the rotation axis $\mathbf{k}$, the unit quaternion is

$$Q = \begin{bmatrix} \cos \frac{\alpha}{2} \\ \mathbf{k} \sin \frac{\alpha}{2} \end{bmatrix}$$

Given the unit quaternion, the corresponding rotation angle and axis are:

$$\mathbf{k} = \frac{\mathbf{q}}{\|\mathbf{q}\|}$$

$$\alpha = 2 \arccos q_0$$

which shows there is not an one-to-one mapping as there are two unit quaternions Q and $-Q$ that represent the same rotation. The rotation matrix R is related to the unit quaternion Q through the Rodrigues formula:

$$R = (q_0^2 - \mathbf{q}^T \mathbf{q})I + 2\mathbf{q}\mathbf{q}^T + 2q_0 S(\mathbf{q})$$

where S is the skew-symmetric cross product matrix:

$$S(\mathbf{q}) = \begin{bmatrix} 0 & -q_3 & q_2 \\ q_3 & 0 & -q_1 \\ -q_2 & q_1 & 0 \end{bmatrix}$$

The unit quaternion can be used to represent the coordinate transform between the inertial frame $\mathcal{I}$ and the body frame $\mathcal{B}$ by defining the multiplication and the inverse quaternion. The multiplication of two quaternions $Q = [q_0, \mathbf{q}^T]^T$ and $P = [p_0, \mathbf{p}^T]^T$ is defined as:

$$Q \otimes P = \begin{bmatrix} q_0 & -\mathbf{q}^T \\ \mathbf{q} & q_0 I + S(\mathbf{q}) \end{bmatrix} \begin{bmatrix} p_0 \\ \mathbf{p} \end{bmatrix} = \begin{bmatrix} q_0 p_0 - \mathbf{q}^T \mathbf{p} \\ p_0 \mathbf{q} + q_0 \mathbf{p} + S(\mathbf{q})\mathbf{p} \end{bmatrix}$$

The inverse unit quaternion is defined as $Q^{-1} = [q_0, -\mathbf{q}^T]^T$ for $Q = [q_0, \mathbf{q}^T]^T$. A vector $\mathbf{x}_{\mathcal{I}} \in \mathbb{R}^3$ in the inertial frame can be expressed as a vector $\mathbf{x}_{\mathcal{B}} \in \mathbb{R}^3$ in the body frame via $\mathbf{x}_{\mathcal{B}} = R^T \mathbf{x}_{\mathcal{I}}$. Using $\bar{\mathbf{x}} = [0, \mathbf{x}^T]^T$, the transformation from the inertial frame to the body frame is expressed as $\bar{\mathbf{x}}_{\mathcal{B}} = Q^{-1} \otimes \bar{\mathbf{x}}_{\mathcal{I}} \otimes Q$.

2.2   Quadrotor Kinematics and Dynamics

Let $\omega = [\omega_x, \omega_y, \omega_z]^T$ be the angular velocity of a quadrotor in the body frame $\mathcal{B}$. The attitude kinematics is: $\dot{Q} = \frac{1}{2} Q \otimes \bar{\omega}$, which is:

$$\begin{bmatrix} \dot{q}_0 \\ \dot{\mathbf{q}} \end{bmatrix} = \frac{1}{2} \begin{bmatrix} -\mathbf{q}^T \omega \\ (q_0 I + S(\mathbf{q})) \omega \end{bmatrix} \tag{1}$$

The attitude dynamics is:

$$J\dot{\omega} = -S(\omega)J\omega - G(\omega) + \tau \tag{2}$$

where J is the 3×3 diagonal matrix representing three inertial moments in the body frame, $G(\omega)$ represents the gyroscopic effect, and $\tau \in \mathbb{R}^3$ is the torque vector applied on the quadrotor.

The translation dynamics of a quadrotor is:

$$\dot{\mathbf{p}} = \mathbf{v}$$
$$\dot{\mathbf{v}} = g\mathbf{e}_3 - \frac{f}{m} R\mathbf{e}_3 \tag{3}$$

where $\mathbf{v} = [v_x, v_y, v_z]^T$ is the linear velocity, $\mathbf{p} = [x, y, z]^T$ is the position vector and the vector $\mathbf{e}_3 = [0, 0, 1]^T$, and the equation of acceleration (3) can be written as:

$$\begin{bmatrix} \ddot{x} \\ \ddot{y} \\ \ddot{z} \end{bmatrix} = \begin{bmatrix} 0 \\ 0 \\ g \end{bmatrix} - \begin{bmatrix} 2(q_1 q_3 + q_0 q_2) \\ 2(q_2 q_3 - q_0 q_1) \\ q_0^2 - q_1^2 - q_2^2 + q_3^2 \end{bmatrix} \frac{f}{m} \tag{4}$$

where $f = f_1 + f_2 + f_3 + f_4$ is the total force applied to the quadrotor. The full mathematical model of the quadrotor can be summarized by equations (1),(2),(3). To control the quadrotor the first step is to stabilize the attitude, which has been already done using H_∞ controller in our work [12]. In this paper, the control aim is to conduct a path tracking of the remaining three position states, $\mathbf{p} = [x, y, z]$, the tracking error can be defined as the position error $\tilde{\mathbf{p}} = [\mathbf{p} - \mathbf{p}_d]$ and the linear velocity error $\tilde{\mathbf{v}} = [\mathbf{v} - \mathbf{v}_d]$, where $\mathbf{p}_d$ and $\mathbf{v}_d$ are the desired position and the desired linear velocity respectively.

Then equations (3) can be rewritten in an error form as:

$$\dot{\tilde{\mathbf{p}}} = \tilde{\mathbf{v}}$$
$$\dot{\tilde{\mathbf{v}}} = g\mathbf{e}_3 - \frac{f}{m} R\mathbf{e}_3 \tag{5}$$

The main goal is to drive asymptotically the quadrotor towards the desired position $\mathbf{p}_d$ from an initial position with the effect of added disturbances tending to disappear and changed parameters tending to be recovered. Now we consider the robust control approach to the path tracking problem. When considering $\mathbf{d}$ as the disturbance, then $\mathbf{d} = [d_x, d_y, d_z]^T$ is applied to the nonlinear system (5).

Let $\mathbf{x} = [\tilde{\mathbf{p}}^T, \tilde{\mathbf{v}}^T]^T$ and $\mathbf{u} = g\mathbf{e}_3 - \frac{f}{m} R\mathbf{e}_3$. The dynamic system (5) with the disturbance $\mathbf{d}$ can be written into an affine nonlinear form:

$$\dot{\mathbf{x}} = f(\mathbf{x}) + g(\mathbf{x})\mathbf{u} + k(\mathbf{x})\mathbf{d} \tag{6}$$

where

$$f(\mathbf{x}) = \begin{bmatrix} \tilde{\mathbf{v}} \\ 0_{3\times 3} \end{bmatrix}; \ g(\mathbf{x}) = \begin{bmatrix} 0_{3\times 3} \\ I \end{bmatrix}; \ k(\mathbf{x}) = \begin{bmatrix} 0_{3\times 3} \\ I \end{bmatrix} \tag{7}$$

3 H_∞ Suboptimal Control Approach

In this section, a brief overview on H_∞ suboptimal control approach is summarized for affine nonlinear systems of the form:

$$\dot{\mathbf{x}} = f(\mathbf{x}) + g(\mathbf{x})\mathbf{u} + k(\mathbf{x})\mathbf{d} \quad ; \quad \mathbf{y} = h(\mathbf{x}) \tag{8}$$

where $\mathbf{x} \in \mathbb{R}^n$ is a state vector, $\mathbf{u} \in \mathbb{R}^m$ is an input vector, $\mathbf{y} \in \mathbb{R}^p$ is an output vector, and $\mathbf{d} \in \mathbb{R}^q$ is a disturbance vector. Detailed information on H_∞ control approach can be found in [9].

We assume the existence of an equilibrium $\mathbf{x}_*$, i.e. $f(\mathbf{x}_*) = 0$, and we also assume $h(\mathbf{x}_*) = 0$. Given a smooth state feedback controller:

$$\mathbf{u} = l(\mathbf{x})$$
$$l(\mathbf{x}_*) = 0 \tag{9}$$

the H_∞ suboptimal control problem considers the L_2-gain from the disturbance $\mathbf{d}$ to the vector of $\mathbf{z} = [\mathbf{y}^T, \mathbf{u}^T]^T$. This problem is defined below.

Problem 1. Let γ be a fixed nonnegative constant. The closed loop system consisting of the nonlinear system (8) and the state feedback controller (9) is said to have L_2-gain less than or equal to γ from $\mathbf{d}$ to $\mathbf{z}$ if

$$\int_0^T \|\mathbf{z}(t)\|^2 dt \leq \gamma^2 \int_0^T \|\mathbf{d}(t)\|^2 dt + K(\mathbf{x}(0)) \tag{10}$$

for all $T \geq 0$ and all $\mathbf{d} \in L_2(0, T)$ with initial condition $\mathbf{x}(0)$, where $0 \leq K(\mathbf{x}) < \infty$ and $K(\mathbf{x}_*) = 0$.

For the nonlinear system (8) and $\gamma > 0$, define the Hamiltonian $H_\gamma(\mathbf{x}, V(\mathbf{x}))$ as below:

$$H_\gamma(\mathbf{x}, V(\mathbf{x})) = \frac{\partial V(\mathbf{x})}{\partial \mathbf{x}} f(\mathbf{x}) + \frac{1}{2} \frac{\partial V(\mathbf{x})}{\partial \mathbf{x}} \left[\frac{1}{\gamma^2} k(\mathbf{x}) k^T(\mathbf{x}) - g(\mathbf{x}) g^T(\mathbf{x}) \right]$$
$$\frac{\partial^T V(\mathbf{x})}{\partial \mathbf{x}} + \frac{1}{2} h^T(\mathbf{x}) h(\mathbf{x}) \tag{11}$$

Theorem 1. *[9] If there exists a smooth solution $V \geq 0$ to the Hamilton-Jacobi inequality:*

$$H_\gamma(\mathbf{x}, V(\mathbf{x})) \leq 0; \ V(\mathbf{x}_*) = 0 \tag{12}$$

then the closed-loop system for the state feedback controller:

$$\mathbf{u} = -g^T(\mathbf{x}) \frac{\partial^T V(\mathbf{x})}{\partial \mathbf{x}} \tag{13}$$

has L_2-gain less than or equal to γ, and $K(\mathbf{x}) = 2V(\mathbf{x})$.

The nonlinear system (8) is called zero-state observable if for any trajectory $\mathbf{x}(t)$ such that $\mathbf{y}(t) = 0, \mathbf{u}(t) = 0, \mathbf{d}(t) = 0$ implies $\mathbf{x}(t) = \mathbf{x}_*$.

Proposition 1. *[9] If the nonlinear system (8) is zero-state observable and there exists a proper solution $V \geq 0$ to the Hamilton-Jacobi inequality, then $V(\mathbf{x}) > 0$ for $\mathbf{x}(t) \neq \mathbf{x}_*$ and the closed loop system (8), (13) with $\mathbf{d} = 0$ is globally asymptotically stable.*

Theorem 1 and Proposition 1 are from [9] and they are cited here for convenience.

4 H_∞ Suboptimal Path Tracking Controller

The H_∞ suboptimal controller will be designed for the path tracking problem in this section. The following form of V is suggested for the path tracking model (6):

$$V(\mathbf{x}) = \frac{1}{2} \begin{bmatrix} \tilde{\mathbf{p}}^T & \tilde{\mathbf{v}}^T \end{bmatrix} \begin{bmatrix} C & K_p \\ K_p & K_d \end{bmatrix} \begin{bmatrix} \tilde{\mathbf{p}} \\ \tilde{\mathbf{v}} \end{bmatrix} \tag{14}$$

where diagonal matrices $K_p > 0$, $K_d > 0$ are the proportional and derivative gain and $C > 0$ is a diagonal matrix. And

$$\frac{\partial V(\mathbf{x})}{\partial \mathbf{x}} = \begin{bmatrix} \tilde{\mathbf{p}}^T C + \tilde{\mathbf{v}}^T K_p & \tilde{\mathbf{p}}^T K_p + \tilde{\mathbf{v}}^T K_d \end{bmatrix} \tag{15}$$

Accordingly the controller is

$$\begin{aligned}
\mathbf{u} &= -g^T(\mathbf{x}) \frac{\partial^T V(\mathbf{x})}{\partial \mathbf{x}} \\
&= -\begin{bmatrix} 0_{3\times3} & I \end{bmatrix} \begin{bmatrix} C\tilde{\mathbf{p}} + K_p\tilde{\mathbf{v}} \\ K_p\tilde{\mathbf{p}} + K_d\tilde{\mathbf{v}} \end{bmatrix} \\
&= -K_p\tilde{\mathbf{p}} - K_d\tilde{\mathbf{v}} \tag{16}
\end{aligned}$$

The following cost variables are chosen with diagonal matrices $H_1 > 0$ and $H_2 > 0$.

$$h(\mathbf{x}) = \begin{bmatrix} \sqrt{H_1}\tilde{\mathbf{p}} \\ \sqrt{H_2}\tilde{\mathbf{v}} \end{bmatrix} \tag{17}$$

which satisfies $h(\mathbf{x}_*) = 0$, where the equilibrium point $\mathbf{x}_* = [0_{1\times3}, 0_{1\times3}]^T$. And we know

$$V(\mathbf{x}_*) = 0 \tag{18}$$

Now the path tracking problem of the quadrotor under the disturbance $\mathbf{d}$ is defined below.

Problem 2. Given the equilibrium point $\mathbf{x}_*$, find the parameters K_p, K_d, c in order to enable the closed-loop system (6) with the above controller $\mathbf{u}$ (16) to have L_2-gain less than or equal to γ.

Next we want to show our main result in the following theorem.

Theorem 2. *If the following conditions are satisfied, the closed-loop system* (6) *with the above controller* $\mathbf{u}$ (16) *has L_2-gain less than or equal to γ. And the closed loop system* (6), (16) *with* $\mathbf{d} = 0$ *is asymptotically locally stable for the equilibrium point* $\mathbf{x}_*$.

$$CK_d \geq K_p^2$$

$$C = K_p K_d \left(1 - \frac{1}{\gamma^2}\right)$$

$$\|K_p\|^2 \geq \frac{\gamma^2 \|H_1\|}{\gamma^2 - 1} \tag{19}$$

$$\|K_d\|^2 \geq \frac{\gamma^2 (\|H_2\| + 2\|K_p\|)}{\gamma^2 - 1} \tag{20}$$

$$\|H_1\| > 0$$

$$\|H_2\| > 0$$

Proof. With the given conditions, we need to show (1) $V(\mathbf{x}) \geq 0$ and (2) the Hamiltonian $H_\gamma(\mathbf{x}, V(\mathbf{x})) \leq 0$. Then the first part of the theorem can be proved by using Theorem 1.

(1) Since

$$V(\mathbf{x}) = \frac{1}{2} \begin{bmatrix} \tilde{\mathbf{p}}^T & \tilde{\mathbf{v}}^T \end{bmatrix} \begin{bmatrix} C & K_p \\ K_p & K_d \end{bmatrix} \begin{bmatrix} \tilde{\mathbf{p}} \\ \tilde{\mathbf{v}} \end{bmatrix}$$

Thus the condition for $V(\mathbf{x}) \geq 0$ is

$$CK_d \geq K_p^2$$

(2)

$$H_\gamma(\mathbf{x}, V(\mathbf{x})) = C\tilde{\mathbf{p}}^T \tilde{\mathbf{v}} + \tilde{\mathbf{v}}^T K_p \tilde{\mathbf{v}} + \frac{1}{2}\left(\frac{1}{\gamma^2} - 1\right) \|K_p \tilde{\mathbf{p}}^T + K_d \tilde{\mathbf{v}}^T\|^2$$

$$+ \frac{1}{2}\|H_1\|\|\tilde{\mathbf{p}}\|^2 + \frac{1}{2}\|H_2\|\|\tilde{\mathbf{v}}\|^2$$

By choosing

$$C = K_p K_d \left(1 - \frac{1}{\gamma^2}\right)$$

Then

$$H_\gamma(\mathbf{x}, V(\mathbf{x})) = \tilde{\mathbf{v}}^T K_p \tilde{\mathbf{v}} + \frac{1}{2}\left(\frac{1}{\gamma^2} - 1\right) \left(\|K_p\|^2\|\tilde{\mathbf{p}}\|^2 + \|K_d\|^2\|\tilde{\mathbf{v}}\|^2\right)$$

$$+ \frac{1}{2}\|H_1\|\|\tilde{\mathbf{p}}\|^2 + \frac{1}{2}\|H_2\|\|\tilde{\mathbf{v}}\|^2$$

By using $|\tilde{\mathbf{v}}^T K_p \tilde{\mathbf{v}}| \leq \|K_p\|\|\tilde{\mathbf{v}}\|^2$, we have

$$H_\gamma(\mathbf{x}, V(\mathbf{x})) \leq \|K_p\|\|\tilde{\mathbf{v}}\|^2 + \frac{1}{2}\left(\frac{1}{\gamma^2} - 1\right) \left(\|K_p\|^2\|\tilde{\mathbf{p}}\|^2 + \|K_d\|^2\|\tilde{\mathbf{v}}\|^2\right)$$

$$+ \frac{1}{2}\|H_1\|\|\tilde{\mathbf{p}}\|^2 + \frac{1}{2}\|H_2\|\|\tilde{\mathbf{v}}\|^2$$

$$= \left(\|K_p\| + \frac{1}{2}\left(\frac{1}{\gamma^2} - 1\right)\|K_d\|^2 + \frac{1}{2}\|H_2\|\right)\|\tilde{\mathbf{v}}\|^2$$

$$+ \left(\frac{1}{2}\left(\frac{1}{\gamma^2} - 1\right)\|K_p\|^2 + \frac{1}{2}\|H_1\|\right)\|\tilde{\mathbf{p}}\|^2$$

Thus, the conditions for $H_\gamma(\mathbf{x}, V(\mathbf{x})) \leq 0$ are

$$\|K_p\| + \frac{1}{2}\left(\frac{1}{\gamma^2} - 1\right)\|K_d\|^2 + \frac{1}{2}\|H_2\| \leq 0$$

$$\frac{1}{2}\left(\frac{1}{\gamma^2} - 1\right)\|K_p\|^2 + \frac{1}{2}\|H_1\| \leq 0$$

i.e.

$$\|K_p\|^2 \geq \frac{\gamma^2\|H_1\|}{\gamma^2 - 1}$$

$$\|K_d\|^2 \geq \frac{\gamma^2(\|H_2\| - 2\|K_p\|)}{\gamma^2 - 1}$$

It is trivial to show that the nonlinear system (6) is zero-state observable for the equilibrium point $\mathbf{x}_*$. Further due to the fact that $V(\mathbf{x}) \geq 0$ and it is a proper function (i.e. for each $\rho > 0$ the set $\{x : 0 \leq V(x) \leq \rho\}$ is compact), the closed-loop system (6), (16) with $\mathbf{d} = 0$ is asymptotically locally stable for the equilibrium point $\mathbf{x}_*$ according to Proposition 1. This proves the second part of the theorem.

Finally from the third row of $\mathbf{u}$, we can have f:

$$\mathbf{u} = g\mathbf{e}_3 - \frac{f}{m}R\mathbf{e}_3$$
$$= -K_p\tilde{\mathbf{p}} - K_d\tilde{\mathbf{v}}$$

$$f = -\left(-K_{p33}\tilde{z} - K_{d33}\tilde{v}_z - g\right)\frac{m}{q_0^2 - q_1^2 - q_2^2 + q_3^2}$$

And from the first and second rows of $\mathbf{u}$, we can have the u_x and u_y respectively.

$$u_x = -\left(-K_{p11}\tilde{x} - K_{d11}\tilde{v}_x\right)\frac{m}{f}$$

$$u_y = -\left(-K_{p22}\tilde{y} - K_{d22}\tilde{v}_y\right)\frac{m}{f}$$

where u_x and u_y are the x and y horizontal position control laws respectively.

5   Simulations

The path tracking problem of a quadrotor using the proposed controller is tested in a MATLAB quadrotor simulater. The quadrotor parameters used in the simulation are described in table 1:

Table 1. Quadrotor Parameters

Symbol	Definition	Value	Units
J_x	Roll Inertia	4.4×10^{-3}	$kg.m^2$
J_y	Pitch Inertia	4.4×10^{-3}	$kg.m^2$
J_z	Yaw Inertia	8.8×10^{-3}	$kg.m^2$
m	Mass	0.5	kg
g	Gravity	9.81	m/s^2
l	Arm Length	0.17	m
J_r	Rotor Inertia	4.4×10^{-5}	$kg.m^2$

Two different paths are tested in order to demonstrate the robustness of the proposed controller. The added disturbances includes $\pm 30\%$ of the model parameter uncertainties (mass and inertia) and a force disturbance.

In the first path, the initial condition of the quadrotor is $\mathbf{p} = [0, 0.5, 0]^T$ meter and $Q = [-1, 0, 0, 0]^T$, and the desired path is $\mathbf{p}_d = [0.5 \sin(t/2), 0.5 \cos(t/2), 1 + t/10]^T$. The constant γ is chosen to be $\gamma = 1.049$ and the norms of the two weighting matrices are chosen to be $\|H_1\| = 1150$ and $\|H_2\| = 10$. Under these parameters the norms of feedback control matrices can be obtained by solving the conditions in (19) and (20) to be $\|K_p\| \approx 112$ and $\|K_d\| \approx 50$ or it can be written in matrix form as

$$K_p = \begin{bmatrix} 0.0417 & 0 & 0 \\ 0 & 0.0417 & 0 \\ 0 & 0 & 112 \end{bmatrix} \quad \text{and} \quad K_d = \begin{bmatrix} 0.417 & 0 & 0 \\ 0 & 0.417 & 0 \\ 0 & 0 & 50 \end{bmatrix}$$

In the second path, the initial condition of the quadrotor is $\mathbf{p} = [0, 0, 0]^T$ meter and $Q = [-1, 0, 0, 0]^T$, and the desired path is a combination of two parts: the first part is $\mathbf{p}_d = [0, 0, 3 - 2\cos(\pi t/20)]^T$ when $0 \leq t < 10$ while the second part is $\mathbf{p}_d = [2 \sin(t/20), 0.1 \tan(t/20), 5 - 2\cos(t/20)]^T$ when $10 \leq t \leq 30$. The constant γ is chosen to be $\gamma = 1.049$ and the norms of the two weighting matrices are chosen to be $\|H_1\| = 1000$ and $\|H_2\| = 20$. Under these parameters the norms of feedback control matrices can be obtained by solving the conditions in (19) and (20) to be $\|K_p\| \approx 105$ and $\|K_d\| \approx 50$ or it can be written in a matrix form as

$$K_p = \begin{bmatrix} 0.48 & 0 & 0 \\ 0 & 0.48 & 0 \\ 0 & 0 & 105 \end{bmatrix} \quad \text{and} \quad K_d = \begin{bmatrix} 0.4 & 0 & 0 \\ 0 & 0.4 & 0 \\ 0 & 0 & 50 \end{bmatrix}$$

The testing results of tracking the first path using the proposed controller are obtained with the conditions (1) no disturbance, (2) force disturbance $d_z = -10N$ at $t = 10s$, $d_x = 10N$ at $t = 20s$ and $d_y = 10N$ at $t = 30s$, (3) $+30\%$ model parameter uncertainty, and (4) -30% model parameter uncertainty. The tracking trajectories, positions, quaternions, Euler angles calculated from quaternions, and rotation speeds are shown in Figures 1,2,3,4,5, respectively.

The testing results of tracking the second path using the proposed controller are obtained with the conditions (1) no disturbance, (2) force disturbance $d_z = -10$N at $t = 10$s, $d_x = 10$N at $t = 15$s and $d_y = 10$N at $t = 25$s, (3) +30% model parameter uncertainty, and (4) −30% model parameter uncertainty. The tracking trajectories, positions, quaternions, Euler angles calculated from quaternions, and rotation speeds are shown in Figures 6,7,8,9,10, respectively.

In Figures 1,2,6,7 the desired path (red) is tracked by the proposed controller (blue) and it is caught with less than 3 seconds. In addition, the controller under disturbance (green) can track the desired path and recovers from the disturbances within less than one second. The controller under +30% model parameter uncertainty (cyan) and −30% model parameter uncertainty (magenta) can track the desired path with very short time. The same result can be found in Figures 3,4,8,9 with even more obvious disturbance rejection.

It can be seen that the proposed controller is able to track the desired trajectories. The expected robustness is demonstrated by the disturbance rejection and the recovery from changes caused by the parameter uncertainties.

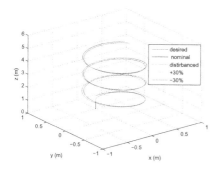

Fig. 1. Path Tracking 1

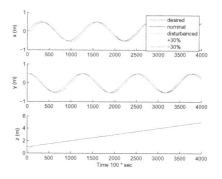

Fig. 2. Positions 1

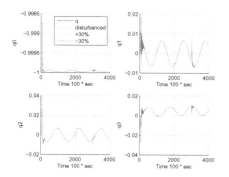

Fig. 3. Quaternion Components 1

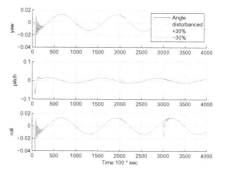

Fig. 4. Angles 1

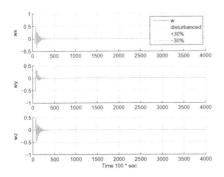

Fig. 5. Angular Velocities 1

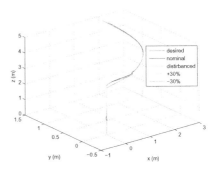

Fig. 6. Path Tracking 2

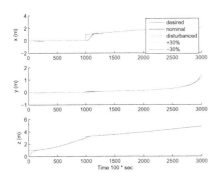

Fig. 7. Positions 2

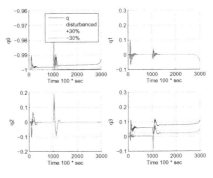

Fig. 8. Quaternion Components 2

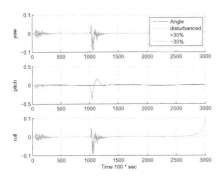

Fig. 9. Angles 2

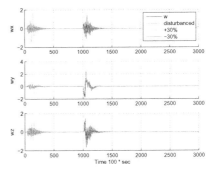

Fig. 10. Angular Velocities 2

6 Conclusions

This paper proposes a robust controller for the path tracking problem of quadrotors. H_∞ approach is used to develop the state feedback controller. A set of conditions is established for the stability of the controller. The details of stability analysis is provided. The proposed controller is tested in simulation and the result shows the proposed controller is able to stabilize the path tracking problem with the added disturbance. Our next step work along this direction of research is to compare this result with popular backstepping controllers to demonstrate how the performance can be improved. We plan to test the proposed controller in real quadrotors to further justify the performance. Taking the uncertainties in position estimation into consideration is also one of our future work.

References

1. Zhao, S., Dong, W., Farrell, J.: Quaternion-based Trajectory Tracking Control of VTOL-UAVs using Command Filtered Backstepping. In: 2013 American Control Conference (ACC), Washington, DC, USA, pp. 1018–1023 (2013)
2. De Monte, P., Lohmann, B.: Trajectory Tracking Control for a Quadrotor Helicopter based on Backstepping using a Decoupling Quaternion Parametrization. In: 2013 21st Mediterranean Conference on Control and Automation (MED), Platanias-Chania, Crete, Greece, pp. 507–512 (2013)
3. Do, K.D., Paxman, J.: Global Tracking Control of Quadrotor VTOL Aircraft in Three Dimensional Space. In: 2013 Australian Control Conference, Perth, Australia, pp. 26–33 (2013)
4. Lee, T., Leok, M., McClamroch, N.H.: Nonlinear Robust Tracking Control of a Quadrotor UAV on SE(3). In: 2012 American Control Conference, pp. 4649–4654. Fairmont Queen Elizabeth, Montreal (2012)
5. Parra-Vega, V., Sanchez, A., Izaguirre, C.: Toward Force Control of a Quadrotor UAV in SE(3). In: 51st IEEE Conference on Decision and Control, Maui, Hawaii, USA, pp. 1802–1809 (2012)
6. Sanchez, A., Parra-Vega, V., Garcia, O., Ruiz-Sanchez, F., Ramos-Velasco, L.E.: Time-Parametrization Control of Quadrotors with a Robust Quaternion-based Sliding Mode Controller for Aggressive Maneuvering. In: 2013 European Control Conference (ECC), Zurich, Switzerland, pp. 3876–3881 (2013)
7. Sanchez, A., Parra-Vega, V., Tang, C., Oliva-Palomo, F., Izaguirre-Espinosa, C.: Continuous Reactive-based Position-Attitude Control of Quadrotors. In: 2012 American Control Conference, pp. 4643–4648. Fairmont Queen Elizabeth, Montreal (2012)
8. Fritsch, O., De Monte, P., Buhl, M., Lohmann, B.: Quasi-static Feedback Linearization for the Translational Dynamics of a Quadrotor Helicopter. In: 2012 American Control Conference, pp. 125–130. Fairmont Queen Elizabeth, Montreal (2012)
9. Van der Schaft, A.: L_2-gain Analysis of Nonlinear Systems and Nonlinear State Feedback H_∞ Control. IEEE Transactions on Automatic Control 37(6), 770–784 (1992)

10. Dalsmo, M., Egeland, O.: State Feedback H_∞ Suboptimal Control of a Rigid Spacecraft. IEEE Transactions on Automatic Control 42(8), 1186–1189 (1997)
11. Raffo, G.V., Ortega, M.G., Rubio, F.R.: Nonlinear H_∞ Controller for the Quad-rotor Helicopter with Input Coupling. In: 18th IFAC World Congress, pp. 13834–13839. Milano, Italy (2011)
12. Jasim, W., Gu, D.: H_∞ Control for Quadrotor Attitude Stabilization. In: UKACC 10th International Conference on Control (CONTROL 2014), Loughborough, UK, July 9-11, pp. 25–30 (2014)

Towards an Ethical Robot: Internal Models, Consequences and Ethical Action Selection

Alan F.T. Winfield[1], Christian Blum[2], and Wenguo Liu[1]

[1] Bristol Robotics Laboratory, UWE Bristol, UK
[2] Cognitive Robotics, Department of Computer Science, Humboldt-Universität zu Berlin, Germany

Abstract. If robots are to be trusted, especially when interacting with humans, then they will need to be more than just safe. This paper explores the potential of robots capable of modelling and therefore predicting the consequences of both their own actions, and the actions of other dynamic actors in their environment. We show that with the addition of an 'ethical' action selection mechanism a robot can sometimes choose actions that compromise its own safety in order to prevent a second robot from coming to harm. An implementation with e-puck mobile robots provides a proof of principle by showing that a simple robot can, in real time, model and act upon the consequences of both its own and another robot's actions. We argue that this work moves us towards robots that are ethical, as well as safe.

Keywords: Human-Robot Interaction, Safety, Internal Model, Machine Ethics.

1 Introduction

The idea that robots should not only be safe but also actively capable of preventing humans from coming to harm has a long history in science fiction. In his short story Runaround, Asimov coded such a principle in his now well known Laws of Robotics [1]. Although no-one has seriously proposed that real-world robots should be 'three-laws safe', work in machine ethics has advanced the proposition that future robots should be more than just safe. For instance, in their book Moral Machines, Wendell and Allen [16] write

> "If multipurpose machines are to be trusted, operating untethered from their designers or owners and programmed to respond flexibly in real or virtual world environments, there must be confidence that their behaviour satisfies appropriate norms. This goes beyond traditional product safety ... if an autonomous system is to minimise harm, *it must also be 'cognisant' of possible harmful consequences of its actions, and it must select its actions in the light of this 'knowledge'*, even if such terms are only metaphorically applied to machines." (italics added).

M. Mistry et al. (Eds.): TAROS 2014, LNAI 8717, pp. 85–96, 2014.

This paper describes an initial exploration of the potential of robots capable of modelling and therefore predicting the consequences of both their own actions, and the actions of other dynamic actors in their environment. We show that with the addition of an 'ethical' action selection mechanism, a robot can sometimes choose actions that compromise its own safety in order to prevent a second robot from coming to harm.

This paper proceeds as follows. First we introduce the concept of internal modelling and briefly review prior work on robots with internal models. In section 3 we outline a generic internal-model based architecture for autonomous robots, using simulation technology, and show in principle how this might be used to implement simple 'Asimovian' ethics. In section 4 we outline an implementation of this architecture with e-puck robots, and in section 5 present experimental results from tests with 1, 2 and 3 robots.

2   Robots with Internal Models

In this paper we define a robot with an internal model as a robot with an embedded *simulation* of itself *and* its currently perceived environment. A robot with such an internal model has, potentially, a mechanism for generating and testing *what-if* hypotheses:

1. *what if* I carry out action x? and, ...
2. ... of several possible next actions x_i, *which* should I choose?

Holland writes: "an internal model allows a system to look ahead to the future consequences of current actions, without actually committing itself to those actions" [4]. This leads to the idea of an internal model as a *consequence engine* – a mechanism for estimating the consequences of actions.

The use of internal models within control systems is well established, but these are typically mathematical models of the plant (system to be controlled). Typically a set of first-order linear differential equations models the plant, and these allow the design of controllers able to cope with reasonably well defined uncertainties; methods also exist to extend the approach to cover non-linear plant [6]. In such internal-model based control the environment is not modelled explicitly – only certain exogenous disturbances are included in the model. This contrasts with the internal simulation approach of this paper which models both the plant (in our case a robot) and its operational environment.

In the field of cognitive robots specifically addressing the problem of machine consciousness [5], the idea of embedding a simulator in a robot has emerged in recent years. Such a simulation allows a robot to try out (or 'imagine') alternative sequences of motor actions, to find the sequence that best achieves the goal (for instance, picking up an object), before then executing that sequence for real. Feedback from the real-world actions might also be used to calibrate the robot's internal model. The robot's embodied simulation thus adapts to the body's dynamics, and provides the robot with what Marques and Holland [8] call a 'functional imagination'.

Bongard *et al.* [2] describe a 4-legged starfish like robot that makes use of explicit internal simulation, both to enable the robot to learn it's own body morphology and control, and notably allow the robot to recover from physical damage by learning the new morphology following the damage. The internal model of Bongard *et al.* models only the robot, not its environment. In contrast Vaughan and Zuluaga [15] demonstrated self-simulation of both a robot and its environment in order to allow a robot to plan navigation tasks with incomplete self-knowledge; they provide perhaps the first experimental proof-of-concept of a robot using self-modelling to anticipate and hence avoid unsafe actions.

Zagal *et al.* [17] describe self-modelling using internal simulation in humanoid soccer robots; in what they call a 'back-to-reality' algorithm, behaviours adapted and tested in simulation are transferred to the real robot. In a similar approach, but within the context of evolutionary swarm robotics O'Dowd *et al.* [11] describe simple wheeled mobile robots which embed within each robot a simulator for both the robot and its environment; a genetic algorithm is used to evolve a new robot controller which then replaces the 'live' robot controller about once every minute.

3   An Internal-Model Based Architecture

Simulation technology is now sufficiently well developed to provide a practical basis for implementing the kind of internal model required to test *what-if* hypotheses. In robotics advanced physics and sensor based simulation tools are commonly used to test and develop, even evolve, robot control algorithms before they are tested in real hardware. Examples of robot simulators include Webots [9] and Player-Stage [14]. Furthermore, there is an emerging science of simulation, aiming for principled approaches to simulation tools and their use [12].

Fig. 1 proposes an architecture for a robot with an internal model which is used to test and evaluate the consequences of the robot's next possible actions. The machinery for modelling next actions is relatively independent of the robot's controller; the robot is capable of working normally without that machinery, albeit without the ability to generate and test *what-if* hypotheses. The *what-if* processes are not in the robot's main control loop, but instead run in parallel to moderate the Robot Controller's normal action selection, if necessary acting to 'govern' the robot's actions.

At the heart of the architecture is the Consequence Engine (CE). The CE is initialised from the Object Tracker-Localiser, and loops through all possible next actions. For each candidate action the CE simulates the robot executing that action, and generates a set of model outputs ready for evaluation by the Action Evaluator (AE). The AE evaluates physical consequences, which are then passed to a separate Safety/ethical Logic (SEL) layer. (The distinction between the AE and SEL will be elaborated below.) The CE loops through each possible next action. Only when the complete set of next possible actions has been tested, does the CE pass weighted actions to the Robot Controller's Action Selection (AS) mechanism.

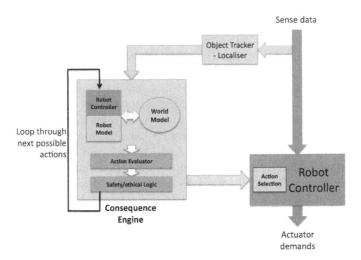

Fig. 1. Internal-model based architecture. Robot control data flows are shown in red (darker shaded); the Internal Model data flows in blue (lighter shaded).

3.1   Towards an Ethical Robot

Consider the scenario illustrated in Fig. 2. Here there are two actors: our self-aware robot and a human. The environment also contains a hole in the ground, of sufficient size and depth that it poses a serious hazard to both the robot and the human. For simplicity let us assume the robot has four possible next actions, each of which is simulated. Let us output *all* safety outcomes, and in the AE assign to these a numerical value which represents the estimated degree of danger. Thus 0 indicates 'safe' and (say) 10 'fatal'. An intermediate value, say 4, might be given for a low-speed collision: unsafe but probably low-risk, whereas 'likely to fall into a hole' would merit the highest danger rating of 10. Secondly,

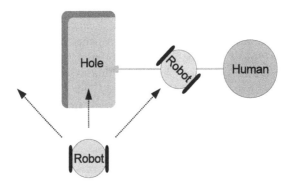

Fig. 2. A scenario with both safety and ethical consequences

we also output, to the AE, the same safety consequence of the other actor(s) in the environment - noting that the way we have specified the CE and its inputs, means that the CE is equally capable of modelling the effect of hazards on *all* dynamic actors in the environment, including itself. The ability to model and hence anticipate the consequences of another dynamic actor's actions means that the CE arguably provides the robot with a very simple artificial theory of mind for that actor. If the actor is a human then we now see the possibility of the robot choosing to execute an unsafe action in order to prevent that human from coming to harm.

Table 1. Safety outcome values for each robot action, for scenario in Fig. 2

Robot action	Robot outcome	Human outcome	Interpretation
Ahead Left	0	10	robot safe, but human falls into hole
Ahead	10	10	both robot and human fall into hole
Ahead Right	4	4	robot collides with human
Stand still	0	10	robot safe, but human falls into hole

Tab. 1 shows the safety outcome values that might be generated by the AE for each of the four possible next actions of the robot, for both the robot and human actors in this scenario. From the robot's perspective, 2 of the 4 actions are safe: *Ahead Left* means the robot avoids the hole, and *Stand Still* means the robot also remains safe. Both of the other actions are unsafe for the robot, but *Ahead* is clearly the most dangerous, as it will result in the robot falling into the hole. For the human, 3 out of 4 of the robot's actions have the same outcome: the human falling into the hole. Only 1 action is safer for the human: if the robot moves *Ahead Right* then it might collide with the human before she falls into the hole.

In order for the AE to generate the action *Ahead Right* in this scenario it clearly needs both a safety rule and an 'ethical' rule, which can take precedence over the safety rule. This logic, in the SEL, might take the form:

```
IF for all robot actions, the human is equally safe
THEN (* default safe actions *)
    output safe actions
ELSE (* ethical action *)
    output action(s) for least unsafe human outcome(s)
```

What we have set out here appears to match remarkably well with Asimov's first law of robotics: *A robot may not injure a human being or, through inaction, allow a human being to come to harm* [1]. The schema proposed here will avoid injuring (i.e. colliding with) a human ('may not injure a human'), but may also sometimes compromise that rule in order to prevent a human from coming to harm ('...or, through inaction, allow a human to come to harm'). This is not

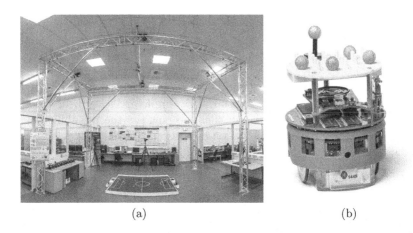

(a)   (b)

Fig. 3. (a) Experimental infrastructure showing Vicon tracking system. (b) An e-puck with Linux board fitted in between the e-puck motherboard (lower) and the e-puck speaker board (upper). Note the yellow 'hat' (which provides a matrix of pins for the reflective spheres which allow the tracking system to identify and track each robot).

to suggest that a robot which apparently implements part of Asimov's famous laws is ethical in any formal sense (i.e. that an ethicist might accept). But the possibility of a route toward engineering a minimally ethical robot does appear to be presented.

4   Implementation

In order to test the ideas set out above we have implemented the internal-model based architecture of Fig. 1 on an e-puck mobile robot [10], equipped with a Linux extension board [7]. We shall refer to this as robot A[1]. A's internal model makes use of the open source simulator Stage [13], and since Stage requires greater resources than are available to the e-puck's Linux board, it is run on an external laptop computer linked to the e-puck via the local WiFi network. Furthermore, object tracking and localisation is not implemented directly using A's onboard sensors, but is implemented as a virtual sensor using the Vicon tracking system. Robot A does, however, use its onboard infra-red proximity sensors for short-range obstacle avoidance. Fig. 3 shows both the experimental infrastructure and an e-puck robot.

The scenario shown in Fig. 2 is implemented experimentally by creating a virtual hole in the ground, of size $60\,cm$ x $60\,cm$ in an arena of size $220\,cm$ x $180\,cm$; this virtual hole is sensed by robot A's virtual sensor. A second e-puck robot (H) acts as a proxy for the 'human' in Fig. 2 (and later, a third robot H2). Robot H does not have the internal-modelling architecture of robot A. It has a simple control system allowing it to move around the arena, avoiding obstacles

---

[1] After Asimov.

with its infra-red proximity sensors, but lacking the virtual sensor of robot A it is unable to 'see' the hole in the arena.

Robot A's virtual sensor allows it to both see the hole in the arena and also track the position and direction of motion of robot H. Robot A is thus able to initialise its CE with both its own position and heading, and that of robot H. Robot A runs its CE every $0.5\,s$, to simulate the next possible actions of both itself and H.

Robots run a stateless controller with a fixed set of pre-programmed sub-actions. Those sub-actions are: `GoStraight(speed)` with a maximum Speed of $1.0\,m/s$, `Avoidance` for Braitenberg [3] style avoidance using IR sensors, `MoveTo(x,y)` using the virtual global position sensors, and `Stop`. Actions are composed of concatenated sub-actions and are executed at $10\,Hz$ within the robots, independently of the CE.

In order to reduce WiFi network traffic and latencies, and facilitate data logging, the CE and AS run on the same laptop computer as the simulation. Furthermore, the set of possible actions is the same in all experiments. Note also that in this implementation the world model is pre-programmed within the simulation and thus robot A is unable to respond to environmental changes.

4.1   Simulation Budget

The CE re-initalises and refreshes at a speed of $\sim 2\,Hz$, allowing $0.5s$ to simulate the set of actions, analyze them and generate the corresponding safety values. In relation to the computational power necessary for the simulation, the other tasks are negligible so, for simplicity, we discount them from this analysis.

In our configuration, Stage runs at about 600 times real time which means a total of about $300\,s$ can be simulated between two runs of the CE. We chose a simulate-ahead time of $10\,s$ which corresponds to $0.7\,m$ movement for robots H and H2 or a maximum of $1\,m$ for robot A. This means we are able to simulate a total of about 30 different next possible actions.

4.2   Real World Safety Outcome Values

In Sec. 3.1 we described how the AE can evaluate the consequence of actions. For simplicity, the example shows only 4 possible actions, tailored to fit the exemplary situation described. In a real robot we can make full use of the simulation budget (see Sec. 4.1) and evaluate more than just a minimal number of tailored actions to generate more flexible robot behaviours.

We generate actions by discretizing the space needed for the experiment into a grid of points to which the robot can move. Trivially one would discretize the whole arena but simulating all these actions would exceed our simulation budget so we chose a smaller area around the virtual hole and the goal. Specifically an area of $1\,m$ x $1\,m$ was discretized into a 6 x 5 grid of points, some of which fall inside the virtual hole.

Since we are now dealing with a larger number of actions, we need an algorithmic way to calculate safety outcome values for all those actions. For this we choose the paradigm of virtual potential functions. We employ one Potential Function (PF) which drives Robot A towards its goal, similar to the second column in Tab. 1. Another, stronger PF is employed if the simulation shows danger for one of the other robots and favours actions which move robot A towards the robot in danger. This second PF is only employed when danger is imminent and is zero otherwise (this PF is not strictly necessary but signifiantly improves the reaction times of robot A). The sum of these PFs is sampled at the grid points and assigned as basic safety values to the actions.

We place no additional penalty on getting too close to other robots during normal operation since we are using the robots' real IR sensors and controllers for basic collision avoidance. If this aspect were to be included, the PF could be used to discourage areas close to other robots. After assigning the basic safety outcome values, robot A's SEL considers the estimated danger for all robots, generating effectively the equivalent to the third column of Tab. 1.

5   Results

We have conducted 3 sets of experimental trials with the setup outlined above. The first trial consists of robot A only, navigating a safe path to its goal destination while using its CE system to safely avoid the hole in the arena. This trial provides us with a baseline test in which A has only to ensure its own safety. The second trial adds robot H, acting as a proxy human, to test the ability of robot A to model both itself and H, and if necessary deliberately interact with H in order to prevent it from reaching the hole. A third trial adds a second proxy human robot H2 in order to present A with a dilemma: can it prevent both H and H2 from coming to harm?

5.1   Trial 1: Baseline with Robot A Only

In this trial the safety values consist only of the original PF driving robot A towards its goal. The starting position and goal are chosen in such a way that the unmodified PF, which is proportional to the distance to the to the goal, would drive robot A straight into the virtual hole. The CE then evaluates all possible actions and penalizes the ones driving robot A into the hole, effectively guiding it around the hole. Overlaid trajectories for this trial are shown in Fig. 4(a) and show that robot A is able to avoid falling into the virtual hole, with 100% reliability.

5.2   Trial 2: Robots A and H

This trial is an extension of the first, with the same goal and initial condition for robot A. To demonstrate our approach, we added the second robot, H, as described in Sec. 4 with its internal Robot Controller (RC) running the simple action (GoStraight(0.7); Avoidance) and initial conditions which point it directly towards the hole. Successive snapshots of a typical experimental run are shown in Fig. 5.

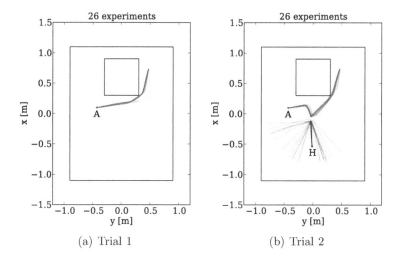

(a) Trial 1   (b) Trial 2

Fig. 4. Superimposed trajectories of robots for trials 1 and 2. Robot A is shown in red, with start position on the left and goal on the upper right; robot H is shown in blue with start position in the lower centre of the arena. Note in trial 2 the near collisions between A and H cause H to be deflected away from the hole.

The run starts with robot A following the same trajectory as in the first trial, but as soon as its CE for robot H shows that H would fall into the hole if not intercepted, A diverts from its normal trajectory to intercept and thus 'rescues' robot H. A then continues back onto its original trajectory and reaches its goal.

Fig. 4(b) shows trajectories for a number of experiments. In all cases robot A succeeds in rescuing robot H by intercepting and hence diverting H. The beginning and end of A's trajectories are exactly the same as in the first trial.

5.3   Trial 3: Robots A's Dilemma

Here a third robot H2 is introduced, presenting robot A with the dilemma of having to decide which of H and H2 to rescue. Both H and H2 start pointing towards, and equidistant from, the virtual hole (see Fig. 6(a)), while the initial and goal positions for robot A remain unchanged.

Fig. 6 shows successive snapshots for one experimental run. Robot A is unable to resolve its dilemma in this particular run since its CE does not favour either H or H2, which results in A trying to rescue both at the same time and failing to rescue either.

Trajectories over a series of 33 runs are shown in Fig. 7(a). The number of robots A actually rescued are shown in Fig. 7(b). Surprisingly and perhaps counter-intuitively, A is able to rescue at least one robot in about 58% of runs, and both robots in 9%. The reason for this is noise. The robots don't start at exactly the same position every time, nor do they start at precisely the same time in every run. Thus, sometimes A's CE for one robot indicates danger first

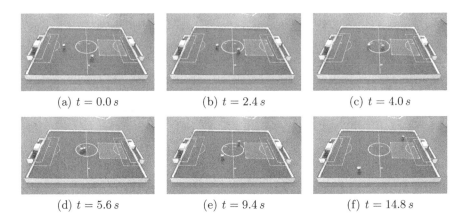

(a) $t = 0.0\,s$ (b) $t = 2.4\,s$ (c) $t = 4.0\,s$

(d) $t = 5.6\,s$ (e) $t = 9.4\,s$ (f) $t = 14.8\,s$

Fig. 5. (a) start (b) Robot A starts normal operation and moves towards its goal. Robot H starts moving towards the rectangular 'hole' (shown shaded). (c) A's CE detects danger for H and moves to intercept it. (d) A intercepts H. (e) Danger for H is averted and A continues towards its goal, avoiding the hole. (f) A reaches its goal. Note, the other markings in the arena have no significance here.

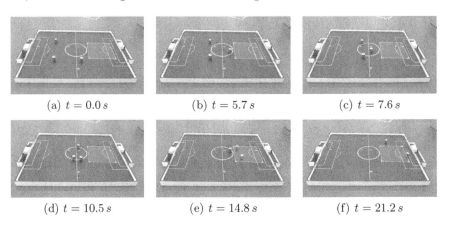

(a) $t = 0.0\,s$ (b) $t = 5.7\,s$ (c) $t = 7.6\,s$

(d) $t = 10.5\,s$ (e) $t = 14.8\,s$ (f) $t = 21.2\,s$

Fig. 6. (a) Initial conditions with H and H2 pointing towards the 'hole'. (b) A detects danger for both H and H2. (c) A cannot decide which of the robots to rescue. (d) A misses the chance to rescue either robot. (e) A turns around to continue towards its goal since it's now too late to rescue the other robots. (f) Robot A reaches its goal.

and since the CE only runs at $2\,Hz$, A by chance rescues this robot. As soon as one robot is rescued, the experiment resembles trial 2 and if physically possible, i.e., A has enough time left to react before the other robot reaches the virtual hole, it also rescues that robot.

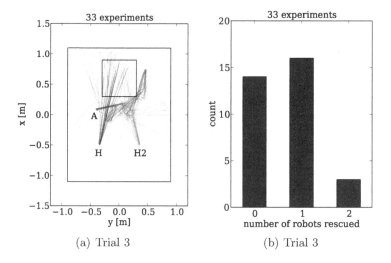

(a) Trial 3   (b) Trial 3

Fig. 7. (a) Trajectories of Robots (b) Success rate

6   Conclusions

In this paper we have proposed an internal-modelling based architecture for a
minimally ethical robot, and – as proof of principle – implemented the architecture on a simple mobile robot we call A. Mobile robot H acts as a proxy human
in a situation hazardous to both robot and human. Since the robots are relatively
simple, then we can run models for both A and H in real-time, with sufficient
simulation budget to be able to simulate ahead and evaluate the consequences
of around 30 next possible actions, for both robots, every 0.5 s. Experimental
trials show that A is able to maintain its own safety, avoid falling into a (virtual)
hole, and if its internal model indicates that H is in danger A will divert from
its course in order to provoke a collision avoidance response from H in order to
deflect H away from danger.

Simulation errors resulting from the reality-gap between real and modelled
robots are mitigated by the periodic memoryless refresh of A's CE, which means
that as A approaches H the error reduces and A is able to reliably encounter
H. A limitation of this implementation is that A assumes H will continue, in a
straight line, at its current heading and velocity. In reality H does not travel in
a perfect line, but again A's periodically refreshed CE compensates for this.

Our 3rd trial, in which A is faced with two robots H and H2 both approaching
danger at the same time, illustrates that even a minimally ethical robot can
indeed face a dilemma. The suprising experimental outcome that A does, in
fact, succeed in 'rescuing' one or more robots in about 58% of runs is a result of
noise, by chance, breaking the latent symmetry in the experimental setup. We
could introduce a rule, or heuristic, that allows A to choose H or H2 (when noise
hasn't already made the choice), but deliberately chose not to on the grounds

that such a rule should be determined on ethical rather than engineering grounds. If ethical robots prove to be a practical proposition their design and validation will need to be a collaborative effort of roboticist and ethicist.

Acknowledgments. We are grateful to the Deutscher Akademischer Austausch Dienst (DAAD) for supporting Christian Blum while visiting researcher at the Bristol Robotics Lab.

References

1. Asimov, I.: I, ROBOT. Gnome Press (1950)
2. Bongard, J., Zykov, V., Lipson, H.: Resilient machines through continuous self-modeling. Science 314(5802), 1118–1121 (2006)
3. Braitenberg, V.: Vehicles: Experiments in synthetic psychology. MIT Press (1984)
4. Holland, J.: Complex Adaptive Systems. Daedalus (1992)
5. Holland, O. (ed.): Machine Consciousness. Imprint Academic (2003)
6. Isidori, A., Marconi, L., Serrani, A.: Fundamentals of internal-model-based control theory. In: Robust Autonomous Guidance. Advances in Industrial Control, pp. 1–58. Springer, London (2003)
7. Liu, W., Winfield, A.F.T.: Open-hardware e-puck Linux extension board for experimental swarm robotics research. Microprocessors and Microsystems 35(1) (2011)
8. Marques, H., Holland, O.: Architectures for functional imagination. Neurocomputing 72(4-6), 743–759 (2009)
9. Michel, O.: Webots: Professional mobile robot simulation. International Journal of Advanced Robotic Systems 1(1), 39–42 (2004)
10. Mondada, F., Bonani, M., Raemy, X., Pugh, J., Cianci, C., Klaptocz, A., Magnenat, S., Zufferey, J.C., Floreano, D., Martinoli, A.: The e-puck, a robot designed for education in engineering. In: Proc. 9th Conference on Autonomous Robot Systems and Competitions, pp. 59–65 (2009)
11. O'Dowd, P.J., Winfield, A.F.T., Studley, M.: The distributed co-evolution of an embodied simulator and controller for swarm robot behaviours. In: Proc. IEEE/RSJ Int. Conf. on Intelligent Robots and Systems (IROS), pp. 4995–5000 (2011)
12. Stepney, S., Welch, P., Andrews, P. (eds.): CoSMoS 2011: Proc. 2011 Workshop on Complex Systems Modelling and Simulation. Luniver Press (2011)
13. Vaughan, R.: Massively multi-robot simulation in stage. Swarm Intelligence 2(2-4), 189–208 (2008)
14. Vaughan, R.T., Gerkey, B.P.: Really reused robot code from the player/stage project. In: Brugali, D. (ed.) Software Engineering for Experimental Robotics, pp. 267–289. Springer (2007)
15. Vaughan, R.T., Zuluaga, M.: Use your illusion: Sensorimotor self-simulation allows complex agents to plan with incomplete self-knowledge. In: Proc. International Conference on Simulation of Adaptive Behaviour (SAB), pp. 298–309 (2006)
16. Wallach, W., Allen, C.: Moral Machines: Teaching Robots Right from Wrong. Oxford University Press, Oxford (2009)
17. Zagal, J.C., Delpiano, J., Ruiz-del Solar, J.: Self-modeling in humanoid soccer robots. Robot. Auton. Syst. 57(8), 819–827 (2009)

"The Fridge Door is Open"–Temporal Verification of a Robotic Assistant's Behaviours

Clare Dixon[1], Matt Webster[1], Joe Saunders[2],
Michael Fisher[1], and Kerstin Dautenhahn[2]

[1] Dept. of Computer Science, University of Liverpool, L69 3BX, UK
[2] Adaptive Systems Research Group, University of Hertfordshire, AL10 9AB, UK

Abstract. Robotic assistants are being designed to help, or work with, humans in a variety of situations from assistance within domestic situations, through medical care, to industrial settings. Whilst robots have been used in industry for some time they are often limited in terms of their range of movement or range of tasks. A new generation of robotic assistants have more freedom to move, and are able to autonomously make decisions and decide between alternatives. For people to adopt such robots they will have to be shown to be both safe and trustworthy. In this paper we focus on formal verification of a set of rules that have been developed to control the Care-O-bot, a robotic assistant located in a typical domestic environment. In particular, we apply *model-checking*, an automated and exhaustive algorithmic technique, to check whether formal temporal properties are satisfied on all the possible behaviours of the system. We prove a number of properties relating to robot behaviours, their priority and interruptibility, helping to support both safety and trustworthiness of robot behaviours.

1 Introduction

Robot assistants are being developed to help, or work, closely with humans in industrial, domestic and health care environments. In these environments the robots will need to be able to act autonomously and make decisions to choose between a range of activities and yet will need to operate close to, or in collaboration with humans. We need to make sure that such robotic assistants are both safe and trustworthy. Safety involves showing that the robot does nothing that (unnecessarily) endangers the person. To this end the International Organization for Standardization (ISO) TC184/SC2 Technical Committee has been working on ISO 13482, a standard relating to safety requirements for non-industrial robots, i.e. non-medical personal care robots[1]. Trustworthiness involves social issues beyond pure safety. It is not just a question of whether the robots are safe but whether they are perceived to be safe, useful and reliable, and will not do anything we would consider unpleasant, unfriendly, or dangerous.

In this paper we consider the application of formal verification to the Care-O-bot® [10], an autonomous robotic assistant deployed in a domestic-type house

---

[1] www.sis.se/popup/iso/isotc184sc2/index.asp

M. Mistry et al. (Eds.): TAROS 2014, LNAI 8717, pp. 97–108, 2014.

at the University of Hertfordshire. Low-level robot actions such as movement, speech, light display, etc., are controlled by groups of high-level rules that together define particular behaviours; for example, a sequence of rules to let the user know that the fridge door is open. We apply *formal verification*, in particular *model-checking* [2], to such behaviours within the Care-O-bot. In model-checking, a mathematical model of *all* the possible executions of the system is constructed (often a finite state transition system) and then *all* possible routes through this model are checked against a required logical formula representing a desired formal property of the system. Model-checking is different from running robot test experiments as it can check that a property holds on *all* paths through the input model. Robot experiments remain useful, allowing the observation of what occurs on *particular* runs of the real system. Together, model-checking the behaviours for the robot alongside real robot experiments within the robot house gives us a stronger assurance that the robot behaves as desired and helps convince users of its trustworthiness.

In preliminary work [18] we modelled the Care-O-bot behaviours using a human-robot teamwork modelling language, known as Brahms [14], and applied formal verification to the resulting Brahms models. Brahms is a multi-agent modelling, simulation and development environment designed to model both human and robotic activity using rational agents and has been used at NASA for modelling astronaut-robot planetary exploration teams. As the Care-O-bot behaviours have similarities with Brahms constructs, and as a tool [15] has been developed to translate from Brahms models into the SPIN model-checker [5], this route was adopted as a preliminary, quick and systematic way to model-check Care-O-bot behaviours. Issues with the approach involve the necessity to first produce a Brahms model, the resulting models being unnecessarily large due to the need to translate aspects of Brahms unnecessary to the Care-O-bot rules, the length of time for verification of simple properties, and issues relating to the modelling of non-determinism. Additionally, as this was a preliminary attempt at model-checking robot behaviours, features such as selecting between alternative behaviours using inbuilt priorities and whether the behaviours could be interrupted were not modelled.

In this paper we provide a translation (by hand) of the behaviours for the Care-O-bot directly into input suitable for a particular model-checker. Whilst a hand crafted translation is both time consuming and potentially error prone it means that we have much greater control over the size of the models (and related verification times) in terms of the level of abstraction taken. Additionally we can allow much more non-determinism in the models (for example allowing the variables relating to sensor information such as "the television being on", "the fridge door being open", "the doorbell ringing", etc., to be set non-deterministically). A longer term aim is to develop an automated, direct translation from sets of Care-O-bot behaviours to one or more model-checkers and this "by-hand" translation gives us useful insight into how to achieve this.

This paper is organised as follows. In Section 2 we describe the Care-O-bot robot assistant and its environment. In Section 3 we give more details about temporal logic and model-checking and explain how the translation into input to the model-checker has been carried out. In Section 4 we give results from formally verifying several properties via the model-checker. In Section 5 we discuss related work, while conclusions and future work are provided in Section 6.

2   The Robot House and the Care-O-bot

The University of Hertfordshire's "robot house" is a typical suburban house near Hatfield. While offering a realistic domestic environment along with typical house furnishings, the robot house is also equipped with sensors which provide information on the state of the house and its occupants, such as whether the fridge door is open and whether someone is seated on the sofa [11,3].

The robot house can be used to conduct Human-Robot Interaction experiments in a setting that is more natural and realistic than a university laboratory (e.g. [12,16]). One of the robots in the house is the (commercially available) Care-O-bot robot manufactured by Fraunhofer IPA [10]. It has been specifically developed as a mobile robotic assistant to support people in domestic environments, and is based on the concept of a robot butler. The Care-O-bot robot, shown in Figure 3, has a manipulator arm incorporating a gripper with three fingers, an articulated torso, stereo sensors serving as "eyes", LED lights, a graphical user interface, and a moveable tray. The robot's sensors monitor its current location, the state of the arm, torso, eyes and tray. The robot can "speak" in that it can express text as audio output using a text-to-speech synthesising module.

The robot's software is based on the Robot Operating System (ROS)[2]. For example, to navigate to any designated location within the house, the robot uses the ROS navigation package in combination with its laser range-finders to perform self-localisation, map updating, path planning, and obstacle avoidance in real-time while navigating along the planned route. High-level rules are sent to the robot via the ROS script server mechanism and these are then interpreted into low-level actions by the robot's software. For example, high-level rules can take the form "lower tray", "move to sofa area of the living room", "say 'The fridge door is open' ", etc. The Care-O-bot's high-level decision making is determined by a set of behaviours which are stored in a database. Behaviours (a set of high level rules) take the form:

`Precondition-Rules -> Action-Rules`

where `Precondition-Rules` are a sequence of propositional statements that are either true or false, linked by Boolean *and* and *or* operators. `Action-Rules` are a sequence rules denoting the actions that the Care-O-bot will perform only if the `Precondition-Rules` hold. The `Precondition-Rules` are implemented as a set of SQL queries and the `Actions-Rules` are implemented through the ROS-based `cob_script_server` package, which provides a simple interface to operate Care-O-bot.

For example the rules for the behaviour `S1-alertFridgeDoor` are provided in Fig. 1. Here, the rule numbers from the database are given, where rules 27 and 31

---

[2] `wiki.ros.org/care-o-bot`

```
27  Fridge Freezer Is *ON* AND has been ON for more than 30 seconds
31  ::514:: GOAL-fridgeUserAlerted is false
    32 Turn light on ::0::Care-o-Bot 3.2 to yellow
    34 move ::0::Care-o-Bot 3.2 to ::2:: Living Room and wait for
          completion
    35 Turn light on ::0::Care-o-Bot 3.2 to white and wait for completion
    36 ::0::Care-o-Bot 3.2 says 'The fridge door is open!' and wait for
          completion
    37 SET ::506::GOAL-gotoCharger TO false
    38 SET ::507::GOAL-gotoTable TO false
    39 SET ::508::GOAL-gotoSofa TO false
    40 ::0::Care-o-Bot 3.2 GUI, S1-Set-GoToKitchen, S1-Set-WaitHere
    41 SET ::514::GOAL-fridgeUserAlerted TO true
```

Fig. 1. The S1-alertFridgeDoor rules

represent the precondition-rules, rules 32, 34-36, 40 provide the (descriptions of the) action-rules while 37-39 and 41 initiate the setting of various flags. Rules 27 and 31 check whether the fridge door is open and GOAL-fridgeUserAlerted is false. If these hold (and the preconditions for no other behaviour with a higher priority hold) then this behaviour will be executed by setting the robot's lights to yellow, moving to the living room, setting the robot's lights to white, saying "The fridge door is open", setting various goals to be false, providing the user several options via the Care-O-bot's interface and finally setting GOAL-fridgeUserAlerted to be true.

The Care-O-bot's database is composed of multiple rules for determining a variety of autonomous behaviours, including checking the front doorbell, telling the person when the fridge door is open, and reminding them to take their medication, etc. The robot house rule database used for this paper (which includes a set of 31 default behaviours) can be obtained from the EU ACCOMPANY projects Git repository[3].

The robot can perform only one behaviour at a time. Each of the behaviours is given a priority between 0 and 90. If the preconditions for more than one behaviour hold at any moment then the behaviour with the highest priority is executed. For example the behaviour S1-alertFridgeDoor has a priority of 60 and S1-gotoKitchen has a priority of 40 so if the preconditions to both were true then the former would be executed. Priorities remain the same throughout the execution. If more than one behaviour has equal priority, then they are loaded in a random order and on execution whatever behaviour is first (in the set of equal priorities) will be executed first. Additionally each behaviour is flagged as interruptible (1) or not (0). In general, behaviours execute to completion, i.e. all the rules that are part of the behaviour are performed, even if the precondition to another behaviour becomes true during its execution. However, behaviours flagged as interruptible are terminated if the precondition of a higher priority behaviour becomes true whilst it is executing. The priorities and interruptible

---

[3] github.com/uh-adapsys/accompany

status (denoted *Int*) of behaviours are given in Table 1. Behaviours with both priority status and interruptible status set to zero are omitted from this table to save space (but are included in the model).

Table 1. Priority Table for Behaviours

Name	Priority	Int	Name	Priority	Int
S1-Med-5PM-Reset	90	0	S1-gotoTable	40	1
checkBell	80	0	S1-kitchenAwaitCmd	40	1
unCheckBell	80	0	S1-sofaAwaitCmd	40	1
S1-remindFridgeDoor	80	0	S1-tableAwaitCmd	40	1
answerDoorBell	70	0	S1-WaitHere	40	1
S1-alertFridgeDoor	60	0	S1-ReturnHome	40	1
S1-Med-5PM	50	1	S1-continueWatchTV	35	1
S1-Med-5PM-Remind	50	1	S1-watchTV	30	1
S1-gotoKitchen	40	1	S1-sleep	10	1
S1-gotoSofa	40	1			

3 Modelling the Care-O-bot Behaviours

Model-checking [2] is a popular technique for formally verifying the temporal properties of systems. Input to the model-checker is a model of all the paths through a system and a logical formula (often termed a *property*) to be checked on that model. A useful feature of model-checkers is that, if there are execution paths of the system that *do not* satisfy the required temporal formula, then at least one such "failing" path will be returned as a counter-example. If no such counter-examples are produced then all paths through the system indeed satisfy the prescribed temporal formula. Here we use the NuSMV [1] model-checker.

The logic we consider is propositional linear-time temporal logic (PTL), where the underlying model of time is isomorphic to the Natural Numbers, $\mathbb{N}$. A model for PTL formulae can be characterised as a sequence of *states* of the form: $\sigma = s_0, s_1, s_2, s_3, \ldots$ where each state, s_i, is a set of proposition symbols, representing those propositions which are satisfied in the ith moment in time.

In this paper we will only make use three of temporal operators: '$\bigcirc$' (*in the next moment in time*), '$\Diamond$' (*sometime in the future*), and '$\Box$' (*always in the future*) in our temporal formulae. The notation $(\sigma, i) \models A$ denotes the truth of formula A in the model σ at state index $i \in \mathbb{N}$, and is recursively defined as follows (where PROP is a set of propositional symbols).

$$(\sigma, i) \models p \quad \text{iff } p \in s_i \text{ where } p \in \text{PROP}$$
$$(\sigma, i) \models \bigcirc A \text{ iff } (\sigma, i+1) \models A$$
$$(\sigma, i) \models \Diamond A \text{ iff } \exists k \in \mathbb{N}. \ (k \geqslant i) \text{ and } (\sigma, k) \models A$$
$$(\sigma, i) \models \Box A \text{ iff } \forall k \in \mathbb{N}. \ (k \geqslant i) \text{ and } (\sigma, k) \models A$$

Note that $\Box$ and $\Diamond$ are duals, i.e. $\Box \varphi \equiv \neg \Diamond \neg \varphi$. The semantics of Boolean operators is as usual.

First we identify the variables to use in the NuSMV representation of the model.

Booleans from the Care-O-bot Rules: Many of the Boolean values from the system can be used directly, for example goals GOAL-fridgeUserAlerted, or GOAL-gotoSofa

Robot Actions: Involving its location, the robot torso position, speech, light colour, the orientation of the tray or providing alternatives on the Care-O-bot display for the person to select between are modelled as enumerated types for example location could have the values livingroom, tv, sofa, table, kitchen, charging.

Scheduling Behaviours: We use a variable schedule with an enumerated type for each behaviour, e.g. if schedule = schedule_alert_fridge_door holds this denotes that the preconditions to the S1-alertFridgeDoor behaviour have been satisfied and this behaviour has been selected to run having the highest priority.

Executing Behaviours: We use a variable called execute with an enumerated type for each behaviour involving more than one step eg execute = execute_alert_fridge_door denotes that the S1-alertFridgeDoor behaviour is executing. An enumerated type execute_step with values step0, step1 etc keeps track of which part of the behaviour has been completed.

Fig. 2 gives the schema showing the changes to variables in subsequent states in the state transition diagram for the behaviour S1-alertFridgeDoor when its precondition holds and this behaviour is scheduled. This corresponds with the behaviour in Fig. 1. The first box shows the preconditions that must hold (i.e. fridge_freezer_on and ¬goal_fridge_userAlerted) and that the behaviour must be scheduled (schedule = schedule_alert_fridge_door) before the other variables are set. This behaviour cannot be interrupted (see Fig. 1) so once it is scheduled it will execute to completion. However other behaviours that are interruptible (eg S1-gotoKitchen) may not complete all their steps if the preconditions of another behaviour with a higher priority become true during its execution.

When the previous behaviour has completed a new behaviour is scheduled. To set the next value of schedule in the NuSMV input file, a list of cases are enumerated as follows

$$condition_1 : schedule_1$$
$$\ldots : \ldots$$
$$condition_n : schedule_n$$

where $condition_i$ represents the preconditions to activate the behaviour and $schedule_i$ is the behaviour selected to execute. The behaviours with higher priorities appear above behaviours with lower priorities and NuSMV selects the first case it encounters where the condition is satisfied.

We need to abstract away from some of the timing details included in the database to obtain a model that is discrete, for example, involving delays or timing constraints of 60 seconds or less. The behaviour S1-watchTV involves checking a goal has been been false for 60 minutes. To achieve this we use an enumerated type goal_watch_tv_time with values for every 15 minutes m0,

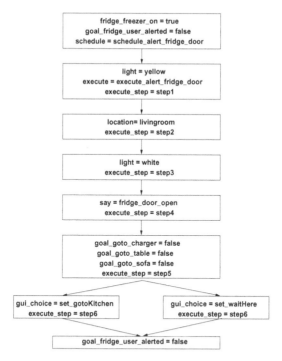

Fig. 2. Schema showing changes to variables for S1-alertFridgeDoor behaviour

m15, m30, m45, m60. We could increase the number of values to represent 5 minute intervals (or even less), for example, but this would increase the size of the model.

4 Verification Using Model-Checking

Here we provide the properties that we checked and the outcome from running these on NuSMV.

1. If the fridge door is open and *goal_fridge_user_alerted* is false then at some-time in the future the Care-O-bot will be in the living room and at some time after that it will say the fridge door is open.

$$\Box((fridge\_freezer\_on \land \neg goal\_fridge\_user\_alerted) \Rightarrow$$
$$\Diamond(location = livingroom \land \Diamond say = fridge\_door\_open))$$

We expect this to be false as, even though the preconditions to the be-haviour S1-alertFridgeDoor are satisfied, preconditions to a behaviour with a higher priority, namely S1-answerDoorBell, might hold and the other behaviour be executed instead of this.

2. If the fridge door is open and *goal_fridge_user_alerted* is false and the `S1-alertFridgeDoor` behaviour has been scheduled then at sometime in the future the Care-O-bot will be in the living room and at some time after that it will say the fridge door is open.

$$\Box((fridge\_freezer\_on \land \neg goal\_fridge\_user\_alerted \land$$
$$schedule = schedule\_alert\_fridge\_door) \Rightarrow$$
$$\Diamond(location = livingroom \land \Diamond say = fridge\_door\_open))$$

We expect this to be true as the `S1-alertFridgeDoor` behaviour is not interruptible so once it is scheduled it should execute to conclusion.

3. If the person selects *goto kitchen* via the Care-O-bot GUI then at sometime in the future its location will be in the kitchen.

$$\Box(gui\_choice = gui\_set\_gotoKitchen \Rightarrow$$
$$\Diamond location = kitchen)$$

We expect this to be false. Selecting *gui_choice = gui_set_gotoKitchen* sets the goal *goal_goto_kitchen* to be TRUE which is the precondition to the behaviour `S1-goToKitchen`. However the behaviour `S1-goToKitchen` may not be scheduled as the preconditions to higher priority behaviours may also be satisfied at the same time. Alternatively it may be scheduled but interrupted before completion.

4. If the person selects *goto kitchen* via the Care-O-bot GUI and this behaviour is scheduled then at sometime in the future its location will be in the kitchen.

$$\Box((gui\_choice = gui\_set\_gotoKitchen \land$$
$$\Diamond schedule = schedule\_goto\_kitchen) \Rightarrow$$
$$\Diamond(schedule = schedule\_goto\_kitchen \land \Diamond location = kitchen))$$

We expect this to be false. By selecting *gui_choice = gui_set_gotoKitchen* (as previously) the goal *goal_goto_kitchen* is set to be TRUE which is the precondition to the behaviour `S1-goToKitchen`. Also, here the behaviour `S1-goToKitchen` has been scheduled but it may be interrupted before completion.

5. If the sofa is occupied, the TV is on and the goal to watch TV has been false for at least 60 minutes, then at some time in the future the Care-O-bot will be located at the sofa and some time after that it will say *shall we watch TV*.

$$\Box((sofa\_occupied \land tv\_on \land \neg goal\_watch\_tv \land goal\_watch\_tv\_time = m60)$$
$$\Rightarrow \Diamond(location = sofa \land say = shall\_we\_watch\_tv))$$

We expect this to be false. Similar to (1) and (3) although the preconditions for behaviour `S1-watchTV` have been satisfied it may not be scheduled as the preconditions to higher priority behaviours may also be satisfied at the same time. Alternatively it may be scheduled but interrupted before completion so that the robot may not have moved to the sofa or may not have said *Shall we watch TV* (or both).

6. If the system attempts to execute a tray raise rule, at some point in the future the physical tray is raised (`physical_tray= raised`) and later the internal flag showing that tray is raised (`tray = raised`) holds.

$$\Box(execute\_raise\_tray \Rightarrow$$
$$\Diamond(physical\_tray = raised \land \Diamond tray = raised))$$

We expect this to be true as `raiseTray` cannot be interrupted.

7. The next property shows the interruption of the behaviour `S1-gotoKitchen` by a higher priority behaviour `S1-alertFridgeDoor`. If the `S1-gotoKitchen` behaviour is executing (and the Care-O-bot is not raising or lowering the tray which is not interruptible) and the preconditions to `S1-alertFridgeDoor` become true then in the next moment the behaviour `S1-gotoKitchen` will not be executing.

$$\Box((execute = execute\_goto\_kitchen \land \neg move\_tray \land$$
$$fridge\_freezer\_on \land \neg goal\_fridge\_user\_alerted) \Rightarrow$$
$$\bigcirc(\neg(execute = execute\_goto\_kitchen)))$$

We expect this to be true due to the priority and interruption settings.

The results from verifying the above properties are given in Fig. 4. The outputs are produced by running NuSMV version 2.5.4 on a PC with a 3.0 GHz Intel Core 2 Duo E8400 processor, 4GB main memory, and 16GB virtual memory running Scientific Linux release 6.4 (Carbon) with a 32-bit Linux kernel. The timings below are carried out running NuSMV with the flags -coi (cone of influence) -dcx (disable generation of counterexamples) -dynamic (enable dynamic variable reordering). The size of the model generated by NuSMV has 130,593 reachable states.

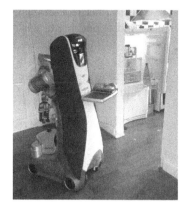

Property	Output	Time (sec)
1	FALSE	11.1
2	TRUE	12.3
3	FALSE	7.7
4	FALSE	9.3
5	FALSE	11.6
6	TRUE	6.4
7	TRUE	6.9

Fig. 3. Care-O-bot in the Robot House   **Fig. 4.** Model-checking Results

The properties (3) and (5) above correspond with properties (1) and (3) from [18]. Note that we cannot prove *any* of the properties that were proved from [18].

One difference between the two models relates to the environment. In [18] a person is explicitly modelled and is the only cause of non-determinism in the model. The person can choose to do one of the following: "move to the kitchen", "move to the living room", "watch television", "send the robot to the kitchen", "send the robot to living room" or "do nothing". The doorbell was assumed not to ring. This means only one behaviour can be triggered at any one time, significantly reducing non-determinism in the previous work. In the current paper a person is not explicitly modelled, although this could be done, and sensors such as the television being on, the doorbell ringing, etc., are allowed to be arbitrarily set to be true or false at any point in the model so several behaviours might be triggered at once. Additionally, in the current work, priorities and interruptions were modelled which were not considered in the previous work. The properties from [18] could not be proved here because the preconditions for higher priority behaviours might also hold and so be executed instead or they might be interrupted by higher priority behaviours during execution.

The size of the model here (130,593 reachable states) is much smaller than that in [18] (with 652,573 or more states). Additionally the verification times for the four properties in [18] took between 20-30 seconds, longer than the properties analysed in this paper. We believe this is because the direct hand-crafted translation we use here avoids the need for the translation of constructs required in the Brahms modelling language whilst still retaining the meaning of the Care-O-bot behaviours. Additionally, as mentioned previously here we have abstracted away from some of the timing details such as "wait for 5 seconds".

5   Related Work

This work applies model-checking to the behaviours of the Care-O-bot robot located in the University of Hertfordshire's robot house. Further details about robot house, robot architecture and control systems and experiments with participants in the robot house can be found, for example, in [3,12,16,11,13]. This paper builds on and extends the work described in [18]. As mentioned in Section 4 we use a different environment model allowing more non-determinism, model behaviour about priorities and interruptions and provide a direct translation into input to the NuSMV model-checker.

In [15] verification of a domestic home care scenario involving a person, a human carer, a robotic assistant and a house agent is considered. The scenario is modelled in Brahms and an automated translation from Brahms models into the SPIN model-checker is provided. However, here the scenario we analyse relates to real code used in practice in real life experiments in the University of Hertfordshire's robot house.

Other research using model-checking to verify aspects of robot behaviour include the verification of safety properties for a core of the Samsung Home Robot (SHR) [6], the analysis of robot motion [4], and the application of model-checking to the control system responsible for slowing down and stopping of a wheeled off-road vehicle [9]. Other formal methods used for robot verification include the application of quantified differential dynamic logic to prove physical location properties for algorithms controlling a surgical robot [7], the use of hybrid automata and statecharts to model and verify a multi-robot rescue scenario [8] and the application of an interactive theorem prover to specify and verify robot collision avoidance algorithms [17].

6 Conclusions and Future Work

We have modelled the behaviours of a robotic assistant in the model-checker NuSMV and proved a number of properties relating to this. The size of models and verification times were less than in previous work [18]. The priorities and interruptibility of the behaviours were modelled so that even if the satisfaction of the preconditions of behaviours such as S1-alertFridgeDoor or S1-gotoKitchen became true these behaviours might not be fully executed because of higher priority behaviours being scheduled instead. Many of the timing details were removed from the models but more detailed timings could be included at the expense of the size of models and verification times. The model of a person in the robot house was not represented but this could be incorporated showing their location for example. In future work we would like to provide an automated translation from a database of rules into a number of different model-checkers and we could experiment with larger databases of behaviours and different model-checkers. Note that we believe that verification of this or more complex systems should be targeted at an appropriate level of abstraction, in particular (as here), the decision making level or high level control rather than low level robot robot movement.

Returning to the issues of safety and trustworthiness, these verification results provide a route towards proving safety requirements and of using proofs to convince users of trustworthiness. These results should be used as a compliment to experiments with real people in the robot house relating to perceptions of trust to give more confidence in robotic assistants.

Acknowledgments. The authors were partially supported by EPSRC grants EP/K006193 and EP/K006509.

References

1. Cimatti, A., Clarke, E., Giunchiglia, E., Giunchiglia, F., Pistore, M., Roveri, M., Sebastiani, R., Tacchella, A.: NuSMV 2: An OpenSource Tool for Symbolic Model Checking. In: Brinksma, E., Larsen, K.G. (eds.) CAV 2002. LNCS, vol. 2404, pp. 359–364. Springer, Heidelberg (2002)
2. Clarke, E., Grumberg, O., Peled, D.A.: Model Checking. MIT Press (2000)
3. Duque, I., Dautenhahn, K., Koay, K.L., Willcock, I., Christianson, B.: Knowledge-driven User Activity Recognition for a Smart House. Development and Validation of a Generic and Low-Cost, Resource-Efficient System. In: Proc. of the Sixth Int. Conf. on Advances in Computer-Human Interactions (ACHI), pp. 141–146 (2013)

4. Fainekos, G., Kress-Gazit, H., Pappas, G.: Temporal Logic Motion Planning for Mobile Robots. In: Proceedings of the IEEE International Conference on Robotics and Automation (ICRA), pp. 2020–2025. IEEE Computer Society Press (2005)
5. Holzmann, G.J.: The Spin Model Checker: Primer and Reference Manual. Addison-Wesley (2003)
6. Kim, M., Kang, K.C.: Formal construction and verification of home service robots: A case study. In: Peled, D.A., Tsay, Y.-K. (eds.) ATVA 2005. LNCS, vol. 3707, pp. 429–443. Springer, Heidelberg (2005)
7. Kouskoulas, Y., Renshaw, D., Platzer, A., Kazanzides, P.: Certifying the safe design of a virtual fixture control algorithm for a surgical robot. In: Proc. of the 16th Int. Conf. on Hybrid Systems: Computation and Control, pp. 263–272. ACM (2013)
8. Mohammed, A., Furbach, U., Stolzenburg, F.: Multi-Robot Systems: Modeling, Specification, and Model Checking, ch. 11, pp. 241–265. InTechOpen (2010)
9. Proetzsch, M., Berns, K., Schuele, T., Schneider, K.: Formal Verification of Safety Behaviours of the Outdoor Robot RAVON. In: Fourth Int. Conf. on Informatics in Control, Automation and Robotics (ICINCO), pp. 157–164. INSTICC Press (2007)
10. Reiser, U., Connette, C.P., Fischer, J., Kubacki, J., Bubeck, A., Weisshardt, F., Jacobs, T., Parlitz, C., Hägele, M., Verl, A.: Care-o-bot®3 - creating a product vision for service robot applications by integrating design and technology. In: IEEE/RSJ Int. Conf. on Intelligent Robots and Systems (IROS), pp. 1992–1998 (2009)
11. Saunders, J., Burke, N., Koay, K.L., Dautenhahn, K.: A user friendly robot architecture for re-ablement and co-learning in a sensorised home. In: Assistive Technology: From Research to Practice (Proc. of AAATE), vol. 33, pp. 49–58 (2013)
12. Saunders, J., Salem, M., Dautenhahn, K.: Temporal issues in teaching robot behaviours in a knowledge-based sensorised home. In: Proc. 2nd International Workshop on Adaptive Robotic Ecologies (2013)
13. Saunders, J., Syrdal, D.S., Dautenhahn, K.: A template based user teaching system for an assistive robot. In: Proceedings of 3rd International Symposium on New Frontiers in Human Robot Interaction at AISB 2014 (2014)
14. Sierhuis, M., Clancey, W.J.: Modeling and simulating work practice: A method for work systems design. IEEE Intelligent Systems 17(5), 32–41 (2002)
15. Stocker, R., Dennis, L., Dixon, C., Fisher, M.: Verifying Brahms Human-Robot Teamwork Models. In: del Cerro, L.F., Herzig, A., Mengin, J. (eds.) JELIA 2012. LNCS, vol. 7519, pp. 385–397. Springer, Heidelberg (2012)
16. Syrdal, D.S., Dautenhahn, K., Koay, K.L., Walters, M.L., Ho, W.C.: Sharing spaces, sharing lives–the impact of robot mobility on user perception of a home companion robot. In: Proc. of 5th Int. Conf. on Social Robotics (ICSR), pp. 321–330 (2013)
17. Walter, D., Täubig, H., Lüth, C.: Experiences in applying formal verification in robotics. In: Schoitsch, E. (ed.) SAFECOMP 2010. LNCS, vol. 6351, pp. 347–360. Springer, Heidelberg (2010)
18. Webster, M., Dixon, C., Fisher, M., Salem, M., Saunders, J., Koay, K.L., Dautenhahn, K.: Formal verification of an autonomous personal robotic assistant. In: Proc. of Workshop on Formal Verification and Modeling in Human-Machine Systems (FVHMS), pp. 74–79. AAAI (2014)

Implementation and Test of Human-Operated and Human-Like Adaptive Impedance Controls on Baxter Robot

Peidong Liang[1,3], Chenguang Yang[1,2,*], Ning Wang[1,4], Zhijun Li[2],
Ruifeng Li[3], and Etienne Burdet[5]

[1] Centre for Robotics and Neural Systems, School of Computing and Mathematics, Plymouth University, UK

[2] The MOE Key Lab of Autonomous System & Network Control, College of Automation Science and Engineering, South China University of Technology, China
cyang@ieee.org

[3] The State Key Laboratory of Robotics and System, Harbin Institute of Technology, China

[4] Department of Computer Science and Engineering, The Chinese University of Hong Kong, China

[5] Department of Bioengineering, Imperial College of Science, Technology and Medicine, London, UK

Abstract. This paper presents an improved method to teleoperate impedance of a robot based on surface electromyography (EMG) and test it experimentally. Based on a linear mapping between EMG amplitude and stiffness, an incremental stiffness extraction method is developed, which uses instantaneous amplitude identified from EMG in a high frequency band, compensating for non-linear residual error in the linear mapping and preventing muscle fatigue from affecting the control. Experiments on one joint of the Baxter robot are carried out to test the approach in a disturbance attenuation task, and to compare it with automatic human-like impedance adaptation. The experimental results demonstrate that the new human operated impedance method is successful at attenuating disturbance, and results similarly to as automatic disturbance attenuation, thus demonstrating its efficiency.

Keywords: EMG, stiffness estimation, adaptive impedance control, teleoperation.

1 Introduction

Human-like robot control has received increasing research attention in recent years e.g. [1,2], boosted in part by the development of serial elastic actuator (SEA) [3] and variable impedance actuator (VIA) [4]. Inspired by the flexibility humans have to adapt visco-elasticity at the endpoint of their arm, impedance control introduced by Hogan in 1985 aims at controlling mechanical impedance,

---

[*] Corresponding author.

M. Mistry et al. (Eds.): TAROS 2014, LNAI 8717, pp. 109–119, 2014.

the relation between force and displacement [5], [6]. Considering how humans can skilfully adapt impedance in various force interactive scenarios, e.g. when using tools, we would like to transfer this flexibility to robots [1]. One possibility to adapt impedance is by using human-like adaptation, e.g. as was developed in [1]. Alternatively, human-like impedance strategies may be achieved using electromyography (EMG) signals to decode impedance in real time, see e.g. [7]. Advantages of teleoperation via EMG include: i) simple and natural control interface which is convenient for teleoperation; ii) low cost sensing EMG signals are easy to be acquired; iii) transparent interaction between human and robot avoiding the delay inherent in providing feedback. In view of these potential advantages, [7] developed a teleimpedance system in which a slave robot is commanded in position from the operator arm position and in impedance from the EMG signals recorded on his or her arm.

This paper proposes to compare these two approaches to adapt impedance in a disturbance attenuation task. A human operator is controlling the position of a robot arm using visual feedback. This robot is subjected to high-frequency external disturbance, that either the robot can attenuate using human-like impedance adaptation [1], or the human operator can attenuate using his EMG. Instead of estimating the stiffness value directly from EMG as in [7], the increment of stiffness value is used to compensate for the potential nonlinear residual error in the linear mapping used in [7]. Furthermore, only the 400-500Hz band of EMG is employed for stiffness estimation [8], which prevent fatigue from affecting the impedance control. An instantaneous amplitude detection algorithm is used to compute stiffness value, which enhances the robustness of EMG processing. The two impedance adaptation methods are tested to teleoperate a Baxter robot arm submitted to external disturbance. The control objective is to maintain a given pose with the Baxter robot in the presence of high frequency external disturbances, for simplicity here in only the shoulder joint.

2   Preliminaries

2.1   Baxter Robot Programming

The Baxter robot used for our experiments, shown in Fig. 1 consists of a torso, 2 DOF head and two 7 DOF arms (shoulder joint: s0, s1; elbow joint: e0, e1; wrist joint: w0, w1, w2), integrated cameras, sonar, torque sensors, and direct programming access via a standard ROS interface. Each joint of the Baxter robot arm is driven by a SEA (illustrated in Fig.1), which provides passive compliance to minimise the force of any contact or impact [9]. ROS (Robot Operating System) SDK is used to control and program the Baxter robot with Baxter RSDK running on Ubuntu 12.04 LTS which is an open source framework with modular tools, libraries, and communications. It simplifies the task of modelling and programming on a wide variety of robotic platforms. In this paper, the robot control is implemented through several sets of nodes and data series of the extracted impedance from human can be sent to the bridge machine for connection between Baxter robot and human. On the human bridge

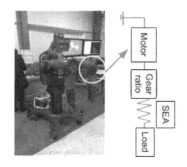

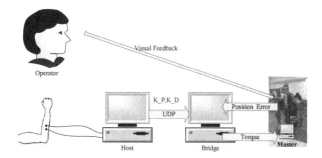

Fig. 1. Experimental setup. Top: Baxter research robot with SEA joints (schematic of SEA shown on the right). Below: Data flow.

machine side, stiffness is computed on the MATLAB 2014a and sent to the bridge machine via UDP, a Python script is set up to receive these data set and build a new node to publish a topic for the robot side via Ethernet. Another interface is built with torque control mode to receive and process human impedance and give feedback to human Data flow is also shown in Fig. 1.

3 Adaptive Impedance Control Methods

3.1 Human-Like Adaptive Impedance Control

Human-like robot control with adaptive impedance was developed in [1] as follows:

$$\tau_u = -\tau_{ff} - \tau_{fb} - L\epsilon + \tau_r \tag{1}$$

where $-\tau_{ff}$ is feedforward torque learned to compensate for the interaction with the environment, $\tau_{fb} = -K_P e - K_D \dot{e}$ plays the role of feedback, with stiffness K_P and damping K_D parameters to be learned during interaction with the environment, and $e = q - q^*, \dot{e} = \dot{q} - \dot{q}^*$ are position and velocity errors relative to the task reference trajectory q^*. $L\epsilon$ denotes the ideal stability margin, and L is a control matrix used to keep robot motion with stability and compliance, whereas $\epsilon = \dot{e} + ke, k > 0$ is the tracking error. The term τ_r is used to counteract the complexity of robot dynamics and disturbances. The adaptation of stiffness is enabled by the following algorithm (originally proposed in iteration domain in [1], and extended to time domain in [10]).

$$\Delta K_P^{t+1} = Q^t (\epsilon^t e^{tT} - \gamma^t K_P^t)$$
$$\Delta K_P^{t+1} = K_P^{t+1} - K_P^t \tag{2}$$

and adaptation of damping K_D is similar and thus omitted here. Notation t is the current time sampling point and $t+1$ next time sampling point, $Q^t = Q^{t^T}$ is a positive definite gain matrix, and the forgetting factor γ can be simply chosen as a constant coefficient. It has been theoretically and experimentally shown that the algorithm of equation (2) is able to adapt robot impedance in a human-like manner, i.e., when there is large unpredictable external disturbances, the stiffness will increase to compensate for the disturbance, while the disturbance decreases stiffness will be reduced to save control effort.

3.2   Improved Stiffness Estimation Using EMG

According to [11,12], we assume that a linear mapping can be employed to approximate the relationship between rectified EMG amplitude and joint torques. We also consider a potential residual term in addition to the linear mapping such that the relationship between rectified EMG amplitude and joint torques is

$$\tau^t = \sum_{i=1}^{n} \alpha_i \cdot A_{i,t}^a - \sum_{i=1}^{n} \beta_i \cdot A_{i,t}^{aa} + r_\tau^t \tag{3}$$

where α_i, β_i are unknown constant coefficients; τ^t is the joint torque generated by agonist/antagonist muscle pairs involved in the joint motion; $A_{i,t}^a$ and $A_{i,t}^{aa}$ represent the rectified EMG amplitude of the agonist and antagonist muscles, respectively (in this work, we will use the identified temporal envelop of the EMG signal extracted from a certain high frequency band to remove effect caused by muscle fatigue); r_τ^t is the potential non-linear residual term. In order to reduce the effect of the residual error and to simplify the non-linearity, incremental stiffness estimation is introduced by taking a first order difference of above equation:

$$\Delta \tau^{t+1} = \sum_{i=1}^{n} \alpha_i \cdot \Delta A_{i,t+1}^a - \sum_{i=1}^{n} \beta_i \cdot \Delta A_{i,t+1}^{aa} + \Delta r_\tau^{t+1} \tag{4}$$

where $\Delta \tau^t = \tau^{t+1} - \tau^t$ is incremental torque; $\Delta A_{i,t+1}^a = A_{i,(t+1)}^a - A_{i,t}^a$; $\Delta A_{i,t+1}^{aa} = A_{i,(t+1)}^{aa} - A_{i,t}^{aa}$ are incremental EMG amplitude; and we assume $\Delta r_\tau^{t+1} = r^{t+1} - r^t \approx 0$. Following [11,12], we assume the joint stiffness can be approximated by the following equation by taking into consideration of a potential residual term

$$K_P^t = \sum_{i=1}^{n} |\alpha_i| \cdot A_{i,t}^a + \sum_{i=1}^{n} |\beta_i| \cdot A_{i,t}^{aa} + r_K^t \tag{5}$$

where K^t denotes the stiffness generated by muscles involved at current time t, and r_K^t is the non-linear residual term. By taking a first order difference of above equation, we propose a method to estimate incremental stiffness of human arm as described in (6). On one side, this method would help to overcome the effect caused by the potential non-linear residual item. On the other side, by transferring estimated incremental stiffness to the robot, possible accident caused by sudden fall of stiffness on robot would be avoided. This may happen when the initial stiffness on robot is higher than the initial stiffness estimated on human operator.

$$\Delta K_P^{t+1} = \sum_{i=1}^{n} |\alpha_i| \cdot \Delta A_{i,t+1}^{aa} + \sum_{i=1}^{n} |\beta_i| \cdot \Delta A_{i,t+1}^{a} \qquad (6)$$

where $A_{i,t}^{aa}$ is the detected amplitude of EMG of the ith antagonistic muscle at the current time instant t, and $A_{i,t}^{a}$ the amplitude of agonistic muscle, and $\Delta A_{i,t+1}^{aa} = A_{i,t+1}^{a} - A_{i,t}^{a}$, $\Delta A_{i,t+1}^{a} = A_{i,t+1}^{a} - A_{i,t}^{a}$. Instead of using amplitude of rectified raw EMG, we will perform signal processing to detect instantaneous amplitude $A_{i,t}^{aa}$ and $A_{i,t}^{a}$, using frequency band decomposition and envelop detection. This will be detailed in section 3.3.

3.3   Signal Decomposition and Amplitude Extraction of EMG Signals

EMG signals are non-invasive and generated in muscle activities, representing muscle tension, joint force and stiffness variation [13], which are a linear summation of a compound of motor-unit action potentials (MUAPs) trains triggered by motor units and correlative with force according to [8]. In this paper, the EMG signal is processed as a multi-component amplitude and frequency modulating (AM-FM) signals due to its sufficiently small-band width [14] described by

$$s^t = \sum_{k=0}^{K} A_k^t \cos \Theta_k^t + \eta^t \qquad (7)$$

where A_k^t denotes the instantaneous amplitude of the kth EMG component, corresponding to A_i^a or A_i^{aa} in Eq. (5) and Θ_k^t denotes its instantaneous phase. η^t denotes residual errors caused by disturbance, modelling error and finite summation etc. The instantaneous amplitude A^t identified using (7) according to [14] can be regarded as temporal envelop of the EMG signal extracted from a certain frequency band (refer to Fig. 3), and the frequency band of 400-500Hz is used in this paper. In this manner, we would have less variation of EMG amplitude in comparison to the amplitude of raw EMG , and avoid muscle fatigue effect which is mainly embedded in low frequency EMG band [8]. In addition, as the instantaneous amplitude detected is always positive [14], the rectification can be omitted. In this work, NI USB6210 (16 inputs 16bit, 250KS/s, and multi-function I/O), and 8 channels EMG Pre-Amplifier are used for detecting EMG signals with MATLAB 2014a 32bit data acquisition toolbox. The sampling rate is 2000Hz with 200 data every 0.1s interval for real time processing.

3.4   Human-Operated Impedance Control Experimental Setup

The whole teleoperation system design is shown in Fig. 2. It consists of the following components: EMG signals acquiring and processing, stiffness estimation, communication between host and master machine, and adaptive controller on Baxter robot. In this paper we only consider controlling a single shoulder (s1) joint of Baxter robot using wrist flexion/extension joint of human operator. EMG signals are gathered through electrodes and filtered, then the stiffness

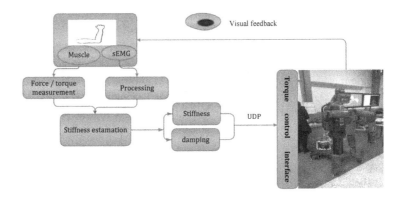

Fig. 2. Overall scheme of teleoperation system

and damping are estimated from EMG with the reference endpoint force measured by a dynamometer. We use UDP to transfer data from host machine with MATLAB2014a 32bit to the master machine which is an interface programmed in Python for interacting with Baxter robot. For simplicity of implementation, only feedback part τ_{fb} in the controller (1) is used in the experiment, where K_P is obtained from (6) and K_D is set as

$$K_D = \sigma \cdot \sqrt{K_P} \tag{8}$$

where $\sigma = 0.2$ is a properly chosen coefficient for the SEA driven joints. The robot will tend to become stable with the increase of impedance. The human operator can adapt impedance using visual feedback, i.e., when the disturbances are increasing, human operator stiffens his/her muscles and relaxes if the robot is little disturbed.

4   Experimental Results

Our experiment compared human-like automatic impedance adaptation and human operated adaptation of impedance. One adult male human operator is performing the experiment. His wrist flexor carpi radialis and extensor carpi radialis were selected to extract stiffness to control Baxter robot impedance in its shoulder (s1) joint.

4.1   Incremental Stiffness Estimation from EMG

Fig. 3 shows an example of EMG signal processing using AM-FM demodulation method described in section 3.3. The bottom panel shows the EMG signal extracted in the 400-500Hz frequency band, and the detected instantaneous amplitude as the temporal envelop of the signal.

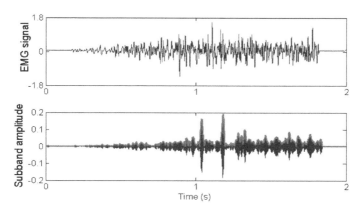

Fig. 3. Top: the raw EMG signal. Bottom: its sub-band of 400-500Hz in blue with identified instantaneous amplitude in pink.

In this preliminary study, only EMG signals of two channels measured from a pair of muscles on wrist were used in the experiments, so equation (6) can be as an equation with only two parameters α and β to identify. For the purpose of calibration, the human operator exerts three different constant forces: $10N$ (wrist extension), $40N$ (wrist flexion) and $50N$ (wrist flexion) against a dynamometer. Before recording EMG data, Labview interface was started to ensure that the signal is at a standard low noisy level, then the human operator's wrist maintains a constant force/torque. When the signals remained stable, recording started using MATLAB2014a 32bit data acquisition toolbox. When the operator maintains the desired force, the lever of force was measured as $0.05m$ and thus torques corresponding to the three forces were $-0.5N \cdot m$, $2N \cdot m$ and $2.5N \cdot m$, respectively.

Fig.4 illustrates the calibration process. The raw EMG signals of the pair of muscles when maintaining the three different levels of torque were measured and shown in 4(a), (b) and (c), respectively. The identified instantaneous amplitudes of these signals extracted from 400-500Hz band are shown in 4(d), (e) and (f). For ease of computation, we took amplitude differences between amplitudes shown in 4(e) and (d), 4(f) and (d), which are shown in 4(g) and (h), instead of taking difference between different sampling points as in (4). The average of the EMG amplitude differences shown in 4(g) and (h) are 0.0112, -0.0040 and $(0.0095, -0.0035)$, shown in Fig. (4)(i). As the torque differences are $2.5N \cdot m$ and $3N \cdot m$, the values of α and β are calculated as $\alpha = -2708.3$ and $\beta = 8208.3$.

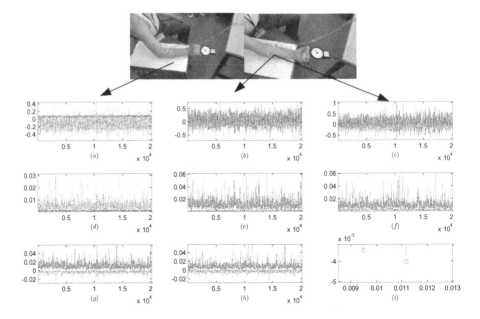

Fig. 4. Calibration process: (a), (b) and (c) denote the raw EMG signals of agnostic and antagonistic muscles of wrist flexion/extension when maintaining -10N (extension), 40N (flexion) and 50N respectively; (d), (e) and (f) are the identified instantaneous amplitude of the 400-500Hz sub-band signals extracted from raw signals shown in (a), (b) and (c); (g) and (h) denote the difference of the amplitudes between (e) and (d), (f) and (d), respectively; (i) denotes the average values of the amplitude difference shown in (g) and (h) (x-axis:flexion muscle, y: extension muscle)

4.2   Comparison between Human-Like Control and Human-Operated Control

Two comparative experiments were carried out to test the proposed methodology in torque control mode of the Baxter robot. Only shoulder joint (s1) was used to implement the adaptive control, while the rest joints stiffness's and damping rate are set with properly chosen values to avoid resonance. Here, the values were: stiffness(N/m)(s0: 400.0; e0: 50; e1: 200; w0: 50; w1:25; w2: 20). Damping rates are set according to Eq. (8). The initial stiffness value for s1 was 15 N/m. Fig. 5 (a) and (b) show the human-like control and human operated control experimental results on Baxter robot under external high frequency disturbances, respectively. Equations (2) and (6) are used to generate the incremental values of stiffness shown in the middle panels of Fig. 5 (a) and (b), respectively. The summation of these values of incremental stiffness along sample points contribute to the variation of the values of stiffness shown in the first panel of Fig. 5 (a) and (b). Note that the summations along the sample points is sampling frequency times the integration along time. From the performance of attenuation of disturbances as shown in the third panels of Fig. 5 (a) and (b), we see that human

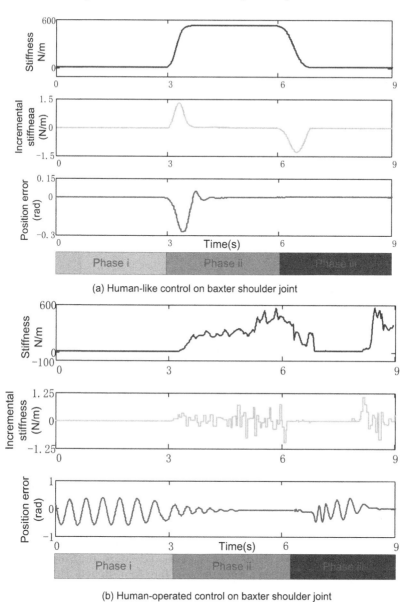

(a) Human-like control on baxter shoulder joint

(b) Human-operated control on baxter shoulder joint

Fig. 5. Comparison between human-like control and human-operated control: (a) Human-like control on Baxter (s1) joint with three phases: i) robot should relaxes in the absence of external disturbance; ii) robot shoulder stiffens up when the high frequency external disturbance is applied; iii) robot shoulder relaxes again when the disturbances are removed. (b) Human-operated control on Baxter (s1) joint with three phases: i) human operator keeps wrist relaxed when of high frequency external disturbances are applied on the Baxter robot; ii) human operator stiffens up his wrist joint to increase impedance on the Baxter robot; iii) human operator relaxes and then stiffens up his wrist joint when the external disturbances are applied.

operated online impedance control via EMG is as good as automatic impedance adaptation to attenuate external perturbations, in turn showing its efficiency.

5 Conclusion

The paper proposed an improved strategy of EMG based incremental stiffness estimation without calibration and nonlinear residual errors for robot teleoperation which requires high dynamic performances in unknown environment. A telecontrol disturbance attenuation experiment was designed to compare this improved algorithm with automatic human-like adaptation of impedance. Results show that the proposed methodology is efficient in teleoperation with human skills transferring to robot in unknown environment, as can be seen in the attached illustrative videos[1]. Using human operated impedance control brings more flexibility than automatic disturbance attenuation, but requires appropriate visual feedback and attention from a human operator.

Acknowledgement. The authors thank Dr. Hai Li who built the amplifier for EMG data collection, Dr Ali Al-Timemy who helped to set up the EMG data capture interface and Zhangfeng Ju who helped to set up the experiment. This work was supported in part by EPSRC grants EP/L026856/1 and EP/J004561/1, Royal Society grants RG130244, IE130681 and JP100992; foundation of Chinese MOE Key Lab of Autonomous Systems and Networked Control grants 2012A04 and 2013A04; and EU-FP7-ICT-601003 BALANCE projects.

References

1. Yang, C., Ganesh, G., Haddadin, S., Parusel, S., Albu-Schaeffer, A., Burdet, E.: Human-like adaptation of force and impedance in stable and unstable interactions. IEEE Transactions on Robotics 27(5), 918–930 (2011)
2. Yang, C., Li, Z., Burdet, E.: Human like learning algorithm for simultaneous force control and haptic identification. In: 2013 IEEE/RSJ International Conference on Intelligent Robots and Systems (IROS), pp. 710–715 (November 2013)
3. Pratt, G.A., Williamson, M.M.: Series elastic actuators. In: Proceedings of the 1995 IEEE/RSJ International Conference on Intelligent Robots and Systems 1995, Human Robot Interaction and Cooperative Robots, vol. 1, pp. 399–406 (August 1995)
4. Vanderborght, B., Albu-Schaeffer, A., Bicchi, A., Burdet, E., Caldwell, D.G., Carloni, R., Catalano, M., Eiberger, O., Friedl, W., Ganesh, G., Garabini, M., Grebenstein, M., Grioli, G., Haddadin, S., Hoppner, H., Jafari, A., Laffranchi, M., Lefeber, D., Petit, F., Stramigioli, S., Tsagarakis, N., Van Damme, M., Van Ham, R., Visser, L.C., Wolf, S.: Variable impedance actuators: A review. Robotics and Autonomous Systems 61(12), 1601–1614 (2013)

---

[1] An illustrative video of the experiment can be viewed at:
Human-operated control: `https://www.youtube.com/watch?v=y5nFrS6gnZQ`;
Human-like automatic control: `https://www.youtube.com/watch?v=C0gU6H2rChs`

5. Hogan, N.: Impedance control: An approach to manipulation. Journal of Dynamic Systems, Measurement, and Control 107(1), 8–16 (1985)
6. Burdet, E., Franklin, D., Milner, T.: Human Robotics - Neuromechanics and motor control. MIT Press (2013)
7. Ajoudani, A., Tsagarakis, N.G., Bicchi, A.: Tele-impedance: Teleoperation with impedance regulation using a body-machine interface. The International Journal of Robotics Research 31(13), 1642–1656
8. Potvin, J.R., Brown, S.H.M.: Less is more: High pass filtering, to remove up to 99% of the surface EMG signal power, improves emg-based biceps brachii muscle force estimates. Journal of Electromyography and Kinesiology 14(3), 389–399 (2004)
9. Baxter-research-robot. Baxter-research-robot profile, http://www.rethinkrobotics.com/products/Baxter-research-robot/
10. Smith, A., Yang, C., Ma, H., Culverhouse, P., Cangelosi, A., Burdet, E.: Bimanual Robotic Manipulation with Biomimetic Joint/Task Space Hybrid Adaptation of Force and Impedance. In: The Proceedings of the 11th IEEE International Conference on Control & Automation to be held in Taichung, Taiwan, June 18-20 (in press, 2014)
11. Osu, R., Franklin, D.W., Kato, H., Gomi, H., Domen, K., Yoshioka, T., Kawato, M.: Short-and long-term changes in joint co-contraction associated with motor learning as revealed from surface EMG. Journal of Neurophysiology 88(2), 991–1004 (2002)
12. Ajoudani, A., Tsagarakis, N.G., Bicchi, A.: Tele-impedance: Preliminary results on measuring and replicating human arm impedance in tele operated robots. In: 2011 IEEE International Conference on Robotics and Biomimetics (ROBIO), pp. 216–222 (December 2011)
13. Ray, G.C., Guha, S.K.: Relationship between the surface e.m.g. and muscular force. Medical and Biological Engineering and Computing 21(5), 579–586 (1983)
14. Wang, N., Yang, C., Lyu, M., Li, Z.: An EMG enhanced impedance and force control framework for telerobot operation in space. In: 2014 IEEE International Conference on Aerospace, pp. 1–12 (March 2014)

Hybrid Communication System for Long Range Resource Tracking in Search and Rescue Scenarios

Michael Boor, Ulf Witkowski, and Reza Zandian

South Westphalia University of Applied Science,
Electronics and Circuit Technology Department, Soest, Germany
{miboo001,witkowski.ulf,zandian.reza}@fh-swf.de

Abstract. This paper focuses on a resource (people or equipment) tracking approach involving radio localization and wireless communication for use in search and rescue scenarios. Two alternative communication options (ZigBee and GSM) are used to facilitate robust wireless data transfers between nodes through either an ad-hoc network or base station. The onboard sensors of the tracking device calculate the speed of the node's movement as well as other environmental parameters such as temperature. The position of each node is regularly uploaded to a map on a server. XBee modules are used for local communication within a range of several hundred meters, while GSM is used for long range communication provided that the infrastructure is available. This paper discusses the hardware platform and network architecture of the developed hybrid communication system. Experiments involving device prototypes are performed to evaluate the feasibility and performance of the system. The results show that a hybrid communication system is extremely practical for search and rescue operations.

Keywords: Hybrid networks, GSM, GPS, ZigBee, XBee, Resource tracking.

1 Introduction

The nature of search and rescue scenarios demands the ability to immediately deploy equipment and forces to an area without any previous planning. Thus no prior infrastructure may exist for communication between people or equipment, nor is there enough time to establish it. An ad-hoc wireless network can solve this problem by directly forming a mesh network between multiple nodes. However, mesh networks face challenges such as network stability, node access range and routing topology. A hybrid type of network that allows for data transfer through multiple mediums is useful in this case, since the optimum medium can be selected based upon the current needs of the system. For example, a communication system based on ZigBee technology is able to communicate between nodes, but the range is limited to some hundred meters in an urban environment [1]. A GSM based communication system may have a longer range, but it does not allow for direct communication between two or more nodes. There would also be data fees for using the GSM network. Therefore, this

M. Mistry et al. (Eds.): TAROS 2014, LNAI 8717, pp. 120–130, 2014.

paper proposes a hybrid type of communication system which can utilize the benefits of multiple networks while fulfilling the requirements of communication systems in search and rescue scenarios.

The advantages of a hybrid communication system make it well-suited for many applications. Venkatakrishnan et al. [3] have employed a hybrid system which includes ZigBee, GSM, GPS and RFID in a public transport ticketing and monitoring application. In their project, the devices are mounted on the buses, bus stations and at the control center. The RFID part is used for ticketing, the GSM system is used for transferring data to the control center, the GPS module is used for determining the position of each bus and the ZigBee system is used for the communication between the buses and the bus stations. Their results show that using the appropriate communication systems for different functions can compensate for the weaknesses of a single communication method, thus creating a stable and extendable communication system.

Stanchev et al. [4] have used GSM and ZigBee to monitor the dynamic parameters of their vehicles and industrial machines. They exploit the benefits of a mixed communication system while developing a new concept involving a multi-point and multi-sensor intelligent system. They attached several sensors to a vehicle which measure vibration, temperature, voltage, speed, position tracking, etc. These parameters are locally retrieved over ZigBee and the complete data is transferred over the GSM network to the control center. The result of their work was a flexible, secure and reliable wireless sensing system with a range equal to that of the GSM network's coverage. B.Amutha et al. [5] have also employed a hybrid communication system (GPS and ZigBee) in their tracking application. They use the wireless system for monitoring and modeling human walking patterns for the purpose of helping blind people navigate safely. They used the Markov chain algorithm to improve the accuracy of their measurements. The hybrid system in this project is used for the localization and tracking of people so that the walking patterns of seeing people can be recorded and compared with the patterns of blind people. Their evaluation results show that the system is accurate and can serve as a path finding device for the blind. Ch.Sandeep et al. [6] created a hybrid communication system for monitoring the health of patients. The ZigBee system is responsible for gathering data from sensors around a patient, while the GSM communication system is used for delivering the processed information to the doctors via SMS packets.

Further sensor integration and tracking work has been performed by Jadhav et al. [7]. They have used the hybrid system (GPS and ZigBee) for monitoring pollution in the air. Multiple sensors are utilized for measuring the pollution level in the air. This data is transmitted to the server via the ZigBee nodes. The GPS system records the position coordinates of the sensors, and these values are transferred to the server so that the corresponding measured sensor values taken at a precise position can be displayed on a map of the area. This system helps regulate pollution by providing the exact locations of pollution sources.

An ad-hoc networking approach based on Wi-Fi, Bluetooth and ZigBee for robotic assistance scenarios has been proposed by Witkowski et al. [2]. The objective was to provide a robust communication system in firefighting scenarios. A limitation of this approach is the system's moderate communication range. A benefit is that in dynamic environments a flexible mesh network can create communication links without the need of a fixed communication infrastructure [2].

In this paper, the hybrid communication system is used in search and rescue scenarios which demand both rapid system deployment and a stable and reliable wireless network. In this type of network, the nodes should be able to communicate with each other as well as with a base station. This paper discusses the hybrid systems architecture, the challenges with such a system, and the design and testing of a prototype.

Within the proposed system architecture, all the end nodes (firefighters, cars, equipment, etc.) are equipped with devices which support communication via both XBee and GSM. The devices can also individually determine their position using GPS. The XBee modules can form a mesh network to create a local connection between the nodes in an ad-hoc networking scheme. In the case that a node is outside an XBee module's range, the node's position and sensor data will be transferred to other nodes and the control center via the GSM network. In future work, strategies will be explored for recovering from link failures in the local network, e.g. active node replacement strategies. Figure 1 shows a simplified diagram of the communication system which includes the nodes and control center.

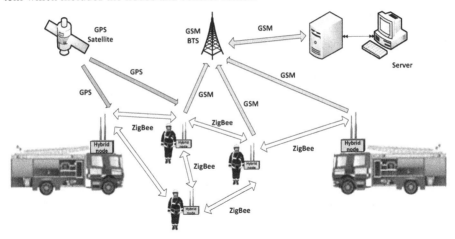

Fig. 1. Overview of the hybrid communication system

Each hardware node is equipped with sensors which collect environmental parameters such as temperature, acceleration, battery level, etc. A few input buttons are also provided which the operator can use for quickly performing actions such as calling for support, raising an alarm, calling for an emergency evacuation, requesting medical aid and so forth.

1.1   Comparison of Available Techniques

All the common communication networks have their own strengths and weaknesses. This is mainly because they are designed for different media and applications. As an example, Wi-Fi is designed for transferring large amounts of data with a high throughput. However, this type of communication system is heavily dependent on its infrastructure. Bluetooth is mainly designed for close range point to point connections, such as the transferring of data between two cooperating devices or in pico-net architectures with one master and up to seven active slaves. The restricted pairing mechanism and limited range are drawbacks to this type of communication interface. ZigBee technology has both a significant signal range and data throughput rate without being dependent on an external infrastructure. The GSM network has a significant signal range, but a network base station is necessary. A list of possible communication systems and their specifications are shown in Table 1.

Table 1. Comparison of common wireless technologies [8]

Technology	Wi-Fi	GSM	Bluetooth	Bluetooth (Low energy)	ZigBee
Selection	Mesh-P2P	Star	P2P-Pico	P2P-P2M	P2P-Mesh
Setup time	<0.1s	~<3s	~3sec	<6ms	<0.1s
Range	< 100m	< 26Km	< 50m	< 100m	~10-4000m
Data rate	54Mbit/s	42.8Kbit/s	2.1Mbit/s	~1Mbit/s	250Kbit/s
Frequency	2.4GHz	900MHz	2.4-2.5GHz	2.4-2.5GHz	2.4-2.5GHz
Cost	High	High	Medium-High	Low-Medium	Low
Required Power	Medium	Low	Medium	Low	Low

1.2 Network Topologies

The ZigBee standard supports multiple network topologies. These topologies include star, tree, and mesh [9]. The star topology has only one coordinator with several other end devices. This type of network is dependent on the coordinator. If this node fails, then the other end nodes will not be able to communicate. Moreover, communication between nodes is only possible through the coordinator [12]. In the tree topology, router nodes help transfer messages between the end nodes and the coordinator. The problem of this topology is if one router fails, then the children of that router node are disconnected from the network. The last possible topology is a mesh network, which is basically a collection of several peer-to-peer connections. In a mesh network, packets can pass through several nodes to reach a target node. This way the range of the network can be increased by adding to its number of nodes. This type of topology is self-healing, which means if that one node fails then the other nodes can transfer the data via another route. Disadvantages of a mesh network include a larger overhead and more complex routing protocols. Figure 2 graphically shows the three possible ZigBee network topologies.

Considering the requirements of search and rescue scenarios, the failure of a single node should not affect the connections between other nodes. Thus the mesh topology is the most suitable network topology. The topology when using the GSM network is star shape, since the nodes cannot contact each other directly, but rather all communication is routed through a base station. Therefore this type of connection will be reserved for situations where certain nodes cannot establish mesh connections with other members of the network. In our case, the coordinator node will be the control station which routinely receives data from all the other nodes. In this project, XBee modules from Digi Company are used as nodes within a Digimesh network. The concept of the Digimesh topology is more or less the same as the mesh network mentioned above. The Digimesh topology however includes some extra features, simplifies the protocol and uses improved routing techniques.

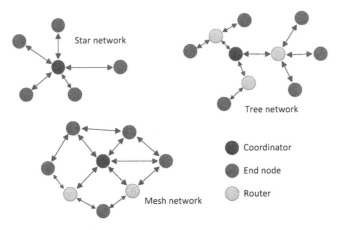

Fig. 2. ZigBee network topologies [10]

2   Communication System Challenges in Search and Rescue Scenarios

The tough conditions of search and rescue scenarios demand for a stable, reliable and redundant network. The environments where these systems will primarily be used will be noisy, the nodes will not be distributed linearly and the signals will likely be passing through thick barriers. Within this paper, these problems are evaluated and potential solutions are provided to improve the overall effectiveness of the network.

2.1   Speed of Data Transmission

Providers of ZigBee modules offer multiple products that operate with different frequencies and data rates. The XBee module used in this project transmits on the 868 MHz frequency band. The XBee module has a data rate of 10 to 80 Kbps. This speed is fast enough for transmission of a device's position and sensor data. In the case which larger amounts of data (such as photos and speech messages) need to be transferred, a Bluetooth connection (and optionally Wi-fi) could be utilized, The pairing process of Bluetooth modules could be done by using ZigBee to find the nearest node in the neighborhood.

2.2   Energy Consumption

Since the nodes run on battery power, it is important that they conserve energy. Therefore the low energy versions of the XBee modules are chosen which consume ~30mA on average while transmitting. The GSM module consumes hundreds of milliamperes while transmitting, and thus is kept in sleep mode and only woken up if the transmission of data is required.

2.3   Range of Signal

The 868MHz XBee modules have a range of up to 8 km outdoors and 100 meters indoors, which is sufficient for sending data inside most buildings. Details of actually measured signal strengths are provided in the results and discussion section.

3   Platform Design

The device is designed around a central microcontroller. Figure 3 shows the various high-level components of the device along with how these components interface to the microcontroller. A GPS module containing a built-in antenna tracks GPS satellites and uses them to compute its position. This location data is then sent to the microcontroller which chooses to store the data on a SD card, send it over the GSM network or transmit it locally via the XBee module. The device can be remotely configured via commands sent to it over the GSM network, XBee mesh network or Bluetooth. Five push buttons attached to the microcontroller allow for the direct control of the device by the operator. The final developed platform is shown in Figure 4.

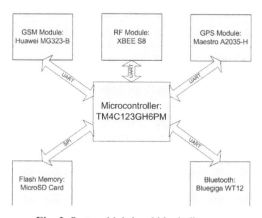

Fig. 3. System high-level block diagram

The TIVA TM4C123GH6PM microcontroller from Texas Instruments was chosen because of its processing power, low power consumption and the versatility of its GPIO ports. The Maestro A2035-H GPS receiver module is a low power GPS receiver integrated with a patch antenna. The module can receive signals from up to 48 GPS satellites at one time. When running, the receiver's position is constantly being calculated and transmitted as serial data over its UART connection. The GPS module can be put into hibernation mode to conserve power.

Fig. 4. Fabricated hardware containing multiple communication modules

The SIM and micro SD card connectors are integrated together into a single space saving device. The SIM card connects directly to the GSM module via four wires. Since the micro SD card will be operated in SPI mode, the SD card requires a clock signal, a chip select signal, a command line and a data line. The Bluegiga WT12 Bluetooth module is a fully integrated device including an antenna. The Bluetooth module interfaces to the microcontroller using a two line UART. A small 3-axis accelerometer is also included on the device. This chip measures acceleration within the range of ±5g. It can measure the static force of gravity as well as acceleration due to motion, shock or vibration.

Host Serial Interface		
UART		SPI
Transparent Mode	AT Command Mode	API Mode
	Command Handler	
Packet Handler		
Network Layer (DigiMesh/Repeater)		
MAC/PHY Layer (Point-Multipoint)		
Antenna		

Fig. 5. XBee Multi-layered firmware [11]

The XBee module interfaces to the microcontroller via its UART port. XBee modules are embedded circuits which provide wireless connectivity between devices. These modules use the IEEE 802.15.4 networking protocol. They can be configured for either point-to-multipoint or peer-to-peer networking. Depending on the environment, the

XBee modules can transmit distances ranging from 100 meters to 8 kilometers. XBee modules are optimal for applications that require both high-throughput and low latency. XBee modules follow the ZigBee standard upon which they add their own added application layer and AT commands. Figure 5 shows how the different layers interact. Blocks that touch signify that the two interfaces can interact. XBee modules employ a LBT (Listen Before Talk) access approach, where they check for activity on a channel prior to transmitting. After transmitting on one channel, the transceiver will then hop to another sub-band. The transceiver will not transmit again on a particular sub-band until after the minimum TX off time has passed.

4   Results and Discussion

Multiple scenarios were evaluated in order to test the performance of the designed hybrid network. In all cases a hybrid network node was maneuvered while it transmitted back to a base station. In the first test scenario, the robot navigated within a single building while transmitting and receiving data. In the second scenario, the robot had to navigate outdoors in an open field. In the third scenario, the hybrid node was taken around an entire city. The analysis of the results is discussed in this paper.

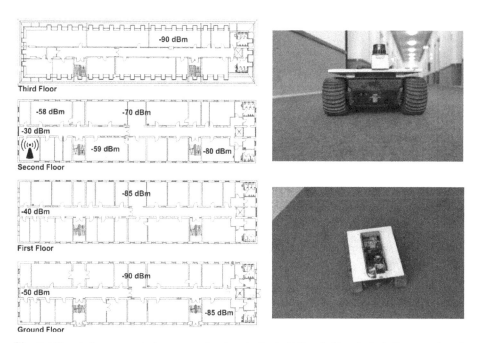

Fig. 6. XBee indoor range test; measured values in the building (left), robot platform equipped with hybrid communication node (right)

In the first experiment the communication range of the XBee modules was eva-luated inside a highly divided and shielded building (Figure 6). One transceiver was kept stationary within a corner room located on the second floor, while a second tran-sceiver was moved about the building. The signal strength of data packets received by the mobile transceiver is marked by location in Figure 6. The XBee modules were able to successfully communicate with one another throughout the entire building. Having multiple devices in a mesh network would only further increase the fidelity of the communication network.

The long rang capabilities of the XBee modules were evaluated by testing the modules in an open outdoor field. One transceiver was kept stationary, while another was moved to greater distances away from it. Figure 7 (left) shows the signal strength of the received data at different distances. The data packets transmitted contained the last recorded GPS position of the mobile transceiver. The field was not completely flat, and the benefits of a slightly higher point of elevation can be seen at around 450 meters out. After about 600 meters, the transceiver goes over a slope and is no longer within line-of-sight. Nevertheless, provided decent elevation and a line of sight be-tween receivers, it is easily conceivable that a transmission could be sent multiple kilometers. A photo of the outdoor experiment is shown in the right of Figure 7, while the effect of distance on the signal strength is shown on the left.

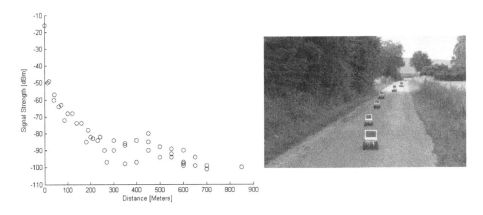

Fig. 7. XBee signal strength vs. distance (left), robot platform in outdoor experiment (right)

A full system test involving the GSM module was performed by taking a bike ride with the hybrid node. The hybrid node was configured so that position data would be logged both to the internet server and to the SD card every 15 seconds. The online map showed that the hybrid node's position was successfully updated in real time. Figure 8 shows the final state of the tracking map after the hybrid node returned back to its start location.

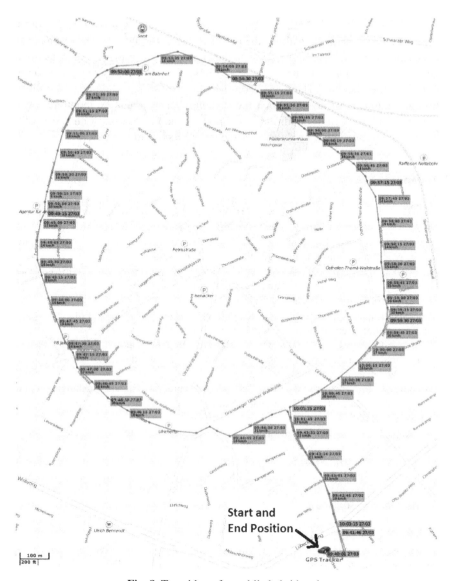

Fig. 8. Travel log of a mobile hybrid node

5   Conclusion

In this paper a hybrid communication system is proposed for use in search and rescue scenarios. According to the experimental results, one can conclude that the performance of a hybrid network is superior to that of networks which utilize only a single communication medium. In hybrid networks, the specific features of one technology may neutralize the weakness of another and improve the overall efficiency and effectiveness of the system. This paper presents a hybrid communication system formed by combining GSM-GPRS, XBee and Bluetooth modules. Several tests were performed in different environments to evaluate the performance of the system. The results prove that multiple technologies cooperating together improve parameters such as signal range, cost of the data transmission, energy consumption, autonomous node routing and dynamic membership. The hybrid communication system should be tested further with higher numbers of nodes so that the system's behavior can be monitored when data traffic is high. A future addition to the node is the integration of an UWB (ultra wide band) communication module to enable indoor localization services.

References

1. Witkowski, U., Herbrechtsmeier, S., El-Habbal, M.: Mobile ad-hoc networking supporting multi-hop connections in multi-robot scenarios. In: Baudoin, Y., Habib, M.K. (eds.) Using Robots in Hazardous Environments: Landmine Detection, De-mining and other Applications. Woodhead Publishing in mechanical engineering (2010)
2. El-Habbal, M., Rückert, U., Witkowski, U.: Topology control in large-scale high dynamic mobile ad-hoc networks. In: Kim, J.-H., et al. (eds.) FIRA 2009. LNCS, vol. 5744, pp. 239–250. Springer, Heidelberg (2009)
3. Venkatakrishnan, V., Seethalakshmi, R.: Public Transport Ticketing and Monitoring System. Journal of Theoretical and Applied Information Technology 38(1) (April 15, 2012)
4. Stanchev, G., Stoyanov, N., Doichev, Z., Balzhiev, P.: Application of Wireless Standards in Dynamic Parameters Monitoring System for Vehicles and Industrial Machines. Proceedings of Manufacturing Systems 7(3) (2012)
5. Amutha, B., Ponnavaikko, M.: Location Update Accuracy in Human Tracking system using ZigBee modules. (IJCSIS) International Journal of Computer Science and Information Security 6(2) (2009)
6. Sandeep, C., Subudhi, K., Sivanandam, S.: Intelligent Wireless Patient Monitoring and Tracking System. International Journal of Interdisciplinary and Multidisciplinary Studies 1(3), 97–104 (2014)
7. Jadhav, D.A., Patane, S.A., Nandarge, S.S., Shimage, V.V., Vanjari, A.A.: Air Pollution Monitoring System Using ZigBee and GPS Module. ISO 9001:2008 Certified Journal 3(9) (2013) ISSN: 2250-2459
8. Tuikka, T., Isomursu, M.: Touch the Future with a Smart Touch. VTT Tiedotteita – Research Notes 2492, 4–7 (2009)
9. Elahi, A., Gschwender, A.: ZigBee Wireless Sensor and Control Network. Prentice Hall communications engineering and emerging technologies series. Pearson Education, Inc. (2010)
10. Khanh Tuan, L.: ZigBee SoCs provide cost-effective solutions, RFIC system architect, ChipCon, online source at:
http://www.eetimes.com/document.asp?doc_id=1273396 (visited on August 11, 2005)
11. Digi International Inc., XBee 865/868 LP User Manuel (February 2012)
12. Cuomo, F., Cipollone, E., Abbagnale, A.: Performance analysis of IEEE 802.15.4 wireless sensor networks: An insight into the topology formation process. Computer Networks 53(18), 3057–3075 (2009)

Wearable Self Sufficient MFC Communication System Powered by Urine

Majid Taghavi,[1,2,3] Andrew Stinchcombe[1], John Greenman[4], Virgilio Mattoli[2], Lucia Beccai[2], Barbara Mazzolai[2], Chris Melhuish[1], and Ioannis A. Ieropoulos[1,*]

[1] Bristol Robotics Laboratory, University of the West of England, Bristol, UK
ioannis.ieropoulos@brl.ac.uk
[2] Center for Micro-BioRobotics, Istituto Italiano di Tecnologia, Viale Rinaldo Piaggio 34, 56025 Pontedera (PI), Italy
[3] BioRobotics Institute, Scuola Superiore Sant'Anna, Pontedera, Italy
[4] Faculty of Health and Applied Sciences, University of the West of England, Bristol, UK

Abstract. A new generation of self-sustainable and wearable Microbial Fuel Cells (MFCs) is introduced. Two different types of energy - chemical energy found in urine and mechanical energy harvested by manual pumping - were converted to electrical energy. The wearable system is fabricated using flexible MFCs with urine used as the feedstock for the bacteria, which was pumped by a manual foot pump. The pump was developed using check valves and soft tubing. The MFC system has been assembled within a pair of socks.

1   Introduction

Demand for ultra-low energy generation is growing in order to develop independent, sustainable, and continuous operation of remote and mobile environmental sensors, micro-electromechanical systems, micro/nanorobotics, portable/wearable personal electronics, and even implantable devices. Furthermore, renewable green energy is one of the international priority areas for addressing the current global warming and energy crises [1]. To this aim, particularly for powering wearable devices, tremendous effort is being exerted to develop systems using alternative energy resources such as solar [2], thermal [3], and mechanical systems [4].

Microbial Fuel Cells (MFC) have been recently reported to utilise urine for energising real electronic devices [5]. In this case, continuous feeding of the MFCs with urine is a necessary condition for improved performance and biofilm community maintenance, which was facilitated with the use of a mains powered peristaltic pump that is in itself consuming valuable energy. The implementation of such MFCs in a self-sustainable manner has been previously studied on-board the EcoBot robots, and in particular EcoBot-III [6]. However, the feasibility of utilising such systems as wearable devices, with no electrical energy input for the pumping, has not been previously attempted. A wearable MFC system that is entirely powered by the human

---

[*] Corresponding author.

M. Mistry et al. (Eds.): TAROS 2014, LNAI 8717, pp. 131–138, 2014.
© Springer International Publishing Switzerland 2014

wearing it is novel and has a great potential in powering a wide range of portable systems, from electronic devices to rehabilitation robots, particularly in specific outdoor conditions. Humans or robots wearing such devices and facilitating fluid-fuel circulation by walking or locomotion may be envisaged, whereby the energy is harnessed from waste and is used to power on-board devices.

In the present study, the first self-sustainable wearable MFC is introduced, which can be used as a promising energy generator for portable electronic devices. Urine is considered as the fuel for the MFC and natural walking as the physical pumping for driving the fuel. Under certain conditions, the pumping is by gaiting, which is enough to replace the existing fuel in the MFCs with fresh urine. The foot pumping part was inspired by the fish circulatory system in terms of structure and material. The wearable generator is made of 24 individual flexible MFCs, integrated in the fabric of a pair of socks. Each pair of MFCs has its own flexible tubing passing under the heel, which facilitates the pumping of sufficient quantities of urine to flow into the cells by means of any type of walking behaviour.

2   Methods

2.1   Preparation of MFC

The single-chamber microbial fuel cells were made of Nafion tubing (TT-110, PermaPure, US) as the membrane, and carbon sleeve (978, Siltex, Julbach, Germany) as the anode and the cathode. The inside of the Nafion tubes was the anode, with anolyte running through, and the open to air outer part, was acting as the cathode. A total of 24 MFCs was developed for this line of experiments.

The Nafion Tubing Membrane was cut to 100mm lengths. These were washed and pre-treated by boiling in H_2O_2 (7% v/v) and deionized water, followed by soaking in 0.5M H_2SO_4 and then deionized water, with each cycle lasting for 1h. The membranes were stored in deionized water prior to being used. Carbon Fibre Sleeves (CFS) were soaked in 100% ethanol for 30min and in 1M HCl for 1h, and then stored in distilled water before use. They were inserted into the tubular membrane and pushed against the two ends of the sleeve in order to make full contact with the inner side of the tube. Two ends of the tubular membrane were connected to the two separate T-shaped fittings. One port of each fitting was used for circulating the anolyte, whereas the other ports were used for running the connection wires through to the anode. Both ends of the sleeve were connected to nickel-chromium (Ni-Ch) wires, which were then connected to the external load. For cathode electrodes, the CFS was positioned around the membrane, and was tightened over the membrane by pulling the two sleeve ends. It was cut at the same length of 100mm, and connected to the Ni-Ch wire. The surfaces of the cathodes were then hydrated by spraying over deionised water. Finally, to keep the cathode moist for longer periods, the electrodes were covered with a waterproof, breathable plaster, manufactured by Masterplast, UK.

Activated anaerobic sludge, collected from Wessex Water, (Cam Valley, Saltford, UK) mixed with 1% (w/v) tryptone and 0.5 % (w/v) yeast extract (Fisher scientific), was used for inoculating the 24 MFCs. The sludge was fed in continuous flow at a slow flow rate, in order to prevent MFCs from blocking. The flow rate was set to 27μL/min, which was facilitated by a Watson Marlow 205U peristaltic pump (Watson Marlow, UK); this allowed for a hydraulic retention time of 1 minute. After one week, the solution was substituted with a mixture of deionized water, 1% (w/v) tryptone, and 0.5% (w/v) yeast extract and the flow rate was increased to 45μL/min. Urine was used as the fuel after a week for feeding the bacteria, which was replaced by a fresh feed every day. The urine was collected from healthy individuals with no known previous medical conditions, and mixed together before being driven into the MFCs. The pH of the pooled urine was measured and found to be 6.3.

All the cells were initially connected to 5kOhm resistors. Polarisation experiments were then performed on each MFC using the resistorstat device [7] every five days; prior to this, the cells were left open-circuit for at least 5 minutes, to reach pseudo steady-state conditions. The value of the external resistance, connected separately to the MFCs, was sequentially reduced (from open circuit) to the lowest possible value, close to short circuit conditions. The total number of external resistors was 38, each one connected to the MFCs for a period of 5min, for establishing quasi steady-state values. At the end of each of the polarisation runs, the resistor producing the maximum power was permanently connected to the MFCs for characterising the real time temporal behaviour of the system. Output voltage was measured and logged using a multi-channel Agilent 34972A, LXI Data Acquisition/Switch Unit (Farnell, UK).

2.2   Stack Assembly

The MFCs, which had reached maturity within 10 days, were connected together electrically as described below. There were 6 lines of tube per sock (foot), each consisting of two MFCs, whose anodes were exposed to the same common anolyte, therefore they were also connected electrically in parallel. Parallel connection was performed using external wiring, and the polarisation experiments were performed on each of the MFC pairs. After 3 days, the 12 pairs (6 pairs per sock) were connected in series, and a new polarisation experiment was carried out in order to evaluate the overall stack performance. A total of 32 different values, starting from 1MΩ down to 50Ω, were connected manually to the stack, with a 4-minute connection time, per value.

2.3   System Design

As described above, the MFCs were matured within 2 weeks and were then disconnected from the apparatus in order to be assembled on the wearable sock support. For the purpose of circulating the anolyte through the MFCs, a simple manual pumping system was developed as shown in the schematic of Figure1a.

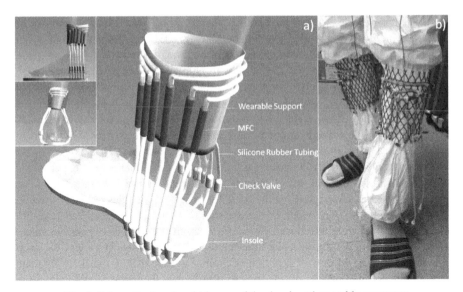

Fig. 1. Schematic drawing; b) image of the developed wearable generator

This manual pumping system includes a silicone tube (pumping-tube) with 1mm inner diameter, placed directly under the heels, and two check valves (SCV21053, The West Group Ltd, UK). The valves were used as connectors between the pumping tube and two other reservoir and carrier silicone tubes. The inner diameter of these tubes was 2mm, which were connected to the tubular MFCs with an inner diameter of 1.8mm. Therefore, in each sole, 12 valves were used to develop 6 separate pumps for feeding 6 pairs of MFC. All the pairs were connected to each other in series to increase the generated voltage. The image of the developed wearable system is illustrated in Figure 1b.

The two phases of heel STRIKE and heel OFF, occurring during each gait cycle, serve the function of foot pumping. Tubes made of silicone rubber were used as the main pumping chamber, reservoirs and carriers. Squeezing the pumping-tube and releasing it provides the two stages required for driving urine, including pumping into the carrier-tube and creating suction from the reservoir-tube, respectively. All the connections were made whilst a foot pressure was applied on the pumping-tube (Figure 2a). Therefore, as shown in Figure 2b, at the first phase of gaiting, i.e. heel OFF, the fluid will flow from the reservoir-tube down to the pumping-tube. Owing to the check valves, it can only flow through the reservoir-tube, which occurs due to the generated differential potential by the released pumping-tube from sustaining the foot pressure. Consequently, urine will move through the MFCs from carrier-tube to the reservoir-tube for compensating the differential pressure in the closed tubing system (Figure 2c). In contrary, as shown in Figure 2d, at the second phase, i.e. heel STRIKE, the tubes are compressed and the fluid is driven into the carrier-tube, through the check valve on the opposite side. The flowing of urine will be continued in the MFCs to refill the reservoir-tube and come back to the first position (Figure 2e). The toe OFF phase from one foot happens at the same time that another heel contacts the ground. Therefore, the identical cycle repeats for the second foot until such time that the first heel strikes the ground again.

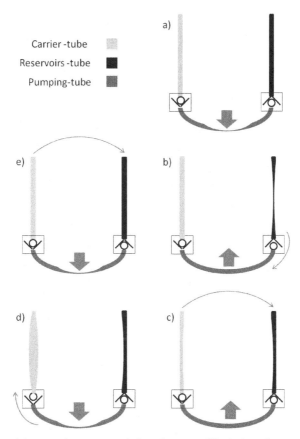

Fig. 2. Schematics of the pumping system; a) the tubes were filled when the pumping-tube was left under the foot pressure; b) urine flows from the reservoir-tube to the pumping-tube as the pressure is released; c) urine flows through MFCs for compensating the differential pressure between reservoir-tube and carrier-tube; d) squeezing pumping-tube flows urine to the carrier-tube; e) The system comes back to the first condition by flowing urine from carrier-tube to the reservoir-tube through the MFCs.

3   Results and Discussion

Figure 3 shows the power curves of the 12 MFC pairs, when fed with fresh urine, under ambient room temperature conditions (22°C), at a flow rate of 45µL/min. Each pair consists of two individual MFCs, fed by a single tube and connected in parallel. Some of the MFC pairs show lower power compared to the average power generated. Several reasons can explain these variations in the performance of the MFCs. Firstly, it may be assumed that the actual volume and surface area of the MFCs was different, since they were all manually made. Secondly, the ends of the Nafion membrane were glued to the fittings, which renders an area of the anode electrode as well as the

Nafion membrane unutilised. Therefore, the effective volume or anode surface area varies for each MFC. The lengths of all MFCs were measured by subtracting the non-utilised parts, which were visible to the naked eye. They vary from 40mm to 70mm with the average length of 56mm. In addition, the variance in power may have resulted from differences in the biofilm growth on the anode of the MFCs, which is subject to the supply of fuel. Finally, it may be related to the leakage of anolyte from the ends of the Nafion tubing, which can affect the performance of the MFCs in terms of insufficient feeding of the bacteria, but also possible 'short circuit' between the negative anode and the positive cathode. Clearly this was an experimental rig with a lot of potential for improvement, which can be the topic of future research into wearable low-power devices.

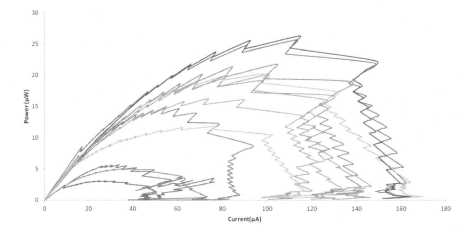

Fig. 3. Power curve of 12 different pairs of MFC

Figure 4 shows the power curves of the whole stack, when the 12 pairs are connected to each other in series. The generated power is plotted against the external load current, produced by sweeping a series of resistors. The power curve confirms that the maximum achievable power is about 110μW, which was produced from the 30KΩ load. These results have been recorded when fresh urine was fed at a flow rate of 45μL/min. The manual pumping system was designed to provide the same flow rate across all the MFC pairs, with 1mm inner diameter chosen to pass through the footwear's insole. The average driven flow by each foot was measured as 100μL/min at a walking speed of 45 steps/min. This number was considered as a normal gaiting for each foot, since only half of the total number of steps contributes to pumping the fuel into the MFCs integrated in each sock. In other words, the average walking speed for a person is 90 steps/min, which results in pumping fluids by means of both legs at the speed of 45 steps/min.

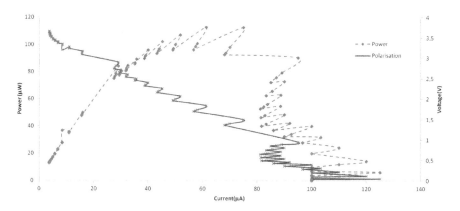

Fig. 4. Power and polarisation curves of the whole MFC stack, where all the 12 MFC pairs were connected to each other in series

As shown in Figure 4, the open circuit output voltage of the stack was 3.66V, which increased to 4V, as the flexible MFCs were been worn and fresh urine (of the same temperature) was fed by the manual foot pump with the same average flow rate. As noted above, there was a small quantity of leakage from the connection points in some MFCs during the bench experiments. It can occur due to the obstructions made by thicker biofilms created on the anodes. However, there was no visible leakage in the same MFCs after assembling them as a wearable system. It is worth noting that although the average flow rates in both conditions have the same value, they provide two different types of flow. The electrical pumping system provides a constant flow. In contrary, the manual pumping system creates a discrete flow with the same flow rate on average. As explained above, it actually acts as a push-pull mechanism for the fluid at the beginning of heel STRIKE and heel OFF phases in each gait cycle, resulting in driving impulsive flow through the MFCs. This will have a direct effect on the output performance, but as shown, it has also prevented the system from leaking. Unlike the mains powered multi-channel pump in which the fluid is only driven into the MFCs, the manual pumping system benefits from differential pressure inside the MFCs and connected tubes for driving the fluid. Therefore, during the circulation, the existing low pressure in the tubes prevents the leakage from happening. Consequently and as explained above, the quasi-discrete fuel supply and no leakage may justify the increase in the open circuit voltage, when the manual pumping system is used.

The inner diameter of the tubes used as reservoir-tube and carrier-tube was 2mm. Since the length of each of these is about 25cm and pumping-tube is 30cm, the whole capacity of the fresh urine stored in each tube can be calculated as 1.8mL. As mentioned above, a normal gaiting produces the flow rate of 45µL/min. Therefore, the total fresh reserved urine will be circulated thoroughly and used within 40 minutes. However, the system can benefit from adding some other reservoirs containing fresh urine in real longer-term applications. A manual subsystem could also be used for rapidly substituting used urine with fresh. The next step in this line of experiments is to evaluate the wearable MFC stack as an energy source for a wireless communication system, and demonstrate its implementation in the context of portable equipment powered by urine.

4   Conclusions

This paper reports for the first time on the successful development of a wearable energy generator based on the microbial fuel cell technology. Flexible materials and structures have been used for fabricating the components for the purpose of making the system wearable. It is powered by urine pumped manually by means of human gaiting. Normal gaiting provides the flow rate of 45μL/min for feeding bacteria by fresh urine. The maximum generated power is approximately 110μW when a 30KΩ resistor is connected as the load. It is envisaged that this will form part of a wearable MFC system capable of useful transmission, which will be powered by urine.

Acknowledgments. Parts of this work were funded by the UK Engineering & Physical Sciences Research Council (EPSRC) grant numbers EP/I004653/1 and EP/L002132/1, and the Bill & Melinda Gates Foundation grant no. OPP1094890.

References

1. Arunachalam, V.S., Fleischer, E.: Harnessing materials for energy. MRS Bulletin, 33 (2008)
2. Leonov, V., Torfs, T., Vullers, R.J., Van Hoof, C.: Hybrid thermoelectric–photovoltaic generators in wireless electroencephalography diadem and electrocardiography shirt. Journal of Electronic Materials 39, 1674–1680 (2010)
3. Leonov, V., Vullers, R.J.M.: Wearable electronics self-powered by using human body heat: The state of the art and the perspective. Journal of Renewable and Sustainable Energy 1, 062701 (2009)
4. Majid, T., Virgilio, M., Ali, S., Barbara, M., Lucia, B.: A novel soft metal-polymer composite for multi-directional pressure energy harvesting. Advanced Energy Materials (2014) doi:10.1002/aenm.201400024
5. Ieropoulos, I., Greenman, J., Melhuish, C.: Urine utilisation by microbial fuel cells; energy fuel for the future. Physical Chemistry Chemical Physics 14, 94–98 (2012)
6. Ieropoulos, I., Greenman, J., Melhuish, C., Horsfield, I.: EcoBot-III: A robot with guts. In: 12th International Conference on the Synthesis and Simulation of Living Systems, Artificial Life XII, pp. 733–740 (2010)
7. Degrenne, N., Buret, F., Allard, B., Bevilacqua, P.: Electrical energy generation from a large number of microbial fuel cells operating at maximum power point electrical load. Journal of Power Sources 205, 188–193 (2012)

Morphogenetic Self-Organization of Collective Movement without Directional Sensing

Ataollah Ramezan Shirazi, Hyondong Oh, and Yaochu Jin

Department of Computing, University of Surrey
Guildford, Surrey, GU2 7XH, UK
{a.ramezanshirazi,h.oh,yaochu.jin}@surrey.ac.uk

Abstract. In this paper, we present a morphogenetic approach to self-organized collective movement of a swarm. We assume that the robots (agents) do not have global knowledge of the environment and can communicate only locally with other robots. In addition, we assume that the robots are not able to perform directional sensing. To self-organize such systems, we adopt here a simplified diffusion mechanism inspired from biological morphogenesis. A guidance mechanism is proposed based on the history of morphogen concentrations. The division of labor is achieved by type differentiation to allocate different tasks to different type of robots. Simulations are run to show the efficiency of the proposed algorithm. The robustness of the algorithm is demonstrated by introducing an obstacle into the environment and removing a subset of robots from the swarm.

Keywords: Collective movement, morphogenetic robotics, pattern formation, self-organizing systems.

1 Introduction

For a robotic swarm, collective movement can be defined as moving in an environment and interacting with objects and obstacles in the environment without losing a predefined form or pattern [1]. Such collective movement can be of great interest in practice in many situations such as rescue or formation flying [2, 3]. One main reason is that multi-robot systems can reduce the chance of complete failure and make the system more adaptable to the environmental changes. However, designing a self-organized multi-robot system is very challenging in that it is not straightforward to develop local interaction rules that can produce the desired global behaviors.

The benefit of multi-robot systems can be fully exploited only if the control mechanism is decentralized. Biological systems are one of the best examples that use the decentralized, self-organizing control mechanisms. In nature, complex living organs develop from a fertilized cell during the embryonic development into a complex morphology without a central controller. Many researchers have attempted to implement principles inspired from biological development in swarm robotics [4-8]. Among others, considerable attention has been dedicated recently to bio-inspired aggregation and patter formation behaviors [9-12], mainly because such aggregation and formation

M. Mistry et al. (Eds.): TAROS 2014, LNAI 8717, pp. 139–150, 2014.
© Springer International Publishing Switzerland 2014

behaviors are the fundamental functions for swarm robotic systems to accomplish more complex tasks.

According to the findings in research on embryonic development, morphogens play a major role in the development of organs and the pattern formation. Diffusing through the tissues, morphogens form gradients that provide cells with relative positional information. Robustness, high complexity, flexibility and adaptability in a changing environment are the most salient features of morphogenesis [13]. Inspired by biological morphogenesis, Jin and Meng [7] introduced the term of "morphogenetic robotics", where genetic and cellular mechanisms in biological morphogenesis are adopted for self-organizing swarm or modular robotic systems.

Inspired by morphogenesis, several self-organizing algorithms for adaptive pattern formation have been developed. Among these, some studies focus on stationary pattern formation only. For instance, Memei et al [9] applied a morphogen gradient to a swarm of robots to achieve polygonal patterns. Shen et al [4] investigated the use of a morphogenetic reaction-diffusion system in swarm robots. Ikemoto et al [10] proposed a method to generate simple polygonal shapes by implementing a reaction-diffusion model containing two morphogens. Other studies explored the use of gene regulatory networks to transform morphogen gradients into robot motions in order to form adaptive patterns [5 7, 8]. However, they are either still stationary [7], or require global position information [5], or rely on the targets' location [8]. Sayama [6] simulated a moving pattern in a swarm of robots, where the patterns do not have a constant shape.

Ho et al [14] proposed an algorithm for navigation of a swarm with predefined shape through a user-defined path. However the initial formation is not self-organizing but user-defined. Moreover, the agents need global position information and are supposed to know the path in advance. There is also no limit on communication between the agents. Navaro and Matia [15] presented a framework for collective movement of a swarm of robots, where robots can maintain a specified distance from each other and form a triangular lattice. The robots in their algorithm are supposed to be capable of directional sensing.

In this paper, we investigate how a swarm of robots can achieve collective movement where each individual robot has no information about the position of others, a short range of communication and no directional sensing. We assume that the only information that is exchanged among the robots is the morphogens' concentrations and the robots' differentiated types. It is also assumed that each robot is able to measure the distance between itself and its neighbors. These assumptions are essential for us because we intend to implement the proposed algorithm in a swarm of very simple robots with limited computational and communication capabilities such as the Kilobot [16].

The rest of the paper is organized as follows. In the next section, we describe in detail the proposed algorithm for self-organized collective movement. In the Section 3, we present the simulation results that demonstrate the effectiveness of the proposed algorithm and its robustness to obstacles and partial robot failures as well. In the last section, we conclude the paper with a brief summary and a discussion of future work.

2   Self-Organized Collective Movement via Diffusion

Robots that constitute a swarm are supposed to be functionally very simple with limited computational, communication and sensing capabilities in order to be affordable in a large number. These constraints make a decentralized control indispensable and impose an immense challenge to the design of effective self-organizing control algorithms. To address these challenges, we turn to nature for getting inspiration because such decentralized control mechanisms are widely seen in nature. One typical example is biological morphogenesis, where cells use the morphogen gradients to find out their position information so that they can take proper actions. These biological mechanisms have been simulated in swarm robotic systems for pattern formation. To meaningfully interpret the gradients information, at least one morphogen should start diffusing from a specified reference position. For example, in a drosophila embryo there are four maternal morphogens that diffuse from specified parts of an egg, including posterior and anterior poles [17]. In our work, the first morphogen diffuses from the robots located at the boundary of aggregated swarm, which are termed edge-robots.

2.1   Edge Differentiation

In an evenly distributed swarm of robots, where the radius of the swarm is significantly larger than the robots' communication range, the number of neighbors the robots at the boundary can have is usually less than half of the number of the robots inside the swarm. As shown in Fig.1, if the distances between the robots and their closest neighbors are the same, the robots will form a hexagonal lattice where each node in this lattice is the location of a robot. Each hexagonal has a node at its center and six nodes on its perimeter. Each node on the perimeter is shared between three hexagonal, i.e., the share of each hexagon from the lattice nodes is three. Hence, for an evenly distributed swarm, if a robot's communication area (see Fig. 1) is completely inside the swarm, the number of robots inside its communication range can be estimated as follows:

$$N = 3 \times \frac{A_{R_c}}{A_{d_e}} = \frac{2\pi R_c^2}{\sqrt{3}d_e^2}, \tag{1}$$

where A_{R_c} is the area of the communication range with a radius R_c and A_{d_e} is the area of a hexagonal shape with a radius d_e. Therefore, the number of neighbors the corresponding robot has is $n = N - 1$.

Consequently, the number of the neighbors of an edge-robot equals approximately to $n/2$. For a disk-like shape swarm, we can estimate the radius of the swarm, S_p as follows by rewriting Equation (1):

$$S_p = \sqrt{\frac{\sqrt{3}}{2\pi}nd_e^2} . \tag{2}$$

We will use this equation in the following to estimate the distance of a robot at the boundary of the swarm from the center of the swarm.

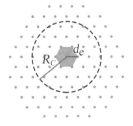

Fig. 1. Estimation of the number of neighbors in an evenly distributed swarm, where the communication area of the robot (dashed circle) is inside the swarm

To have an evenly distributed swarm, we define a simple task for the robots before edge differentiation occurs. Each robot should adjust its distance to its nearest neighbor to be equal to the predefined value d_e. We assume that each robot diffuses a morphogen gradient that decreases as the distance to it increases. Using these gradients, the robot can avoid collision with others by autonomously adjusting their distance to each other. The mechanism of the movement will be explained in Sections 2.3 and 2.4. As the specified distance d_e to the nearest neighbor cannot be achieved for the all robots in one shot, it will take some time for the robots to stabilize. While the robots are fine tuning to distribute evenly, they count the number of neighbors and calculate the average over the time. A better estimation can be obtained when a robot slightly moves and counts the number of neighbors. At the end of this stage if the average number of neighbors of a robot is equal to or less than $n/2$, the robot considers itself at the edge of the swarm and will differentiate to an edge-robot and remains unchanged in its type. Fig. 2(b) shows the status of the swarm after distributing evenly and the edge differentiation stage.

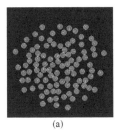
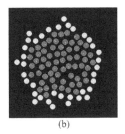

(a)   (b)

Fig. 2. Edge differentiation. a) Randomly distributed swarm. b) Edge differentiation after the robots distribute evenly. The edge-robots are indicated in yellow.

2.2   Morphogen Diffusion

In our algorithm, there are two additional types of morphogens: edge morphogen and center morphogen. The edge morphogen is secreted by the edge-robots. After the edge differentiation is complete, the concentration of the edge morphogen is initialized to its highest amount (100) in the edge-robots, and sets to zero in non-edge robots. Then the diffusion of the edge morphogen begins. A robot asynchronously changes its concentration of the edge morphogen by injecting it into its all neighbors according to the following equation:

$$V_{ij} = c_d \frac{M_i - M_j}{n_i d_{ij}}, \tag{3}$$

$$\text{if } V_{ij} > 0 \begin{cases} M_j(t+1) = M_j(t) + V_{ij} \\ B_i = B_i + 1 \end{cases}, \tag{4}$$

$$M_i(t+1) = M_i(t) - \sum_{j=1}^{B_i} V_{ij}, \tag{5}$$

where M_i and M_j are the morphogen concentration in robot i and robot j, respectively, d_{ij} is distance between robot i and robot j, c_d is the diffusion rate, and B_i is the number of robot i's neighbors whose concentration of the morphogen is lower than that of robot i.

Fig. 3 shows the edge morphogen gradient after ten asynchronous morphogen injections iterations. The closer a robot is to the center of the swarm, the lower the edge morphogen it receives. Obviously, the center-robot has the lowest concentration of the edge morphogen. A robot identifies itself as the center-robot if all of its neighbors have a higher concentration of the edge morphogen.

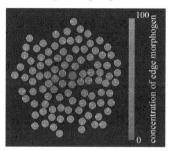

Fig. 3. The edge morphogen gradient. The center-robot is indicated by a red border.

Once the center-robot is determined, another type of morphogen, named center morphogen, is secreted by the center-robot. For the center-morphogen, a slightly different diffusion rule will be used to have a more precise sensing of the distance from the center. The center morphogen is set to 100 in the center-robot and remains constant. The other robots update their concentration of the center morphogen using following form:

$$M_i(t+1) = \max_j \{M_j(t) - d_{ij}\}. \tag{6}$$

Using this morphogen all other robots can estimate their distance from the center-robot. This morphogen will serve as a glue to keep the disc-like shape of the swarm while the swarm moves in the environment.

2.3   Movement Direction Estimation

In this section, we explain how the robots can move collectively following the morphogen gradient. We assume that all robots have an omnidirectional sensor, i.e., they are not able to sense the direction of the messages received and therefore, they cannot distinguish which of neighbors the received message is sent from. Consequently, the robots cannot understand the orientation of the morphogen gradients unless they perform some random walks and measure the received morphogen concentration at different places. The orientation of the morphogen gradient can be estimated using the following equations:

$$\Delta C_{t,i} = \frac{P_i - P_{i-1}}{\|P_i - P_{i-1}\|} (C_{P_i} - C_{P_{i-1}}),$$ (7)

$$\Delta C_t = \sum_{i=t-N_s}^{t} \Delta C_{t,i},$$ (8)

$$\Delta \widehat{C}_t = \frac{\Delta C_t}{\|\Delta C_t\|},$$ (9)

where P_i represents the position of robot i at time t in its local reference frame, C_{P_i} is the morphogen concentration measured at position P_i, $\Delta \widehat{C}_t$ is a unit vector indicating the orientation of the morphogen concentration and N_s is the number of previous positions taken into account to evaluate the morphogen gradient.

At each iteration of the simulation, the next position of a robot is calculated by:

$$P_{t+1} = P_t + \widehat{\theta}_t V dt,$$ (10)

where P_t is the current position, P_{t+1} is the next position, V is the velocity of the robot, and $\widehat{\theta}_t$ is a unit vector determining the direction of the movement. At each iteration, a robot corrects its current direction using Eq. (9):

$$\Delta \theta_t = \Delta \widehat{C}_t - \widehat{\theta}_{t-1},$$ (11)

$$sat(\Delta \theta_t) = \begin{cases} \|\Delta \theta_t\| = \Delta \theta_{max}, & if \ \|\Delta \theta_t\| > \Delta \theta_{max} \\ \Delta \theta_t, & otherwise \end{cases},$$ (12)

$$\widehat{\theta}_t = \widehat{\theta}_{t-1} + sat(\Delta \theta_t).$$ (13)

To consider the rotation speed limitations of real robots, if the direction change in one step of the simulation is greater than $\Delta \theta_{max}$, then the direction change will be limited, refer to Eq. (12).

2.4   Collective Movement

The main objective of this work is to accomplish collective movement by following the center-robot and maintaining a dis-like shape of the swarm. The movement trajectory of the center-robot is a randomly generated spline function. It is assumed that the center-robot's position is available to none of the other robots.

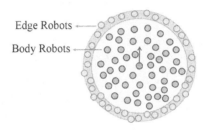

Fig. 4. An illustration showing the two types of robots and their distribution in the swarm

As illustrated in Fig. 4, all other robots, which are neither an edge-robot nor the center-robot, differentiate to body-robots. Here, the body-robots play the role of keeping the edge-robots communicationally connected to the center-robot as the communication range of each robot is very much limited.

The edge-robots have two tasks. The first task is to stay inside the grey ring indicated in the Fig. 4 to keep the disc-like shape of the swarm. Using the concentration of the center morphogen an edge-robot has a sense of its distance to the center. Once the concentration of the received center-morphogen falls in a certain range associated with the grey ring, an edge-robot performs its second task. To enclose the body-robots perfectly, the edge-robot should distribute inside the boundary ring. We have tried different strategies for distributing the edge-robots inside the gray ring and the most efficient one is to minimize the following function:

$$f = \frac{n_i}{(\sum_{j=1}^{n_i} d_{ij})/n_i} = \frac{n_i^2}{\sum_{j=1}^{n_i} d_{ij}}, \tag{14}$$

where n_i is the number of neighbors of robot i, and d_{ij} is the distance between robot i and its neighbor j. In this way, a robot goes in a direction that decreases the number of neighbors and increases the average distance to its neighbors.

More details of the algorithm can be found in the pseudo code below. The definition and description of the functions and parameters in the code can be found in Tables 1 and 2. Here, C_{pi} indicates the concentration in Eq. (7).

```
if type= body_bot
    if disToObastale()< D_ob
        status= "close_to_obstale";
        C_pi = -disToObastale();
    else if Density(body) < DEN_1 OR Density(edge) > DEN_2
        status= "go_inside";
        C_pi = -centerMorpho();
    else if nearestBot(edge)< NE_min
        status= "close_to_edge ";
        C_pi = nearestBot(edge);
    else if nearestBot(center)< D_cent
        status= "close_to_center ";
        C_pi = nearestBot(center);
    else
        status= "distribution";
        C_pi = -Density(body);
if type= edge_bot
    if disToObastale()< D_ob
        status= "close_to_obstale";
        C_pi = -disToObastale();
    else if centerMorpho()< R_min OR centerMorpho()> R_max
        status= "outside_ring";
        C_pi = - abs(centerMorpho()- (R_min+ R_max)/2);
    else
        status= "inside_ring";
        C_pi = -Density(body);
```

Table 1. Functions executed by a robot in the algorithm

Function	Description
disToObastale()	Return the distance from the robot to an obstacle
density(T)	Evaluate the density of the neighbors of type T using Eq. (14)
centerMorpho()	Return the concentration of the center morphogen
nearestBot(T)	Return the distance from the nearest robot of type T
amICenter()	Differentiate to the center-robot if all neighbors have a lower concentration of the edge-morphogen
amIEdge	Differentiate to an edge-robot if the number of neighbors is lower than $n/2$ (see Eq. (1)).
secreteMorpho(M)	Secrete morphogen type M to neighbors using Eqs. (3)-(6).
move(dir, Vdt)	Move one step forward using Eq. (9). 'Vdt' is the step length and 'dir' is the direction of the morphogen gradient calculated by Eqs. (11)-(13).

Whenever a robot performs a task and changes its status, it records the number of its last vectors of $\Delta C_{t,i}$ (N_s). The next time when the robot switches to the same task, it resets all $\Delta C_{t,i}$ and reloads the vectors of $\Delta C_{t,i}$ stored after successfully performing the last task. This can help the robot find the correct gradient direction faster.

To prevent a robot from being lost from the swarm, a robot makes a turn once the number of its neighbors drops to one. If the number of neighbors drops to zero, the robot moves in a random direction. If it cannot find the swarm after a specified number of steps then it takes another random direction.

3   Simulation and Discussion

The proposed algorithm is implemented under the environment of Visual Studio 2012, and the SDL library is used for visualization. In order to simulate the asynchronous status changes, each robot is selected randomly to update the received information from its neighbors and change its status. To quantitatively assess the performance of the algorithm, we calculate the percentage of the robots that are correctly positioned in the required movement. After type differentiation, a randomly generated spline is generated as the trajectory of the center-robot. In the simulations, 100 robots with a diameter of three units and a communication radius of nine units are used. We have performed a large number of experiments to tune the parameters of the algorithm. Table 2 lists the parameter values found to produce good performance.

The motion speed of the center-robot is set to 15 times slower than others, which is critical for the success of the system. The reason is that all robots, apart from the center-robot, do not have a directional sensing and they have to perform many random walks in different directions to learn about the morphogen gradient. They are also very likely to collide with others, both of which reduce the speed of movement. So if the center-robot moves too fast, other robots can easily get lost.

Table 2. Parameter values used in the simulations

Parameter	Description	Chosen value
S_P	Swarm size	100
Vdt	Displacement of a robot at each step of simulation	0.1
$V_c$dt	Displacement of a center-robot at each step of simulation	0.0066
B_D	Diameter of each robot	3
R_C	Radius of communication	9
c_d	Diffusion rate in Eq. (3)	0.5
d_e	Distance between robots in an evenly distributed swarm	1
$\Delta\theta_{max}$	Maximum change in motion direction at each step of simulation	$\pi/10$
D_{ob}	Threshold distance from an obstacle	2
NE_{min}	Threshold distance from an edge-robot	3
D_{cent}	Threshold distance from the center-robot	1
N_s	Number of gradient history used in Eq. (8)	5
R_{min}	Inner radius of the edge-robot ring	S_p
R_{max}	Outer radius of the edge-robot ring	$1.1S_p$
DEN_1	Density of neighbor body-robot that triggers "go-inside" status	
DEN_2	Density of neighbor edge-robot that triggers "go-inside" status	

To understand the impact of variation on the number of concentration histories, N_s, on the algorithm's performance, we conducted experiments to measure the performance when N_s varies from 1 to 12. For each N_s, we ran the simulation for 40 times and calculated the mean and standard deviation. The results shown in Fig. 5 indicate that the best performance is obtained for N_s=5. For a value greater than five, the performance starts to degrade because the concentration pattern changes over the time and older data become obsolete.

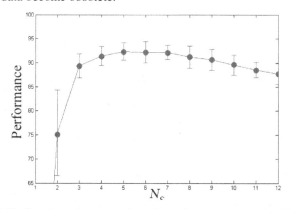

Fig. 5. Finding the optimum number of morphogen concentration histories

Fig. 6 shows a few snapshot of one simulation using the parameters given in Table 2 when the center-robot moves along a randomly generated spline. As a whole, the swarm is able to move collectively along the center-robot without explicitly knowing the movement speed and direction of the center-robot. This is of great interest given that the robots do not have directional sensing and only local information is available to the robots. It can be noticed, however, that the density of the edge-robots is slightly higher in the rear of the swarm than in the front with respect to the movement direction. The main reason is that, due to the lack of directional sensing, the edge-robots need to perform a lot of random walks to find the right direction to move and to keep moving in the boundary area. Collisions also happen and reduce their speed of movement. Nevertheless, there are always a sufficient number of edge-robots in the front of the swarm to enclose the body-robots and keep the shape of the swarm.

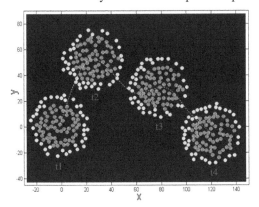

Fig. 6. Snapshots of a collective movement

We conducted two additional case studies to assess the robustness of the proposed algorithm. In the first case study, we introduce an obstacle into the environment. As it is described in the pseudo code of the algorithm in Section 2, where there is an obstacle, the robots try to avoid it whenever the distance to the obstacle is less than a predefined value D_{ob}. As seen in Fig. 7, the swarm successfully passes an obstacle with the collective movement being maintained. We note however, that a small number of robots do get lost from the swarm because of the obstacle, which causes a big problem when the obstacle is in the center of the swarm. Fortunately, the swarm can recover its shape after passing through the obstacle.

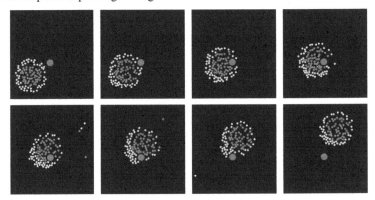

Fig. 7. Obstacle avoidance. Time increases from left to right, top to bottom

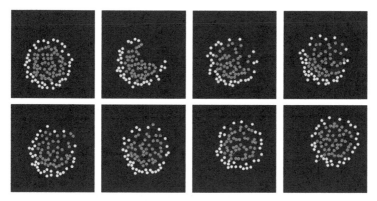

Fig. 8. Self-repairing after removal of the robots located between angles 0 and 90 in the swarm. Time increases from left to right, top to bottom.

In the second case study, we demonstrate that the swarm is capable of self-repairing after removing part of the robots in the swarm. As shown in Fig. 8, the robots between angles 0 and 90 are suddenly removed during the simulation. Immediately after that the rest robots autonomously adjust their position and move towards the space where the robots are removed and reform the disc-like shape.

The movies of the simulation can be downloaded from the link below:
https://www.dropbox.com/sh/tcmylfimeo7cqby/WIXZXM7NZy

4   Conclusion and Future Work

Collective movement while maintaining a predefined shape is very important for swarm systems to achieve various tasks. In this paper, we presented a morphogenetic approach to collective movement, assuming that the robots (agents) in the swarm do not have directional sensing. Note that not much work has been reported on self-organization of collective movements without directional sensing. The proposed approach first identifies the robots located on the boundary of the swarm using a diffusion mechanism. Then, the robot sitting in the center of the swarm is located with the help of the morphogen gradients emitted by the edge robots. A method for detecting the desired movement direction to follow the center robot has also been proposed according to the previous concentration records of morphogens. Empirical studies have been performed to find out the optimum number of the gradient histories for an accurate estimation of the movement direction. Simulation results showed that the robots can follow the center-robot while maintaining the disc-like shape. It also has been demonstrated that the algorithm is able to avoid obstacles and robust to a sudden removal of some robots in the swarm.

Our future work includes extending the proposed algorithm to form a more complex shape while moving in a dynamic environment. We are also interested in applying more complex biological mechanisms in biological morphogenesis such as gene regulatory networks to deal with more complex environments. Instead of fine tuning the parameters in the algorithm, future research will also be dedicated to optimize the parameters using global optimization algorithm e.g., evolutionary algorithms.

Acknowledgement. This research is funded by the European Commission 7th Framework Program, Project No. 601062, SWARM-ORGAN.

References

1. Navarro, I., Matia, F.: A Survey of Collective Movement of Mobile Robots. Int. J. Adv. Robot. Sys. 10, 1–9 (2013)
2. Couceiro, M.S., Portugal, D., Rocha, R.P.: A Collective Robotic Architecture in Search and Rescue Scenarios. In: 28th Annual ACM Symposium on Applied Computing, pp. 64–69. ACM Press, New York (2013)
3. Stirling, T., Wischmann, S., Floreano, D.: Energy-efficient indoor search by swarms of simulated flying robots without global information. Swarm Intell. 4, 117–143 (2010)
4. Shen, W.-M., Will, P., Galstyan, A.: Hormone-Inspired Self-Organization and Distributed Control of Robotic Swarms. Auton. Robot. 17, 93–105 (2004)
5. Guoa, H., Meng, Y., Jin, Y.: A Cellular Mechanism For Multi-Robot Construction Via Evolutionary Multi-Objective Optimization of a Gene Regulatory Network. BioSystems 98, 193–203 (2009)
6. Sayama, H.: Robust Morphogenesis of Robotic Swarms. IEEE Comput. Intell. Mag. 5, 43–49 (2010)
7. Guo, H., Jin, Y., Meng, Y.: A Morphogenetic Framework for Self-Organized Multirobot Pattern Formation and Boundary Coverage. ACM Trans. Auton. Adap. Syst. 7, Article No. 15 (2012)
8. Jin, Y., Guo, H., Meng, Y.: A Hierarchical Gene Regulatory Network for Adaptive Multirobot Pattern Formation. IEEE Trans. Syst., Man, Cybern.,Syst. 42, 805–816 (2012)
9. Mamei, M., Vasirani, M., Zambonelli, F.: Experiments of Morphogenesis in Swarms of Simple Mobile Robots. Appl. Artif. Intell. 18, 903–919 (2004)
10. Ikemoto, Y., Hasegawa, Y., Fukuda, T., Matsuda, K.: Gradual Spatial Pattern Formation of Homogeneous Robot Group. Inf. Sci. 171, 431–445 (2005)
11. Sayama, H.: Swarm Chemistry. Artif. Life. 15, 105–114 (2009)
12. Eyiyurekli, M., Bai, L., Lelkes, P.I., Breen, D.E.: Chemotaxis-based Sorting of Self-Organizing Heterotypic Agents. In: 25th ACM Symposium on Applied Computing, Sierre, Switzerland, pp. 1315–1322 (2010)
13. Davies, J.: Mechanisms of Morphogenesis. Elsevier Academic Press, Amsterdam (2005)
14. Ho, C.S., Nguyen, Q.H., Ong, Y.-S., Chen, X.: Autonomous Multi-agents in Flexible Flock Formation. In: Boulic, R., Chrysanthou, Y., Komura, T. (eds.) MIG 2010. LNCS, vol. 6459, pp. 375–385. Springer, Heidelberg (2010)
15. Navarro, I., Matia, F.: A Framework for the Collective Movement of Mobile Robots Based on Distributed Decisions. Robot. Auton. Syst. 59, 685–697 (2011)
16. Rubenstein, M., Christian, A., Radhika, N.: A Low Cost Scalable Robot System for Collective Behaviors. In: IEEE International Conference on Robotics and Automation (ICRA), pp. 3293–3298. Computer Society Press of the IEEE, Saint Paul (2012)
17. Shvartsman, S.Y., Coppey, M., Berezhkovskii, A.M.: Dynamics of Maternal Morphogen Gradients in The Drosophila Embryo. Curr. Opin. Genet. Dev. 18, 342–347 (2008)

The Pi Swarm: A Low-Cost Platform for Swarm Robotics Research and Education

James Hilder, Rebecca Naylor, Artjoms Rizihs, Daniel Franks, and Jon Timmis

York Robotics Laboratory
University of York, Heslington, YO10 5DD, UK
james.hilder@york.ac.uk
http://www.york.ac.uk/robot-lab

Abstract. The paper introduces the Pi Swarm robot, a platform developed to allow research and education in swarm robotics. Motivated by the goals of reducing costs and simplifying the tool-chain and programming knowledge needed to investigate swarming algorithms, we have developed a trackable, sensor-rich and expandable platform which needs only a computer with internet browser and no additional software to program. This paper details the design and use of the robot in a variety of settings, and we feel the platform makes for a viable, low-cost alternative for development of swarm robotic solutions.

Keywords: Swarm robotics, Tracking.

1 Introduction

Swarm robotics is a research field that allows for the study of complex swarm behaviours and emergent behaviours ("swarm-intelligence") as applied to robotic systems. Simple robotic platforms are ideal tools for teaching a wide range of engineering disciplines, encompassing digital electronics, control theory, signal processing, energy management and embedded programming; when the scope of the individual robot is expanded to allow communication and interaction with other similar robots, their potential value as a tool for education and research expands to allow investigation of swarm intelligence. What is required is a system which is flexible to multiple problems, robust to environmental changes and faults, and scalable enough to permit large swarms without impacting performance. However, such systems often come at a cost which is prohibitive to education and research, both in pure monetary terms, and also in the context of the time needed to learn how to use and program a system and general maintenance.

Historically, a number of platforms have been developed which attempt to allow swarm robotics research at a low unit cost, many of which have been highly popular amongst academic institutions for education and research. Unfortunately, such systems still often cost several-hundred GBP per unit; this can make their use for research, and especially as an educational tool, prohibitive, particularly in the context of a swarm of tens- to hundreds- of robots. Additionally, there is an inevitable compromise to be made in such systems between cost,

M. Mistry et al. (Eds.): TAROS 2014, LNAI 8717, pp. 151–162, 2014.

processing capability and complexity of use. Those platforms which offer the lowest cost tend to use very limited microcontrollers; those which offer more processing capability have higher associated costs and often also challenging learning-curves to be able to program the devices.

The motivation for this research was to have a platform that could achieve a number of objectives: be able to sense and react to its environment, be able to communicate and interact as part of a swarm, be able to be accurately tracked by a infrared tracking camera system, and be able to be expanded at a future date to add additional capability. In addition to these basic requirements, to allow for the platform to be used by a wider variety of researchers and students with varying levels of programming experience and expertise, it was desirable for it to have a simple tool-chain and an easy and concise Application Programming Interface (API) to control all basic functions.

1.1   Alternative Swarm Robotic Platforms

Whilst there exist a wide variety of low-cost robotic platforms, only a limited number of these have been designed with swarm robotics in mind. Swarm systems generally need to be able to interact, so some form of communication, the ability to detect neighbours, and some method of identification are prerequisites. Also of significant importance is ease-of-use; the ease of (re-)programming and charging robots are factors which become increasingly important as the size of swarm increase. One of the largest obstacles is the cost of the whole system; this is particularly pressing in the educational field, where multiple swarms may be needed to support classes. Here we discuss a number of the more popular and relevant platforms which have been developed for swarm research.

The e-puck is one of the most widely used swarm robotic platforms, a small 2-wheeled stepper-motor driven robot developed at EPFL [10]. A significant benefit of being widely used in research is that is has accurate simulator models, including in Webots, V-REP and Player-Stage. The retail price of a e-puck is around £650[1]. To overcome the limited processing capabilities, an open-source Linux development board has been developed for the E-Puck [7]. Whilst this creates a powerful platform ideal for research, the overall cost and complexity of the tool-chain needed to program it present an obstacle to its use as a teaching platform.

The Khepera series of robots developed by K-Team are highly successful platforms that mix a reliable, sensor-rich base with a compact size and extensive simulator support. Whilst they have been promoted and used as a teaching platform [5], their high retail cost (in excess of £2000 for a single Khepera III model) prohibit their use for swarm research in many establishments.

The SWARM-BOT project developed fully autonomous platforms that had the ability to communicate and physically connect with other SWARM-BOTs to allow task completion that would not be possible with single robots, such as dragging heavy objects in rough terrain [4]. The electronic subsystem of the

---

[1] Price correct at time of writing: 840 CHF plus taxes from official retailers.

SWARM-BOT linked an ARM based main microcontroller running a lightweight Linux operating system to twelve auxiliary PIC microcontrollers which controlled the sensors and actuators [11].

The Jasmine is an extremely small (approximately 3cm cube), low cost (around 100 Euro for components) open-source robot designed for large scale swarm experiments. Communication of short messages between robots uses infrared [6]. The low-cost and small size make it an attractive prospect for research into lightweight distributed swarm intelligence algorithms on larger swarms, but with this comes limited processing power on an ATMega168 microcontroller and a lack of high-level communication or environmental sensors. Such a compact size also limits the effectiveness of tracking systems, due to the resolution required to discriminate between such small robots.

The K-Team developed Kilobot is a widely used, low cost platform designed to be scalable up to hundreds and even thousands of platforms. It uses vibrating legs to allow directed motion. Many aspects of the platform, such as the ability to perform mass-reprogramming and large scale charging, make it well suited to large swarms. Rubenstein et al claim a raw parts cost of $14 USD for each platform [12], however the retail cost for 10 assembled robots is in excess of £1000 in the UK at the time of writing [2].

The r-one robot, created at Rice University, was developed with very similar goals to the Pi-Swarm robot: a platform that is low-cost, scalable and with a simple to use development environment to make it more accessible to students and outreach. It manages to include a wide array of sensors and high-speed communication system onto an accurate two-wheeled platform. The cost of parts for the robot are approximately $250 [8].

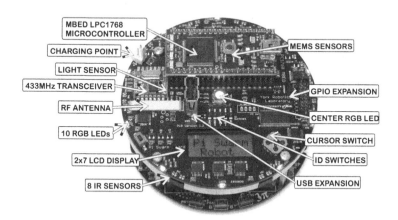

Fig. 1. The Pi Swarm robot annotated with the core features

---

[2] Prices taken from official UK distributor, Rahal Technology, www.rahalco.co.uk, at time of writing.

1.2   The 3-Pi and m3-Pi Platforms

Whilst researching possible platforms that would fit the teaching and research requirements for a new swarm system, it was observed that there exist a number of low-cost robotic platforms already developed which are designed for educational use, without explicitly being part of a swarm based system. One of these platforms, the Pololu 3-Pi, had a lot of desirable characteristics: a simple 2-wheeled design with a circular chassis is a proven platform for development of swarms, allowing simple motion, on-the-spot turning and minimising problems that can occur when robots collide or get very close to each other. The 3-Pi is widely available at a low-cost (approximately £70) and with the use of base-mounted infrared proximity sensors, designed to excel at line-following and maze-solving competitions [1].

The power source, 4xAAA Ni-Mh cells, is low in cost, readily available and easy to charge, and whilst not providing as high a charge density as Li-Ion type batteries, is generally safer. The built-in display is a useful addition not found on many other platforms, allowing debug information or user options to be shown as text on an individual robot. A set of five simple, but effective, infrared sensors on the base of the platform allow line-following tasks to be completed accurately, which provides an excellent introduction into control systems programming. The limiting factors of the basic 3-Pi platform as a swarm robotics platform are its limited processing capability, based around a ATmega328 microcontroller with 2KB of RAM and 32MB of FLASH memory, its lack of communication systems, a lack of proximity sensing and the use of standard geared motors, without wheel-encoders or stepping-motor capability, which limits the accuracy to which position can be estimated. However, it was observed that many of these potential issues could be overcome with the development of an expansion board, and given the low basic cost of the platform this could prove a very cost-effective solution.

Pololu themselves have developed an expansion board, the m3-Pi, for the 3-Pi robot which includes a socket for a MBED LPC1768 rapid prototyping board. The MBED is a small 40-pin module based around an ARM Cortex-M3 microcontroller. One of the core features of the MBED board is it is designed for use with a special set of online compiler tools, which allow for the development of programs from any machine with an internet connection and compatible browser; this eliminates the need for purchasing and installing a specific tool chain [3]. The use of the MBED microcontroller adds a significant boost in raw processing power over the ATmega328 microcontroller on the 3-Pi and provides a wide array of IO expansion buses. Whilst the m3-Pi in itself offers little additional hardware to the base robot, it was observed that it would be possible to add a number of sensors and actuators suited to swarm robotics to the basic PCB, and this forms the basis of the Pi-Swarm system.

---

[3] Projects created in the online tools can also be exported to a number of different tool-chains designed for ARM microcontrollers, such as Keil μVision, IAR Systems and GCC if desired.

2   Hardware Design

The basic Pi-Swarm robot, as shown in Figure 1, comprises a standard 3-Pi base, onto which a custom PCB: the "Pi Swarm Extension Board", is attached. It is this PCB that contains most of the sensors and actuators that convert the platform from a standard robot into one specifically engineered with swarm robotics as the core function. The total components costs for the Pi Swarm Extension Board are under £45 [4]. The PCB costs around £10 to manufacture; connecting hardware and plastic parts cost approximately £5 and a MBED board costs £38, bringing the total parts cost for a robot to around £165.

2.1   Pi Swarm Extension Board

The Pi Swarm Extension board is a 95mm diameter circular PCB, which connects to the 3-Pi base using a 14-pin 0.1" pitch dual row peripheral connector, with a pair of additional 2-pin connectors allowing duplication of the reset switch and the recharging pins. It is secured in place using a set of 4xM2.5 bolts and spacers. The top of the board contains a socket for attaching the MBED and a number of sensors and actuators. On the underside of the board, a set of 8 IR optocouplers which act as proximity detectors are arranged around the edge. A socket allowing the connection of a separate ultrasonic range detector is present at the front. The Pi Swarm boards are designed in Eagle PCB design software as 2-layer boards. The board is equipped with the following sensors:

Sensor	Description
Infrared (IR)	A set of 8 IR-proximity detectors on the underside of the board, placed at $\pm 15°, \pm 45°, \pm 90°$ and $\pm 144°$. The optical component is a TCRT1000 manufactured by Vishay Semiconductor, which combines a phototransistor and infrared emitter. The phototransistors all feed into an Analog Devices AD7997 8-channel, 10-bit analogue to digital converter.
Accelerometer	A MEMS 3-axis accelerometer (ST Micro LIS332AX), which produces an output voltage of 1.25V at zero G and has a sensitivity of 363mV/G over its acceleration range of ± 2.0G.
Magnetometer	A MEMS 3-axis magnetometer (Freescale MAG3310FCR1), capable of measuring magnetic fields with an output data rate of up to 80Hz. It has a full scale range of $\pm 2000\mu$T and a sensitivity of 0.1μT.
Gyroscope	A MEMS 3-axis accelerometer (ST Micro LIS332AX), which produces an output voltage of 1.25V at zero G and has a sensitivity of 363mV/G over its acceleration range of ± 2.0G.
Temperature	A linear digital temperature (MCP9701) which provides an output of 400mV at 0°C and ± 19.5mV from this value for each 0°C difference, within the operating limits.
Light	An ambient light sensor (APDS-9005), facing upwards on the expansion PCB. The sensor produces an analogue output of 1V at its peak value, at approximately 1000 Lux of light.

---

[4] The component are based on purchasing components for 20 boards, and are sourced from major UK component retails, Farnell and RS Electronics.

A number of additional hardware components are included, which allow for communication, visual feedback, user input and data storage:

Component	Description
RF Transceiver	A self-contained 433MHz RF transceiver (Alpha TRX-433) module forms the primary communication system for the Pi Swarms and provides up to 115.2kbps data rate. A chip antenna is mounted on the PCB providing a typical range of 10 meters at the higher data rates.
Outer LEDs	A set of 10 RGB LEDs equally spaced 36° apart in a ring around the edge of the board. These LEDs can be enabled/disabled individually with the colour selected by setting three PWM values.
Center LED	A high-power RGB LED in the middle of the board, facing up. This has independent colour control from the edge LEDs.
ID Switch	A set of 5-DIL switches, for the setting of robot ID within the swarm. This allows a swarm of up to 31 robots (the ID 0 is to be reserved).
Cursor	A 5-way directional switch which may be used to trigger interrupts and control the robot
EEPROM	A 64 kilobit EEPROM (24AA64T) provides non-volatile storage on the Pi Swarm, which can be used to storage calibration values. Additional FLASH memory storage is available on the MBED board.

2.2   Additional Hardware

To increase the number of peripherals which can be attached to the MBED, the Pi Swarm PCB makes use of a PCA9505 I/O Expansion IC manufactured by NXP. This device connects to the I^2C interface of the MBED, and contains 40 general purpose Input/Output pins. Of these, 24 are used by the Pi Swarm hardware, and the remaining 16 are available on a 2mm pitch expansion port on the top of the PCB for connecting additional hardware.

(a) HC-SR04 ultrasonic module   (b) PC Radio Modem   (c) Hand-held Controller

Fig. 2. Additional hardware for the Pi Swarm robot

A set of plastic shims have been designed which are attached beneath the expansion board in normal use. The shims are laser-cut from 3mm Perspex, and provide protection to the optocouplers from collisions with obstacles. They also help balance the weight of the platform, provide an opaque layer to allow robots to detect other robots, and set the correct height above the 3-Pi to allow the use of 20mm PCB spacers. An additional plastic disc has been designed to attach above the expansion board, which allows unique patterns of reflective tracking balls to be fixed to the robot.

A connector on the underside of the expansion board allows a low-cost HC-SR04 ultrasonic sensor to be attached, facing to the front of the robot. This sensor module is a self-contained unit which can detect and range obstacles which are approximately 10cm to 100cm in front of the sensor. Originally developed for the Arduino range of controllers, it is widely available for less than £2. A Pi Swarm robot with the ultrasonic sensor attached can be seen in Figure 2a.

Whilst communication is primarily handled using the 433MHz transceivers, other communication systems are possible. A USB connector is included on the board and libraries are available for the MBED to use a number of low cost BlueTooth dongles; it should be noted that BlueTooth protocols can impose restrictions on swarm size. Additionally, the infrared sensors can be programmed to send and receive simple messages from adjacent robots.

2.3   Radio Modem

To facilitiate communication between the robot swarm and external sources, two different systems have been developed. The first is a simple radio-modem, shown in Figure 2b, which combines the Alpha-TRX433 transceiver with an MBED board and a 2-line LCD display. The MBED board can be connected to a computer using a standard mini-USB cable, and contains the hardware and drivers to allow it to act as a serial-USB interface. Using this, messages can be sent to-and-from the computer to the MBED, and in turn to the swarm of robots using the 433MHz transceiver.

Additional, a stand-alone handheld controller has been developed, as can be seen in Figure 2c, which includes a multi-directional controller switch and software to allow the remote control of one or many robots simultaneously. A special debug routine has been added to robot firmware to allow the controllers to remotely read the values of the robot's sensors, and control the various actuators. The use of MBEDs in the controllers simplifies code development and the addition of extra sensor inputs.

2.4   Tracking the Robots

A specially designed plastic disc to facilitate the use of a tracking camera system has been designed and tested. The tracking hat is laser cut from 3mm Perspex sheet, and contains 21 regularly spaced 3mm diameter holes arranged in a grid. Short M3 machine screws can bolted through these holes to provide secure mounting points for placing reflective Scotchlite balls for use with the

OptiTrack commercial tracking camera system. A set of distinct patterns of different placements for between 4 and 6 balls on each hat has been created using an evolutionary algorithm based on NSGA-II [3]; these placements ensure that all the patterns are maximally unique in both translation and rotation. More information on the tracking system, the algorithm to evolve patterns and its use in practice can be found in Millard et al [9].

2.5   Battery Life

A drawback of the use of the MBED board is it requires a relatively high current draw and does not provide easy access to the lower power states of its Cortex microcontroller, as can be seen in the table below showing the average power consumption when using a 5V bench power supply. When using high-capacity 1000mAH cells, a battery life of around 2 hours can be expected for most tasks with intermittent motor use.

Device	Power	Description
3-Pi	400mW	Robot switched on but idle
MBED	800mW	MBED board only in idle loop
Pi Swarm	1700mW	Complete Pi Swarm in idle loop
Pi Swarm	2450mW	Performing obstacle avoidance, motors at 25% power, all LEDs blinking
Pi Swarm	2650mW	Performing obstacle avoidance, motors at 50% power
Pi Swarm	11500mW	Motors stalled at 100% power

3   Software Design

One of the core principles for the Pi Swarm is that it should be easy for someone with limited programming experience in embedded and robotic systems to program. To achieve this, an API has been written which simplifies the process of interacting with all of the sensors and actuators. This is implemented as a published C++ library that can be imported into a new project in the online MBED compiler. The programmer is given a shell *"Hello World"* program which performs all the code necessary to set up the expansion board, start communication with the 3-Pi base and initialise the RF communication stack.

3.1   Communication Stack

The API includes a communication stack for use with the 433MHz RF transceiver which simplifies many of the core communication tasks. A 4-byte header encodes the sender and target IDs, an identifying flag for the message and a set of predefined functions. These functions include operations which allow for most of the sensors and actuators on the robot to be read and set by the external sender. The stack also allows for user-defined functions to be augmented to the pre-defined ones, so more complex interactions can be implemented if desired. Messages can be defined as broadcast, meaning they are to be handled by all members of the swarm, and a TDMA-system based on the robot ID is used to minimise the risk of collisions in acknowledgements and replies. For brevity the API and communication stack is not discussed in detail here, the reader is referred to the manual available online for more in-depth details [5].

3.2   Simulation

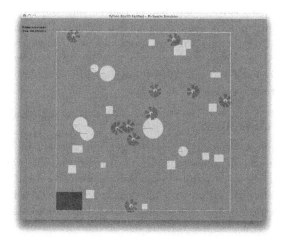

Fig. 3. Pi-Swarm simulator with robots as blue circles with an ID number and IR sensor cones shown in purple. Obstacles are shown in white.

The Pi-Swarm simulator is a 2D top-down environment being developed in Python. It uses **pygame** for graphics and **pyBox2D** as a physics engine [6]. The simulated Pi-Swarms use a simplified model of the physical hardware, they are shown as light blue circles in Figure 3 with a unique ID number and a centre line indicating their heading. The robots move using two wheels on the perimeter of the robot body. A small force is applied at each time-step to represent the motor drive in the direction of movement (either forwards or backwards).

The environment is a square arena with obstacles placed at random locations. The Pi-swarms use 8 IR sensors, at the same positions as the physical robots, to perform obstacle avoidance and navigate the arena. The sensors' fields of view are represented as cones surrounding the robots. When running the robots draw 0.33C from their battery, based on an average use case for the real Pi-Swarm robots. They recharge at a rate of 0.5C in power areas on the ground (bottom left of Figure 3).

---

[5] Manual and videos available at www.york.ac.uk/robot-lab/piswarm
[6] See http://pygame.org and http://code.google.com/p/pybox2d

4   Case Studies

Despite the short-development time, the Pi Swarm platforms have already been used successfully in a number of taught-courses, research projects and outreach demonstrations. Some examples are outlined below; example videos of some of these projects are available for viewing online [5].

4.1   Maze-Solving Tasks

As an introduction to the Pi Swarm platform and embedded programming, students were required to solve a number of tasks using the platform. These included line-following, making the robots solve a small walled-maze, and locating the warmest or brightest zone within an arena using the on-board sensors and an optimisation algorithm. Some students were starting this task with very limited programming experience and managed to successfully complete all the tasks in 3 x 3-hour laboratory sessions.

4.2   Block-Counting Tasks

As a taught module assessment, single Pi Swarm robots were programmed to autonomously navigate a number of walled-courses. On one side of the wall a number of black lines were painted; the robots were to count the number of lines, then automatically switch to a line-following task when the walls ended. The task was designed to test the students ability to implement effective PID controllers and make use of the various sensors and actuators on the platform to provide feedback from the counting task.

4.3   Creative Demonstrations

Students were assigned the task of performing a creative task over two minutes with multiple robots. This resulted in a diverse set of work, examples of which can be viewed on the website. Some students made use of the lights and high speed of the robots to produce creative synchronised dances; others created novel examples of human-robot interaction with playable games, including a version of the popular "Angry Birds" game in which one robot behaved as the bird, with trajectory and power determined by hand-actions behind the IR sensors, and others as the pigs (the targets for the bird).

4.4   Implementation of a Swarm Taxis Algorithm

A implementation of Bjerknes' ω algorithm, which demonstrates behaviours of aggregation and taxis towards a beacon [2], was implemented on the Pi Swarms. The goal was to replicate prior examples of the algorithm implemented on the e-puck robots. The algorithm was implement using only the IR sensors, which uses signal processing to descriminate between self-, other robots and an IR beacon. The algorithm lets an individual to predict the average heading of the swarm based on the strength of IR signals received, which allows the taxis behaviour to emerge.

5   Future Work

The Pi Swarm platform is still in active development, with new features planned to improve the robots reliability, ease of use and longevity. Some of the current work in progress and future plans are discussed here.

5.1   Scalability

The Pi Swarm has been designed for small swarms of up to 32 robots, based on the budget available and size of arena we currently have, and the limitation of the tracking system that is being used. Whilst the hardware ID switch and the TDMA protocol used in the communication system are limited to 5-bits, these limitations could be adapted with minimal efforts for larger swarms. As observed by Kornienko et al [6], RF-based communication can be problematic when scaled to large numbers. Also, the practicalities of reprogramming and charging become an issue; a system to allow dynamic reprogramming of multiple robots would be a useful solution if scaling to larger swarms was desired.

5.2   Dynamic Recharging

Currently in development is a system to allow the platform to recharge whilst it is in use. The system works by collecting power from the base of the arena, on which a chequerboard arrangement of copper pads allows wide regions of power distribution. The power is rectified by a small daughterboard sandwiched between the two main PCBs which contains a small charging IC, allowing the batteries to be charged whilst the robot is still drawing power. The current prototype model for the recharging PCB and components costs approximately £15.

5.3   Improved Base

One of the most biggest limitations of the platform is the lack of feedback about wheel positions; many similar platforms employ the use of either wheel-encoders or stepper motors which allow turning angles to be more accurately known. The 3-Pi base does use a stable, regulated supply for power to the motors, which allows a good level of repeatability of turns, but this relies on smooth, flat surfaces to avoid wheel slip. With careful arena design it is possible to use the magnetometer or either infrared system to provide feedback on the robots approximate heading; this has been successfully achieved in experiments using a programmable infrared beacon at one end of a 2.5m square arena. However, it is planned to investigate a redesign of the base-portion of the robot (the 3-Pi) with the use of stepper motors to improve its ability to move controlled distances and turns on multiple terrains. Replacing the 3-Pi base will also allow a redesign of the power source; replacing the AAA cells with a combination of a Lithium Polymer battery and low internal resistance supercapicitor will allow for a longer running time and the ability to rapidly recharge the platform.

6   Conclusions

This paper has discussed the design and implementation of a low-cost platform, designed for the teaching and research of swarm-robotics, built on an existing commercial base and easy to use rapid-prototype board. As with all designs, there are limitations and trade-offs the occur between versatility and scalability of the platform and cost; the authors believe, however, that the broad array of sensors and actuators provided coupled with the simplicity of the programming interface of the MBED and the API written for the platform make it a unique prospect for teaching and research. Further information about the Pi Swarm, including the reference manual, schematics, software API and demonstration videos can be found at `www.york.ac.uk/robot-lab/piswarm`.

Acknowledgements. Rebecca Naylor is funded by a grant from the Biological and Physical Sciences Institute. Jon Timmis is part funded by the Royal Society and the Royal Academy of Engineering.

References

1. Benitez, A., de la Colleha, J., et al.: Reactive behavior for mobile robotic through graphic simulation environment. Procedia Engineering 23, 728–732 (2011)
2. Bjerknes, J.D.: Scaling and Fault Tolerance in Self-organized Swarms of Mobile Robots. PhD thesis, University of West of England (2009)
3. Deb, K., Pratap, A., et al.: A fast and elitest multiobjective genetic algorithm: NSGA-II. IEEE Trans. on Evolutionary Computation 6(2), 182–197 (2002)
4. Groß, R., Bonani, M., Mondada, F., Dorigo, M.: Autonomous self-assembly in swarm-bots. IEEE Transactions on Robotics 22(6), 1115–1130 (2006)
5. Harlan, R.M., Levine, D.: The khepera robot and the kRobot class. SIGCSE Bull. 33(1) (February 2001)
6. Kornienko, S., Levi, P., et al.: Minimalistic approach towards communication and perception in microrobotic swarms. In: Intelligent Robots and Systems (August 2005)
7. Liu, W., Winfield, A.: Open-hardware e-puck linux extension board for experimental swarm robotics research. Microprocessors and Systems 35(1), 60–67 (2011)
8. McLurkin, J., Lynch, A.J., Rixner, S., Barr, T.W., Chou, A., Foster, K., Bilstein, S.: A low-cost multi-robot system for research, teaching, and outreach. In: Martinoli, A., Mondada, F., Correll, N., Mermoud, G., Egerstedt, M., Hsieh, M.A., Parker, L.E., Støy, K. (eds.) Distributed Autonomous Robotic Systems. STAR, vol. 83, pp. 597–609. Springer, Heidelberg (2013)
9. Millard, A.G., Hilder, J.: A low-cost real-time tracking infrastructure for ground-based robot swarms. In: ANTS: 9th Int. Conf. on Swarm Intelligence (2014)
10. Mondada, F., Bonani, M., et al.: The e-puck, a robot designed for education. In: Proc. 9th Conf. on Autonomous Robot Systems and Competitions (2009)
11. Mondada, F., Pettinaro, G.C., et al.: SWARM-BOT: A swarm of autonomous mobile robots with self-assembling capabilities. In: Proc. of the Int. Workshop on Self-organisation and Evolution of Social Behaviour, pp. 307–312 (2002)
12. Rubenstein, M., Ahler, C., et al.: Kilobot: A low cost scalable robot system for collective behaviors. In: Int. Conf. on Robotics and Automation (2012)

Tactile Features: Recognising Touch Sensations with a Novel and Inexpensive Tactile Sensor

Tadeo Corradi, Peter Hall, and Pejman Iravani

University of Bath, Bath, UK
t.m.corradi@bath.ac.uk

Abstract. A simple and cost effective new tactile sensor is presented, based on a camera capturing images of the shading of a deformable rubber membrane. In Computer Vision, the issue of information encoding and classification is well studied. In this paper we explore different ways of encoding tactile images, including: Hu moments, Zernike Moments, Principal Component Analysis (PCA), Zernike PCA, and vectorized scaling. These encodings are tested by performing tactile shape recognition using a number of supervised approaches (Nearest Neighbor, Artificial Neural Networks, Support Vector Machines, Naive Bayes). In conclusion: the most effective way of representing tactile information is achieved by combining Zernike Moments and PCA, and the most accurate classifier is Nearest Neighbor, with which the system achieves a high degree (96.4%) of accuracy at recognising seven basic shapes.

Keywords: Haptic recognition, tactile features, tactile sensors, supervised learning.

1 Introduction

The aim of this paper is to find an accurate low-dimensional representation of a tactile image perceived by a novel tactile sensor developed by us, these representations are from now on referred to as 'encodings'. Tactile sensors and tactile information encoding have been focus of much research lately [5]. Whilst numerous standards exist in Computer Vision, there is no consensus on the best approach to encoding tactile sensing information [5], and the only tactile database known to us [24] is limited to a single sensor type. Unlike visual information, haptic information can be distributed over a potentially unknown geometry [5] (for example a single robotic hand can be fitted with many different combinations of tactile sensors), so the equivalent problem to 'camera calibration' is a significantly more difficult task. Whilst the majority of efforts have gone to low resolution sensor pads [2], [16], [19], [20], [26], a new biologically inspired sensor design, called the TacTip [3] aims to provide higher resolution whilst remaining inexpensive. This paper presents a similar, simplified, low cost tactile sensor and evaluate its accuracy recognising 7 basic tactile shapes (Corner, Cylinder, Edge, Flat-to-Edge, Flat, Nothing, Point), comparing a selection of encodings and a range of supervised classifiers.

M. Mistry et al. (Eds.): TAROS 2014, LNAI 8717, pp. 163–172, 2014.

2   Related Work

Tactile sensors can be designed using a variety of techniques, perhaps the most popular being resistive sensors [28]; but also including magnetic, piezo-electric, capacitive and others [5]. A large amount of effort has been put into texture recognition [7], [11], [14], [25], since texture is usually difficult to capture from vision alone. The most direct approach to tactile feature classification is to use the tactile images with no encoding and use a simple distance metric [23]. Recently, there have been several projects involving recognition by grasping using Pattern Recognition techniques to find the best dimensionality reduction function for tactile information. Early approaches focused on tailored designs [1] or classical Artificial Neural Networks (ANNs) [27]. More recently, PCA, moment analysis and binary (contact/no contact) have been compared in a system that integrates tactile and kinesthetic information for object recognition [8], finding that the use of central moments outperforms other encodings. A variation on Self-Organizing Maps (SOMs) [13] has been developed and applied to fusing proprioceptive and tactile input for object recognition [12]. PCA and SOMs have been used to extract tactile features which were then used for object recognition [16]. Novel recursive gaussian kernels have been used to encode the various stages of contact during grasping leading to a robust online classifier [26].

2.1   The TacTip

Most previous studies are based on pressure sensor arrays. An innovative biologically inspired sensor was proposed recently [3] which uses a flexible hemispherical membrane with internal papillae which move as the membrane deforms whenever it touches an object. A digital camera records and transmits the image of the displaced papillae (see right side of Fig. 1). This sensor, called the TacTip, was shown to achieve a high degree of accuracy in sensing edges [4] to a point where a small object is clearly identifiable by a human from its tactile image and has been theoretically shown to have potential in tele-surgery [21]. More recently it has also been successfully used to identify textures [29]. The new sensor presented by this paper is an adaptation of the TacTip. No papillae nor internal gel is needed (significantly simplifying the sensors manufacture process and cost) and the shading pattern of light is used as input, instead of the papillae locations. This paper shows that the new sensor is effective at recognising tactile shapes.

3   Sensor Specification

3.1   Design

The new sensor consists of an opaque silicone rubber hemispherical membrane of radius 40mm and thickness 1mm, mounted at the end of a rigid opaque cylindrical ABS tube. At the base of the tube, there is a PC web-cam equipped with 8 white LEDs. The LEDs illuminate the rubber, the shading pattern of the image changes as the rubber makes contact with various surfaces (see Fig. 1).

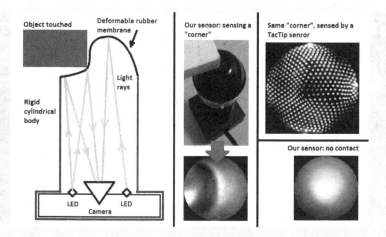

Fig. 1. The new tactile sensor design (left). The main body is 3D printed in ABS. The tip is a 1mm thick silicone rubber hemisphere. At the base (not visible) there is a USB eSecure©web-cam with 8 LEDs illuminating the inside of the silicone hemisphere (bottom right). As the tip makes contact with an object, it deforms resulting in a specific shading pattern (middle). As a comparison, the same tactile shape as perceived by a TacTip is shown (top right).

4   Methods

4.1   Preprocessing: Discrete Derivative

The shading pattern is related to the angle between the membrane's normal and the light rays going to the camera. Therefore drastic changes in luminosity are to be expected whenever the discrete spatial derivative of the normal of the surface is highest, that is where the rubber is most sharply bent (see Fig. 2). This concept motivates the analysis of the images' discrete derivative's magnitude matrix $D(I)$, defined, for any square image matrix $I \in \mathbb{R}^{w \times w}$, as:

$$D(I)_{i,j} := +\sqrt{(I_{i-1,j} - I_{i+1,j})^2 + (I_{i,j-1} + I_{i,j+1})^2}, \ \forall i, j \in [1, w-1] \quad (1)$$

In the experiments described below, encodings will be applied to the raw image received by the camera, and to the magnitude of its discrete derivative, $D(I)$.

4.2   Rotationally Invariant Encodings

Due to the circular geometry of the sensor image, a rotation invariant encoding was required. Five alternatives were explored: Hu moments [10], Zernike Moments [30], Principal Component Analysis (with regularized rotation), Zernike-PCA (PCA applied to the Zernike moments), and image scaling.

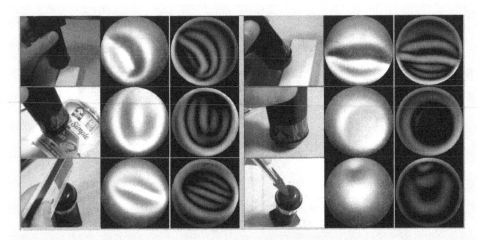

Fig. 2. Examples of occurrences of 6 of the 7 basic tactile shapes (the 7th is "nothing", in Fig. 1) (left columns), and their corresponding shading pattern (middle columns) and the magnitude of its first spatial derivative $D(I)$ (right columns). From the top left, downwards: Corner, Cylinder, Edge, Flat-to-Edge, Flat, Point.

Hu Moments. Hu moments are special combination of central moments which aim to be invariant to rotation, translation and scale (for details see [10]). The implementation used here was the one by [15], who have demonstrated the use of Hu moments in effective feature extraction on edge images for object recognition.

Zernike Moments. A Zernike Moment is the element-wise product of an image with a Zernike polynomial evaluated at the locations of the pixels of the image, rescaled to circumscribe a unit disk.

Definition 1. *Let $m \geq n$ be non-negative integers, and let $0 \leq \phi \leq 2\pi, 0 \leq \rho \leq 1$ define a polar coordinate system. Then the even and odd Zernike polynomials are defined as:*

$$Z_n^m(\rho, \varphi) = R_n^m(\rho) \cos(m\varphi) \tag{2}$$

$$Z_n^{-m}(\rho, \varphi) = R_n^m(\rho) \sin(m\varphi), \tag{3}$$

Which can be indexed by:

$$Z_j = Z_{n(j)}^{m(j)} \tag{4}$$

Where $m(j), n(j)$ are Noll's indices (See Table 1) of Zernike polynomials [17], and

$$R_n^m(\rho) = \sum_{k=0}^{(n-m)/2} \frac{(-1)^k (n-k)!}{k! ((n+m)/2 - k)! ((n-m)/2 - k)!} \rho^{n-2k} \tag{5}$$

Table 1. First ten Noll indeces [17] to compose a linear sequence of Zernike polynomials

j	1	2	3	4	5	6	7	8	9	10
n(j)	0	1	1	2	2	2	3	3	3	3
m(j)	0	1	-1	0	-2	2	-1	1	-3	3

Now, the dth Zernike Moment of an image M is given by:

$$Zer_d(M) = \left| \sum_{i,j \in \{i^2 + j^2 \leq n^2/2\}} M(i,j) Z'_d(i,j) \right| \tag{6}$$

Where,

$$Z'_d(i,j) := Z_j \left(\frac{\sqrt{(i^2 + j^2)}}{\frac{\sqrt{2}}{2} n}, arctan \left(\frac{j - n/2}{i - n/2} \right) \right) \tag{7}$$

PCA and Zernike-PCA. In the third encoding, the orientation of each image was computed (from central moments) and the image was rotated so as to regularize its orientation. Then PCA was performed on the vectorised images. The fourth encoding, Zernike-PCA, was simply applying PCA to the Zernike Moments of all images. In both of these, the dimensionality reduction matrix was computed on training data and used for both the training dataset and the testing dataset.

Scaling (Vectorized). For the fifth encoding, image orientations are regularised first, then images are resized by averaging pixel intensities, into a much smaller resolution (up to 13 by 13 pixels, from an original resolution of 300 by 300). The resulting images are vectorized, so for example, a 13-by-13 image, is converted into a 1-by-13^2 vector, by concatenating the pixel columns.

4.3   Encoding Evaluation

Each of these encodings was applied to a training dataset of 175 images, labelled from 1 to 7, corresponding to the tactile shapes they represented (see Fig. 2). Each encoding will produce a different set of data clusters. Good encodings will result in clusters which are spatially conglomerate: vectors corresponding to images of equal label will be close together and those with different labels will be far apart. One way of measuring this property is the Davies-Bouldin index in L^2 [6], defined below. Lower values of this index represent more distinctive clusters.

Definition 2. *Let $d(a, b)$ represent the euclidean distance metric. Let X be a set of vectors of dimension d, partitioned into k disjoint clusters, $X = \bigcup_{i=1}^{k} X_i$. Let c_i be the centroid of cluster X_i. The Davies−Bouldin index is given by:*

$$D = \frac{1}{k} \sum_{i=1}^{k} \max_{j:i \neq j} \left(\frac{\sigma_i + \sigma_j}{d(c_i - c_j)} \right) \tag{8}$$

Where,

$$\sigma_i := \sqrt{\frac{1}{|X_i|} \sum_{x \in X_i} d(x - c_i)^2} \tag{9}$$

Classifiers Cross Validation. As a second way of judging the suitability of a particular encoding is to train a supervised classifier given the known labels and to test their accuracy at predicting the labels of the encoded data. The measure used here is the 5-fold cross validation accuracy, defined as the average percentage of correct classifications performed by a given classifier trained with $\frac{4}{5}$ of the labelled data and tested on the remaining $\frac{1}{5}$ of the data. The process is repeated 5 times so that all data is used for testing. This method was applied to the following classifiers:

- Nearest Neighbor classifier
- Artificial Neural Network with a single 7 neuron hidden layer, trained using backpropagation.
- A group of seven binary Support Vector Machines (one per label) used in conjunction, arbitrarily choosing the largest label id, if more than one returned a positive classification.
- A simple Naive Bayes classifer, using Kernel Density Estimation (KDE) [18], [22].

For the implementation of these four algorithms, and for the simulations described in this paper, MATLAB[1] was used.

5   Results

Seven basic tactile shapes were defined: Corner, Cylinder, Edge, Flat-to-Edge, Flat, Nothing, and Point. Using the new sensor images were manually captured, resulting in 70 sample frames of each one (see Fig. 2). Data was split: 175 images were used for training (selecting the optimum encoding vector size), and the remaining 175 images for validation. Each one of the encodings defined in Section 4.2 was applied to each training image and the magnitude of its discrete derivatives (as described in Section 4.1). Then two tests were performed: cluster evaluation and classifier evaluation.

---

[1] MATLAB©, Statistics Toolbox and Neural Network Toolbox Release 2013b, The MathWorks, Inc., Natick, Massachusetts, United States.

5.1   Cluster Evaluation

First, the Davies-Bouldin index was computed on the training data data (175 images) to find the optimum number of components to use in each encoding (number of principal components, number of zernike polynomials, etc.). This parameter (number of components) is then fixed and the Davies-Bouldin index is computed on the remaining 175 images (the validation dataset). Table 2 shows the result. Zernike moments combined with PCA seem to produce the most distinct clusters under this criteria. Cluster formation using the new sensor seems superior with respect to the TacTip using this measure. This is possibly due to the fact that papillae displacements mean that small perturbations in the object surface translate into significant non-linear changes in the image.

Table 2. Davies–Bouldin index (described in Section 4.3) computed for the clusters resulting from the different encodings. They represent the distinctiveness of a cluster, smaller numbers represent better defined clusters.

Encoding	Applied to Image (Our sensor)	Applied to Image (TacTip sensor)	Applied to D(Image) (Our sensor)	Applied to D(Image) (TacTip sensor)
Hu Moments	5.3	10.4	5.1	13.2
Zernike M.	2.0	2.5	1.9	3.8
PCA	2.6	5.9	1.8	5.4
ZernikePCA	1.5	2.6	1.4	2.9
Scale (Vect.)	37.1	37.9	10.4	1378.8

5.2   Classifier Evaluation

Each one of the classifiers described in Section 4.3 is now trained. Using 20 iterations of randomized 5-fold cross validation on the training dataset the optimal vector sizes for each encoding and classifier are obtained. Then, the process is repeated on the validation dataset, but only using these optimum vector sizes. Figure 3 shows the accuracy of each encoding/classifier pair.

Zernike PCA applied directly to the image outperforms other encodings in general. In terms of classifiers, Nearest Neighbor is the overall best for both sensors, reaching an accuracy on the validation dataset of 96.4%. It must be born in mind that Nearest Neighbor classifiers using cross validation are prone to data twinning (bias if similar data are present in a dataset). To reduce the effects, a small value for k (5) was used in k-fold cross validation, together with randomisation and multiple trials; furthermore, separate dataset were used for training and validation. Nevertheless, if data twinning is likely to be an issue in further applications, it may be advisable to use Naive Bayes (KDE).

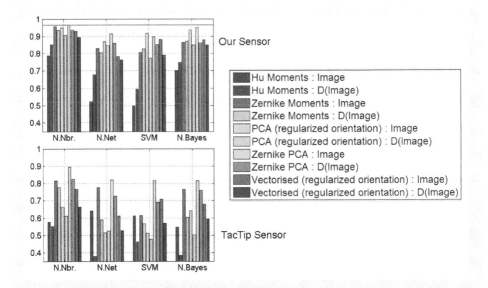

Fig. 3. Randomized 5-fold cross validation accuracy for the 7 basic tactile shapes (higher is better, 1 is 100% perfect recognition). Input set of 175 labelled tactile images, corresponding to 7 clusters. Comparison between our sensor and the TacTip, using four different encodings as classified by four different supervised algorithms.

There is no significant difference between the performance of any encoding/classifier pairing when comparing their use on the image and on its derivative. This may be due to the fact that the discrete derivative only loses base intensity information, which is a single degree of freedom over images which are 90000-dimensional. The accuracy achieved with our sensor is slightly higher to the one with the TacTip, for these particular choices of encodings and classifiers. Once again, the non-linearity introduced by papillae is potentially a factor, and so the comparison is by no means exhaustive in scope.

6   Conclusions

This paper presented a novel, simple and inexpensive tactile sensor based on shading resulting from the deformation of a rubber membrane. Various encodings were tested on the input images and on their discrete derivatives. For each encoding, the accuracy of a selection of classifiers was tested, by performing tactile shape recognition. The new sensor is capable of distinguishing between these shapes, the most accurate encoding is Zernike Moments combined with PCA, applied directly to the input image. The most accurate classifier is Nearest Neighbor, which reaches a classification accuracy of 96.4%. Our sensor performed slightly better than the TacTip in these tests, which is remarkable considering the simplicity of our sensor's design. However it must be stressed that other approaches may very well favor the TacTip. The discrete localization of the papillae may be a disadvantage in linear encodings, but it can be an advantage in general, as it is more resilient to image noise and less dependent on calibration of camera parameters. Only pattern recognition was discussed in this paper, it may be of interest to use "shape from shading" [9] to reconstruct the exact shape of the deformed hemisphere. Further work should also focus on this sensor's potential for object recognition.

Acknowledgments. This work was supported by the Engineering and Physical Sciences Research Council (EPSRC), UK. We would like to thank Bristol Robotics Lab[2] for lending us the TacTip sensor.

References

1. Allen, P.K.: Integrating vision and touch for object recognition tasks. The International Journal of Robotics Research 7(6), 15–33 (1988)
2. Barron-Gonzalez, H., Prescott, T.: Discrimination of social tactile gestures using biomimetic skin. In: IEEE International Conference on Robotics and Automation, Karlsruhe, Germany (2013)
3. Chorley, C., Melhuish, C., Pipe, T., Rossiter, J.: Development of a tactile sensor based on biologically inspired edge encoding. In: International Conference on Advanced Robotics, ICAR 2009, pp. 1–6 (2009)
4. Chorley, C., Melhuish, C., Pipe, T., Rossiter, J.: Tactile edge detection. In: 2010 IEEE Sensors, pp. 2593–2598 (2010)
5. Dahiya, R., Mittendorfer, P., Valle, M., Cheng, G., Lumelsky, V.: Directions toward effective utilization of tactile skin: A review. IEEE Sensors Journal 13(11), 4121–4138 (2013)
6. Davies, D.L., Bouldin, D.W.: A cluster separation measure. IEEE Transactions on Pattern Analysis and Machine Intelligence PAMI-1(2), 224–227 (1979)
7. Decherchi, S., Gastaldo, P., Dahiya, R., Valle, M., Zunino, R.: Tactile-data classification of contact materials using computational intelligence. IEEE Transactions on Robotics 27(3), 635–639 (2011)
8. Gorges, N., Navarro, S., Goger, D., Worn, H.: Haptic object recognition using passive joints and haptic key features. In: 2010 IEEE International Conference on Robotics and Automation (ICRA), pp. 2349–2355 (2010)
9. Horn, B.K.P., Brooks, M.J. (eds.): Shape from Shading. MIT Press, Cambridge (1989)
10. Hu, M.K.: Visual pattern recognition by moment invariants. IRE Transactions on Information Theory 8(2), 179–187 (1962)
11. Jamali, N., Sammut, C.: Majority voting: Material classification by tactile sensing using surface texture. IEEE Transactions on Robotics 27(3), 508–521 (2011)
12. Johnsson, M., Balkenius, C.: Sense of touch in robots with self-organizing maps. IEEE Transactions on Robotics 27(3), 498–507 (2011)
13. Kohonen, T.: Self-organized formation of topologically correct feature maps. Biological Cybernetics 43(1), 59–69 (1982)
14. Liu, H., Song, X., Bimbo, J., Seneviratne, L., Althoefer, K.: Surface material recognition through haptic exploration using an intelligent contact sensing finger. In: 2012 IEEE/RSJ International Conference on Intelligent Robots and Systems (IROS), pp. 52–57 (2012)

---

[2] www.brl.ac.uk

15. Mercimek, M., Gulez, K., Mumcu, T.V.: Real object recognition using moment invariants. Sadhana 30(6), 765–775 (2005)
16. Navarro, S., Gorges, N., Worn, H., Schill, J., Asfour, T., Dillmann, R.: Haptic object recognition for multi-fingered robot hands. In: 2012 IEEE Haptics Symposium (HAPTICS), pp. 497–502 (2012)
17. Noll, R.J.: Zernike polynomials and atmospheric turbulence. Journal of the Optical Society of America 66(3), 207–211 (1976)
18. Parzen, E.: On estimation of a probability density function and mode. Annals of Mathematical Statistics 33, 1065–1076 (1962)
19. Pezzementi, Z., Plaku, E., Reyda, C., Hager, G.: Tactile-object recognition from appearance information. IEEE Transactions on Robotics 27(3), 473–487 (2011)
20. Ratnasingam, S., McGinnity, T.: A comparison of encoding schemes for haptic object recognition using a biologically plausible spiking neural network. In: 2011 IEEE/RSJ International Conference on Intelligent Robots and Systems (IROS), pp. 3446–3453 (2011)
21. Roke, C., Melhuish, C., Pipe, T., Drury, D., Chorley, C.: Deformation-based tactile feedback using a biologically-inspired sensor and a modified display. In: Groß, R., Alboul, L., Melhuish, C., Witkowski, M., Prescott, T.J., Penders, J. (eds.) TAROS 2011. LNCS, vol. 6856, pp. 114–124. Springer, Heidelberg (2011)
22. Rosenblatt, M.: Remarks on some nonparametric estimates of a density function. The Annals of Mathematical Statistics 27(3), 832–837 (1956)
23. Schneider, A., Sturm, J., Stachniss, C., Reisert, M., Burkhardt, H., Burgard, W.: Object identification with tactile sensors using bag-of-features. In: IEEE/RSJ International Conference on Intelligent Robots and Systems, IROS 2009. pp. 243–248 (2009)
24. Schopfer, M., Ritter, H., Heidemann, G.: Acquisition and application of a tactile database. In: 2007 IEEE International Conference on Robotics and Automation, pp. 1517–1522 (2007)
25. Sinapov, J., Sukhoy, V., Sahai, R., Stoytchev, A.: Vibrotactile recognition and categorization of surfaces by a humanoid robot. IEEE Transactions on Robotics 27(3), 488–497 (2011)
26. Soh, H., Su, Y., Demiris, Y.: Online spatio-temporal gaussian process experts with application to tactile classification. In: 2012 IEEE/RSJ International Conference on Intelligent Robots and Systems (IROS), pp. 4489–4496 (2012)
27. Taddeucci, D., Laschi, C., Lazzarini, R., Magni, R., Dario, P., Starita, A.: An approach to integrated tactile perception. In: 1997 IEEE International Conference on Robotics and Automation, vol. 4, pp. 3100–3105 (1997)
28. Weiss, K., Worn, H.: The working principle of resistive tactile sensor cells. In: 2005 IEEE International Conference on Mechatronics and Automation, vol. 1, pp. 471–476 (2005)
29. Winstone, B., Griffiths, G., Pipe, T., Melhuish, C., Rossiter, J.: TACTIP - tactile fingertip device, texture analysis through optical tracking of skin features. In: Lepora, N.F., Mura, A., Krapp, H.G., Verschure, P.F.M.J., Prescott, T.J. (eds.) Living Machines 2013. LNCS, vol. 8064, pp. 323–334. Springer, Heidelberg (2013)
30. Zernike, V.: Beugungstheorie des schneidenver-fahrens und seiner verbesserten form, der phasenkontrastmethode. Physica 1(7-12), 689–704 (1934)

Polygonal Models for Clothing

Jan Stria, Daniel Průša, and Václav Hlaváč

Center for Machine Perception, Department of Cybernetics,
Faculty of Electrical Engineering, Czech Technical University
Karlovo nám. 13, 121 35 Prague 2, Czech Republic
{striajan,prusapa1,hlavac}@cmp.felk.cvut.cz
http://cmp.felk.cvut.cz

Abstract. We address the problem of recognizing a configuration of a piece of garment fairly spread out on a flat surface. We suppose that the background surface is invariant and that its color is sufficiently dissimilar from the color of a piece of garment. This assumption enables quite reliable segmentation followed by extraction of the garment contour. The contour is approximated by a polygon which is then fitted to a polygonal garment model. The model is specific for each category of garment (e.g. towel, pants, shirt) and its parameters are learned from training data. The fitting procedure is based on minimization of the energy function expressing dissimilarities between observed and expected data. The fitted model provides reliable estimation of garment landmark points which can be utilized for an automated folding using a pair of robotic arms. The proposed method was experimentally verified on a dataset of images. It was also deployed to a robot and tested in a real-time automated folding.

Keywords: clothes folding, robotic manipulation.

1 Introduction

We present a solution for identifying an arrangement of a piece of garment spread out on a flat surface. Our research is motivated by the needs of the European Commission funded project Clothes Perception and Manipulation (CloPeMa) [16]. This project focuses on a garments manipulation (sorting, folding, etc.) by a two armed industrial robot which is shown in Fig. 1. The general aim is to advance the state of the art in the autonomous perception and manipulation of limp materials like fabrics, textiles and garments, placing the emphasis on universality and robustness of the methods.

The task of clothes state recognition has already been approached by Miller et al. [9]. They consider a garment fairly spread on a green surface, which allows to segment images using simple color thresholding. The obtained garment contour is fitted to a parametric polygonal model specific for a particular category of clothing. The fitting procedure is based on iterative estimation of numeric parameters of the given model. The authors report quite accurate results. However, the main drawback is a slow performance. It takes 30–150 seconds for a single contour and a single model. This makes the algorithm practically unusable for a

M. Mistry et al. (Eds.): TAROS 2014, LNAI 8717, pp. 173–184, 2014.

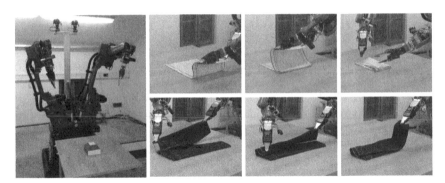

Fig. 1. Our robotic test platform utilizes two industrial hollow-wrist welding manipulators Motoman MA1400 with mounted cameras, Kinect-like rangefinders and dedicated grippers. We use it as the experimental platform for folding of various types of clothes.

real-time operations. The authors also use the parametric model for recognition and fitting of pairs of socks [15]. Information about texture of socks was utilized here as well. Another successful application is an automated folding of towels based on a robust visual detection of their corner points [8]. The two-armed robot starts with a towel randomly dropped on a table and folds it in a sequence of manipulations performed both on the table and in the air.

Kita et al. use single-view image [5] and stereo image [6] to estimate state of the hanging clothes being held by a gripper. Their approach is based on matching a deformable model to the observed data. Hata et al. [3] solve the problem of lifting a single towel from a pile of highly wrinkled towels and grasping it for its corner. The solution is based on detection of the highest point of the pile followed by corner detection in stereo data. Ramisa et al. [12] are also interested in determination of the grasping point. They combine features computed from both color and range images in order to locate highly wrinkled regions. The research area of cloth modeling is explored mainly by the computer graphics community. Hu et al. [4] give an overview of the known methods.

In this work, we propose a complete pipeline for clothes configuration recognition by estimating positions of the most important landmark points (e.g. all four corners of a towel). The identified landmarks can be used for automated folding performed by robotic arms. We introduce our own polygonal models describing contours of various categories of clothing and develop a fast, dynamic programming based methods for an efficient fitting of an unknown contour to the models. Moreover, we have modified the grabcut image segmentation algorithm to work automatically without being initialized by a user input, utilizing a background model learned in advance from training data. The recognition pipeline can be summarized as follows:

1. *Capturing input:* The input is a single color image of a piece of garment spread on a table. We assume that type of the clothing (e.g. towel, pants, shirt) is known in advance. The image is taken from a bird's eye perspective

by a camera attached to the robot. Since the relative position of the table
and the camera is known, all pixels not displaying the table and the garment
lying on it can be cropped.

2. *Segmentation:* The goal is to segment the garment and its background which
is a wooden table in our case. We assume that the table and the garment
have statistically dissimilar colors. We also assume that the table is invariant
and thus its color properties can be learned from data. These assumptions
make it possible to modify the grabcut segmentation algorithm [13] in a way
that it does not require manual initialization.

3. *Contour detection:* The binary mask obtained from the segmentation is processed by Moore's algorithm [2] for tracing 8-connected boundary of a region.
This gives a bounding polygon of the garment. Vertices of the polygon are
formed by individual contour pixels.

4. *Polygonal approximation:* The dense boundary is then approximated by a
polygon having fewer vertices. Their exact count depends on a model of
garment which we want to fit in the following step. Generally, the number of
vertices is higher than the number of landmark points for a specific model.

5. *Model fitting:* The polygonal approximation of the garment contour is matched to a polygonal model defined for the corresponding type of garment. The
matching procedure employs dynamic programming approach to find correspondences between approximating vertices and landmark points defining
the specific polygonal model. The matching considers mainly local features
of the approximating polygon. As there are more vertices than landmarks,
some of the contour vertices remain unmatched.

2   Contour Extraction

2.1   Learning the Background Color Model

The background color model is a conditional probabilistic distribution of RGB
values of background pixels. The distribution is represented as a mixture of K
3D Gaussians (GMM):

$$p(z) = \sum_{k=1}^{K} \pi_k \, \mathcal{N}(z; \mu_k, \Sigma_k) = \sum_{k=1}^{K} \pi_k \, \frac{\exp\left(-\frac{1}{2}(z - \mu_k)^T \Sigma_k^{-1}(z - \mu_k)\right)}{\sqrt{(2\pi)^3 |\Sigma_k|}} \qquad (1)$$

Here π_k is a prior probability of k-th component and $\mathcal{N}(z; \mu_k, \Sigma_k)$ denotes 3D
normal distribution having a mean vector μ_k and a covariance matrix Σ_k.

The mixture is learned from training data, i.e. from a set $Z = \{z_n = (z_n^R, z_n^G, z_n^B)^T \in [0, 1]^3\}$ of vectors representing RGB intensities of $|Z|$ background pixels.
The number of GMM components K is determined empirically based on the
number of visible clusters in RGB cube with visualized training data. E.g. for
a nearly uniform green background one component should be sufficient, for the
table in our experiments we choose three components.

To train the GMM probabilistic distribution, we split the training data to K
clusters $C_1 \ldots C_K$ at first, employing the binary tree algorithm for the palette

design [10]. The algorithm starts with all training data Z assigned to a single cluster and it iteratively constructs a binary tree like hierarchy of clusters in a top-bottom manner. In each iteration, the cluster having the greatest variance is split to two new clusters. The separating plane passes through the center of the cluster and its normal vector is parallel with the principal eigenvector of the cluster. The algorithm stops after $K - 1$ iterations with K clusters.

Prior probability π_k, mean vector μ_k and covariance matrix Σ_k for the k-th GMM component is computed using the maximum likelihood principle [1] from data vectors contained in the corresponding cluster C_k:

$$\pi_k = \frac{|C_k|}{|Z|}, \quad \mu_k = \frac{1}{|C_k|} \sum_{z_n \in C_k} z_n, \quad \Sigma_k = \frac{1}{|C_k|} \sum_{z_n \in C_k} (z_n - \mu_k)(z_n - \mu_k)^T \quad (2)$$

2.2   Unsupervised Segmentation

The segmentation is based on the grabcut algorithm [13] which is originally a supervised method. It expects an RGB image $Z = \{z_n \in [0,1]^3 : n = 1 \dots W \times H\}$ of size $W \times H$. Moreover, the user is expected to determine a trimap $T = \{t_n \in \{F, B, U\} : n = 1 \dots W \times H\}$. The value t_n determines for the n-th pixel whether the user considers it being a part of the foreground ($t_n = F$), background ($t_n = B$) or whether the pixel should be classified automatically ($t_n = U$). The trimap T is usually defined via some interactive tool enabling to draw a stroke over foreground pixels, another stroke over background pixels and leave the other pixels undecided.

In the proposed method, the input trimap is created automatically using the learned background GMM probabilistic model from Eq. 1 and two predetermined probability thresholds P_B and P_F:

$$t_n = \begin{cases} F, & p(z_n) < P_F \\ U, & P_F \le p(z_n) \le P_B \\ B, & P_B < p(z_n) \end{cases} \quad (3)$$

The thresholds P_B and P_F are set based on the training data so that 3% training pixels have the probability lower than P_F and 80% training pixels have probability higher than P_B in the learned background model. The foreground component of the trimap is thus initialized by lowly probable pixels while the background component by highly probable pixels.

The core part of the grabcut [13] algorithm is an iterative energy minimization. It repeatedly goes through two phases. First, GMM models for the foreground and the background color are reestimated. And second, the individual pixels are relabeled based on finding the minimum cut in a special graph. To estimate the GMM color models we utilize the binary tree algorithm [10] described in Sec. 2.1 followed by the maximum likelihood estimation introduced in Eq. 2. We use three components both for background and foreground GMM which is sufficient in our case of not so varying table and garment. The grabcut algorithm iterates until convergence which usually takes 5–15 cycles. However, the segmentation

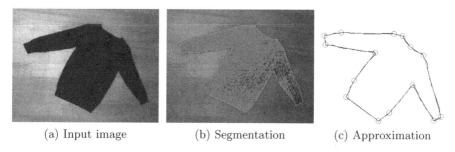

| (a) Input image | (b) Segmentation | (c) Approximation |

Fig. 2. (a) Input is formed by a single RGB image. (b) The input trimap for grabcut algorithm is automatically initialized with foreground (plotted in yellow), background (blue) and unknown (red) pixels. The resulting segmentation gives a garment contour (green). (c) The contour is simplified by approximating it by a polygon (magenta).

mask is being changed only slightly in the later cycles. Since we need to get the segmentation as fast as possible, we stop the optimization after three cycles.

2.3 Contour Simplification

The segmentation algorithm proposed in the previous section is followed by Moore's algorithm [2] for contour tracing. The result is a closed contour in the image plane, i.e. a list $(q_1 \ldots q_L)$ of 2D coordinates $q_i = (x_i, y_i)$. The number of distinct points L depends on the image resolution as well as on the piece of garment size. Typically, L has an order of hundreds or thousands.

To be able to fit out polygonal model to the contour effectively, we need to simplify the contour by approximating it with a polygon having N vertices where $N \ll L$. More precisely, we want to select a subsequence of N points $(p_1 \ldots p_N) \subseteq (q_1 \ldots q_L)$. Additionally, we want to minimize the sum of Euclidean distances of the original points $(q_1 \ldots q_L)$ to edges of the approximating polygon $(p_1 \ldots p_N)$ as seen in Fig. 3a.

The simplification procedure is based on the dynamic programming algorithm for the optimal approximation of an open curve by a polyline [11], [7]. It iteratively constructs the optimal approximation of points $(q_1 \ldots q_i)$ by n vertices from previously found approximations of $(q_1 \ldots q_j)$ where $j \in \{n-1 \ldots i-1\}$ by $n-1$ points. The construction is demonstrated in Fig. 3b.

Time complexity of the algorithm is $O(L^2 N)$. Since the algorithm works with an open curve, it would have to be called L times for every possible cycle breaking point q_i to obtain optimal approximation of the closed curve. However, we only call it constantly many times to get a suboptimal approximation which is sufficient for our purpose.

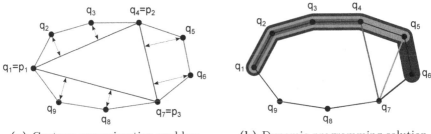

(a) Contour approximation problem   (b) Dynamic programming solution

Fig. 3. (a) The original contour $(q_1 \ldots q_L)$ (plotted in red) is simplified by approximating it with a polygon $(p_1 \ldots p_N)$ (blue) while minimizing distances of the original points q_i to polygon edges. (b) Dynamic programming algorithm for polygonal approximation utilizes previously constructed approximations of points $(q_1 \ldots q_4, q_5, q_6)$ by $n-1$ vertices (plotted in various colors) to obtain approximation of points $(q_1 \ldots q_7)$ by n vertices.

3   Polygonal Models

3.1   Models Definition and Learning

To be able to recognize the configuration of a piece of garment, we describe contours of various types of clothing by simple polygonal models. The models are determined by their vertices. Inner angles incident to the vertices are learned from training data. Additional conditions are defined in some cases to deal with inner symmetries or similarities of distinct models. We use the following models:

1. *Towel* is determined by 4 corner vertices as shown in Fig. 4. All inner angles incident to the vertices share the same probability distribution. There is an additional condition that the height of the towel (distance between the top edge and the bottom edge) is required to be longer that its width (distance between the left edge and the right edge).
2. *Pants* are determined by 7 vertices. There are 3 various shared distributions of inner angles as shown in Fig. 4.
3. *Short-sleeved shirt* is determined by 10 vertices and 4 shared distributions of inner angles as shown in Fig. 4. There is an additional condition that the distance between the armpit and the inner corner of the sleeve is required to be maximally 50% of the distance between the armpit and the bottom corner of the shirt.
4. *Long-sleeved shirt* is similar to the short-sleeved model. The distance between the armpit and the inner corner of the sleeve should be minimally 50% of the distance between the armpit and the bottom corner of the shirt.

The probability distributions for inner angles incident to vertices of polygonal models are learned from manually annotated data. We assume that the angles have normal distributions. This seems as a reasonable assumption, since e.g.

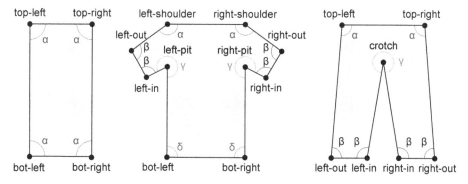

Fig. 4. Polygonal model for towel, short-sleeved shirt and pants. Angles sharing one distribution are denoted by the same letter and plotted with the same color.

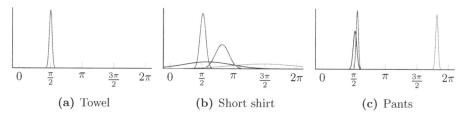

Fig. 5. Angle distributions learned for various types of clothes models. Colors of the plotted distributions correspond to angles in Fig. 4.

a corner angle of a towel should be approximately 90° with a certain variance caused by deformations of the contour. The mean and the variance of each normal distribution is estimated using a maximum-likelihood principle similarly to Eq. 2. Various vertices in a model can share the same angles distribution because of obvious symmetries, e.g. all 4 corners of a towel should be statistically identical.

3.2   Problem of Model Matching

We described in Sec. 2.3 how to approximate a contour by N points $(p_1 \ldots p_N)$. Each polygonal model defined in Sec. 3.1 is determined by M vertices $(v_1 \ldots v_M)$ where M is specific for the particular model. See examples of models in Fig. 4. We show how to match an unknown simplified contour onto a given model.

We assume that $N \geq M$, i.e. the simplified contour contains more points than is the number of vertices of the model to be matched. The problem of matching can be then defined as a problem of finding a mapping of simplified contour points to model vertices $f \colon \{p_1 \ldots p_N\} \to \{v_1 \ldots v_M\} \cup \{s\}$. Symbol s represents a dummy vertex which corresponds to a segment of the polygonal model. It makes it possible to leave some of the contour points unmapped to a real vertex. Additionally, a proper mapping f has to satisfy several conditions:

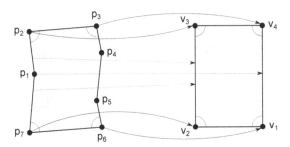

Fig. 6. Points of the simplified contour $(p_1 \ldots p_7)$ are matched (plotted in blue) to vertices of the polygonal model $(v_1 \ldots v_4)$. Some of them remain unmatched (green), i.e. they are mapped to a dummy vertex s representing a segment. The energy of the particular matching is based on a similarity of corresponding inner angles (red).

1. There exists a point p_i mapped to it for each vertex v_m. More formally, $\forall v_m \exists p_i : f(p_i) = v_m$.
2. No two points p_i and p_j are mapped to the same vertex v_m. However, many points can be mapped to segments represented by a dummy vertex s. Formally, $\forall p_i \neq p_j : f(p_i) = f(p_j) \Rightarrow f(p_i) = f(p_j) = s$.
3. The mapping preserves the ordering of points on the polygonal contour and the ordering of vertices of the polygonal model in the clockwise direction. For example of such a proper mapping see Fig. 6.

The number of mappings f satisfying the aferementioned conditions for N contour points and M model vertices is given by the combinatorial formula:

$$N \binom{N-1}{M-1} \geq N \left(\frac{N-1}{M-1} \right)^{M-1} \tag{4}$$

The interpretation is that we can choose 1 of N points to be mapped to the first vertex v_1. From the remaining $N-1$ points, we select a subset of $M-1$ points which are mapped to vertices $v_2 \ldots v_M$. All other points are mapped to the dummy vertex s representing all segments of the polygonal model.

We introduce an energy function $E(f)$ associated with a matching f. Let us denote ϕ_i the inner angle adjacent to the point p_i of the simplified contour. Let us also denote μ_m the mean value and σ_m^2 the variance of the normal distribution of inner angles $\mathcal{N}(\phi; \mu_m, \sigma_m^2)$ learned for the vertex v_m of a particular polygonal model. We recall that the same distribution can be shared by several vertices of one polygonal model as seen in Fig. 4. The energy function is then given by:

$$E(f) = - \sum_{f(p_i)=v_m} \log \mathcal{N}\left(\phi_i; \mu_m, \sigma_m^2\right) - \sum_{f(p_i)=s} \log \mathcal{N}\left(\phi_i; \pi, \frac{\pi^2}{16}\right) \tag{5}$$

It can be seen that we force angles of unmatched points (p_i such that $f(p_i) = s$) to be close to π, i.e. we want the unmatched parts of the contour to resemble straight segments. We set the variance $\pi^2/16$ for unmatched points empirically. Since the energy is inversely proportional to a probability, the optimal mapping f^* is obtained as $f^* = \arg\min_f E(f)$.

3.3 Matching Algorithm

Eq. 4 shows that the count of all admissible mappings is exponential in the number of vertices M. Thus it would be inefficient to evaluate the energy function for each mapping. We have rather developed an algorithm employing a dynamic programming approach which has a polynomial time complexity.

The dynamic programming optimization procedure seen in Alg. 1 is called for every shifted simplified contour $(p'_1 \ldots p'_N) = (p_d \ldots p_N, p_1 \ldots p_{d-1})$, where the shift is $d \in \{1 \ldots N\}$. The reason is that Alg. 1 finds a mapping f such that $f(p_{i_m}) = v_m$ for $m \in \{1 \ldots M\}$ and $1 \le i_1 \le i_2 \le \ldots \le i_M \le N$, i.e. one of the first points is mapped to the vertex v_1, some of its successors along the contour to the vertex v_2 and so on. Thus we have to try various shifts in order to be able to map any point to vertex v_1 as seen in Fig. 6.

Alg. 1 does not work with points and vertices directly. It expects a precomputed matrix $V \in \mathbb{R}^{N \times M}$ and a vector $S \in \mathbb{R}^N$ instead. The value $V_{i,m}$ is a cost of matching the inner angle ϕ_i associated with the point p_i to the learned angle distribution for vertex v_m, i.e. $V_{i,m} = -\log \mathcal{N}(\phi_i; \mu_m, \sigma_m^2)$ as in Eq. 5. The value S_i is a cost of matching ϕ_i to the angle of a dummy vertex s, i.e. $S_i = -\log \mathcal{N}(\phi_i; \pi, \pi^2/16)$ as in Eq. 5.

Both minimizations in Alg. 1 can be performed incrementally in $O(N)$ time by remembering the summation value for previous j. The first minimization is performed N times, the second one $O(NM)$ times. Thus the time complexity of Alg. 1 is $O(N^2M)$. The Alg. 1 is called N times for variously shifted contour, i.e. for $d \in \{1 \ldots N\}$. Thus the overall complexity of contour matching is $O(N^3M)$.

Algorithm 1. Contour matching algorithm

Input: $V_{i,m}$ = cost of mapping point p_i to vertex v_m
$\qquad\quad S_i$ = cost of mapping point p_i to segment s
Output: $T_{i,m}$ = cost of mapping sub-contour $(p_1 \ldots p_i)$ to vertices $(v_1 \ldots v_m)$
for all $i \in \{1 \ldots N\}$ **do**

$$T_{i,1} \leftarrow \min_{j \in \{1 \ldots i\}} \left(\sum_{k=1}^{j-1} S_k + V_{j,1} + \sum_{k=j+1}^{i} S_k \right)$$

end for
for all $m \in \{2 \ldots M\}$ **do**
$\quad$ **for all** $i \in \{m \ldots N\}$ **do**

$$T_{i,m} \leftarrow \min_{j \in \{m \ldots i\}} \left(T_{j-1,m-1} + V_{j,m} + \sum_{k=j+1}^{i} S_k \right)$$

$\quad$ **end for**
end for
return $T_{N,M}$

---

4   Experiments

The proposed methods were tested on a dataset of spread garments collected at the Czech Technical University [14]. The dataset contains color images (as in Fig. 2) and depth maps taken by Kinect-like device from a bird's eye perspective. All images were manually annotated by specifying positions of landmark points which correspond to vertices of the proposed polygonal models in Fig. 4. The resolution of images is 1280×1024. The edge of 1 pixel approximately corresponds to 0.09 cm in real world coordinates. We used 158 testing images (29 towels, 45 pants, 45 short-sleeved shirts and 39 long-sleeved shirts). The algorithms were implemented mainly in Matlab. Some of the most time-critical functions were reimplemented in C++. The performance was evaluated on a notebook with 2.5 GHz processor and 4 GB memory.

The input images were downsampled to the resolution 320×256 for the purpose of segmentation. The smaller resolution preserves all desired details and significantly improves the time performance of the segmentation algorithm. Totally 153 of 158 input images were correctly segmented which gives 97% success ratio. The incorrectly segmented images were excluded from the further evaluation. The time spent by segmenting one image is on average 0.87 seconds.

The contour simplification algorithm is the most time consuming operation of the proposed pipeline. The running times can be seen in Tab. 1. They highly depend on the length of the contour which is induced mainly by the shape complexity of the particular category of clothing. The subsequent model matching procedure is working with the already simplified contour and thus it is very fast as seen in Tab. 1. The whole pipeline including also segmentation and contour simplification runs around 5 seconds in the worst case. This is a significant improvement compared to 30–150 seconds required just for model fitting which is reported by Miller et al. [9].

Table 1. Time performance (in seconds) of contour simplification phase and polygonal model matching phase for various categories of clothing

Phase	Towel	Pants	Short-sleeved	Long-sleeved
Contour	1.33	3.95	0.64	1.88
Matching	0.01	0.01	0.03	0.03

Table 2. Displacements (in centimeters) of the identified vertices to ground-truth vertices found by polygonal model matching for various categories of clothing

Error	Towel	Pants	Short-sleeved	Long-sleeved
Median	0.41	0.52	0.53	0.59
Mean	0.43	0.69	1.07	1.26
Std. dev.	0.23	1.00	1.40	1.79

Fig. 7. Displacements of the vertices found by model matching (plotted in green) and the manually annotated landmarks (red). The displacements were computed for various configurations of garments and then they were projected to the canonical image

Tab. 2 summarizes displacements of vertices found by the proposed algorithm compared to the manually annotated landmark points. These errors are similar to those reported by Miller et al. [9]. They are small enough to determine the configuration of a piece of garment reliably and then use this information to manipulate the garment with robotic arms. Fig. 7 visualizes the displacements for the selected representatives of clothing. The errors were computed for various configurations of the same piece of garment and then they were projected to a canonical image. The biggest source of displacements are shoulders as seen in Fig. 7 for the green long-sleeved sweater. However, estimation of their position can be ambiguous even for a human. Moreover, their exact position is rather unimportant for automated manipulation. A few other significant errors were made while estimating armpits of a shirt with very short sleeves as seen in Fig. 7 for the white shirt. They are caused by indistinguishable shape of the sleeves on the contour. The contour resembles a straight line around the armpits.

The proposed algorithms were deployed to a real robot and successfully tested in several folding sequences of various garments, as seen in Fig. 1. The folding procedure succeeds approximately in 70% attempts. However, observed folding failures were almost never caused by the described vision pipeline. Main source of these failures lies in an unreliable grasping mechanism and in occasional inability to plan move of robotic arms.

5   Conclusion

We have fulfilled our goal and proposed a fast method allowing to recognize the configuration of a piece of garment. We have achieved a good accuracy, comparable to those of known approaches, despite the usage of a more challenging nonuniform background. The presented model has proved to be sufficient for the studied situation. The recognition procedure was deployed to a real robot and successfully tested in fully automated folding.

In the future, we would like to strengthen power of the model by introducing more global constraints. Our intention is to generalize the method to folded pieces of garment. We would also like to learn the robot how to detect folding failures and how to recover from them.

Acknowledgment. The authors were supported by the European Commission under the project FP7-ICT-288553 CloPeMa (J. Stria), by the Grant Agency of the Czech Technical University in Prague under the project SGS13/205/OHK3/ 3T/13 (J. Stria, D. Průša) and by the Technology Agency of the Czech Republic under the project TE01020197 Center Applied Cybernetics (V. Hlaváč).

References

1. Duda, R.O., Hart, P.E., Stork, D.G.: Pattern Classification, 2nd edn. Wiley (2000)
2. Gonzalez, R.C., Woods, R.E., Eddins, S.L.: Digital Image Processing Using MAT-LAB, 2nd edn. Gatesmark (2009)
3. Hata, S., Hiroyasu, T., Hayashi, J., Hojoh, H., Hamada, T.: Robot system for cloth handling. In: Proc. Annual Conf. of IEEE Industrial Electronics Society (IECON), pp. 3449–3454 (2008)
4. Hu, X., Bai, Y., Cui, S., Du, X., Deng, Z.: Review of cloth modeling. In: Proc. SECS Int. Colloquium on Computing, Communication, Control and Management (CCCM), pp. 338–341 (2009)
5. Kita, Y., Kita, N.: A model-driven method of estimating the state of clothes for manipulating it. In: Proc. IEEE Workshop on Applications of Computer Vision (WACV), pp. 63–69 (2002)
6. Kita, Y., Ueshiba, T., Neo, E.S., Kita, N.: Clothes state recognition using 3D observed data. In: Proc. IEEE Int. Conf. on Robotics and Automation (ICRA), pp. 1220–1225 (2009)
7. Kolesnikov, A., Fränti, P.: Polygonal approximation of closed discrete curves. Pattern Recognition 40(4), 1282–1293 (2007)
8. Maitin-Shepard, J., Cusumano-Towner, M., Lei, J., Abbeel, P.: Cloth grasp point detection based on multiple-view geometric cues with application to robotic towel folding. In: Proc. IEEE Int. Conf. on Robotics and Automation (ICRA), pp. 2308–2315 (2010)
9. Miller, S., Fritz, M., Darrell, T., Abbeel, P.: Parametrized shape models for clothing. In: Proc. IEEE Int. Conf. on Robotics and Automation (ICRA), pp. 4861–4868 (2011)
10. Orchard, M., Bouman, C.: Color quantization of images. IEEE Trans. on Signal Processing 39(12), 2677–2690 (1991)
11. Perez, J.C., Vidal, E.: Optimum polygonal approximation of digitized curves. Pattern Recognition Letters 15(8), 743–750 (1994)
12. Ramisa, A., Alenyà, G., Moreno-Noguer, F., Torras, C.: Using depth and appearance features for informed robot grasping of highly wrinkled clothes. In: Proc. IEEE Int. Conf. on Robotics and Automation (ICRA), pp. 1703–1708 (2012)
13. Rother, C., Kolmogorov, V., Blake, A.: Grabcut – interactive foreground extraction using iterated graph cuts. ACM Trans. on Graphics 23(3), 309–314 (2004)
14. Wagner, L.: Krejčová, D., Smutný, V.: CTU color and depth image dataset of spread garments. Tech. Rep. CTU-2013-25, Center for Machine Perception, Czech Technical University (2013)
15. Wang, P.C., Miller, S., Fritz, M., Darrell, T., Abbeel, P.: Perception for the manipulation of socks. In: Proc. IEEE Int. Conf. on Intelligent Robots and Systems (IROS), pp. 4877–4884 (2011)
16. CloPeMa project – clothes perception and manipulation, http://www.clopema.eu/

Design of a Multi-purpose Low-Cost
Mobile Robot for Research and Education

Sol Pedre[1], Matías Nitsche[2], Facundo Pessacg[2],
Javier Caccavelli[2], and Pablo De Cristóforis[2]

[1] Centro Atómico Bariloche, Comisión Nacional de Energa Atómica, Argentina
[2] Departamento de Computación, Facultad de Ciencias Exactas y Naturales,
Universidad de Buenos Aires, Argentina
sol.pedre@cab.cnea.gov.ar,
{mnitsche,fpessacg,jcaccav,pdecris}@dc.uba.ar

Abstract. Mobile robots are commonly used for research and education. Although there are several commercial mobile robots available for these tasks, they are often costly, do not always meet the characteristics needed for certain applications and are very difficult to adapt because they have proprietary software and hardware. In this paper, we present the design principles, and describe the development and applications of a mobile robot called ExaBot. Our main goal was to obtain a single multi-purpose low-cost robot -more than ten times cheaper than commercially available platforms- that can be used not only for research, but also for education and public outreach activities. The body of the ExaBot, its sensors, actuators, processing units and control board are described in detail. The software and printed circuit board developed for this project are open source to allow the robotics community to use and upgrade the current version. Finally, different configurations of the ExaBot are presented, showing several applications that fulfill the requirements this robotic platform was designed for.

1 Introduction

Mobile robots can be found in many fields, ranging from missions in environments that are hostile for human beings (such as in space exploration), to home service robots (such as autonomous vacuum cleaners). To develop new applications, it is necessary to have test platforms. Thus, mobile robots are commonly used at research laboratories as well as universities. Nowadays, there are many commercial mobile robots available for this purpose.

The most popular commercial research robots are those of Adept MobileRobots and K-Team companies, in particular the Adept's Pioneer [1] robots and K-Team Kheperas [2]. However, many times commercial robots do not quite fit the necessary characteristics and are difficult to adapt since they have proprietary software and hardware. Moreover, a big drawback is their cost: the basic Pioneer 3-DX academic price is around $4,500, the basic Khepera III academic price is around $3,000 and the basic Koala II (also from K-Team) around $9,000.

M. Mistry et al. (Eds.): TAROS 2014, LNAI 8717, pp. 185–196, 2014.

Extra sensors, processing elements or part replacements are also quite expensive. An attempt to provide a low-cost platform is TurtleBot 2 [3]. TurtleBot 2 is a mobile robot with open-source software based on ROS (Robot Operation System). It is designed for 3D vision applications as its main sensor is a Microsoft Kinect. Although its creators claim that it is inexpensive, the market price for this robot is around $2,000.

In addition of them being expensive, it is often the case that commercial robots do not meet your needs. These problems have led some universities to develop their own robotic platforms to lower costs and/or tackle particular tasks for which commercial robots are not well suited. Even universities with extensive experience in robotics have started to propose cheaper robots for research and/or education. Some works present designs of mini robots similar to the Kheperas. That is the case of Rice University's r-one platform [4] and Harvard's miniature robot Kilobot [5].

Educational robots are also a growing field. The most commonly used commercial robots for education are the Lego kits [6]. Although these kits are widely used in K-12 education, they are not suitable for research or undergraduate and graduate education. Some universities have also developed robots for K-12 education. That is the case of Miniskybot, a 3D printable mobile robot proposed by the Autonomous University of Madrid [7].

In this paper we present our approach to develop a mobile robotic platform, more than ten times cheaper than similar commercial research robots like those of Adept MobileRobots and K-Team companies. Instead of lowering costs by tailoring the design to a particular task -like other academical designs-, we have built a highly reconfigurable robot. Our goal was to design a single robotic platform that could be tuned for different research experiments, outreach activities and undergraduate education. For these reasons, we decided to build a small size robotic platform, with reconfigurable sensing capabilities and reconfigurable processing power. In this paper, we present the design ideas, development and applications of the ExaBot, a new multi-purpose low-cost mobile robot.

The rest of the paper is organized as follows: section 2 presents the main design goals and ideas, section 3 presents the resulting system design; section 4 presents results, showing the different configurations used for a wide set of applications; and section 5 outlines conclusions and future work.

2   General Design Considerations

The goal of the ExaBot is to have one single robotic platform that allows to carry out research, education and outreach activities, focusing on the low cost compared to its commercial counterparts and with similar functionalities. Therefore, we address a three way general design that trades off between size, cost and functionality.

Size: The body of the ExaBot should be small enough to be transported around easily, but also big enough to support many sensors and different processing units. On the other hand, the dimensions of the chassis should be large enough to allow a single-board computer inside. Also, the robot should be able to carry a laptop or a mini external PC on top of it. The relation of robot cost to robot size indicates that off-the-shelf components are the best option. If the platform is too small it gets expensive due to the advanced technologies and fabrication techniques required (e.g. micro-electromechanical systems, micro assembly, etc.), and to the lack of off-the-shelf components. On the other hand, if the platform is too big, the cost of the chassis increases considerably and it also becomes more difficult to use because it requires a large workspace. Based on these considerations, we decided that the ExaBot was going to be a medium size robot (for more details see Section 3.1).

Cost: The cost is one of the main constraints of the robot design. It should be in the order of ten times less compared to its commercial counterparts. Thus, the chassis, sensors and actuators of the robot should be inexpensive, but allow a wide application spectrum. Using a pre-built body, off-the-shelf components and developing the electronics of ExaBot on our own, it is possible to achieve this goal. Because the robot should be small we decided to use small, cheap and still readily available sensors. On the other hand, as the ExaBot has the ability to carry a laptop mounted on it, we can use it as the main processing unit, decreasing significantly the cost of the robot.

Functionality: When considering functionality one must consider the environment that the robot is expected to work in, and what it is expected to do in that environment. As we want a multipurpose robot that can be used for a variety of activities, the ExaBot should be designed to support many different sensors and processing units, and to be easily reconfigured with a particular subset of them for a given task. Regarding locomotion, the robot should be able to operate in both indoor and outdoor environments (see details in section 3.1). Regarding sensors, the ExaBot should have a variety of removable sensors for different applications (see section 3.2). Finally, the robot should have different processing capabilities depending on each task (see section 3.3).

2.1   Design Flow

From the point of view of design, a mobile robot can be thought of as an embedded system that deals with real world interactions and control. Thus, it is necessary a design methodology that supports the cooperative and concurrent development of hardware and software (co-specification, co-development, and co-verification) in order to achieve shared functionality and performance goals for the system. To develop the ExaBot, we adapted a traditional hardware-software co-design flow to the particular field of mobile robot design. Although it would be interesting to go over each design stage, this exceeds the length and aim of this work. In this paper we will only outline the first stage, i.e. goal definition and body, locomotion and sensors definition, and show an overview of the final system with the particular angle of reconfigurability. In Fig. 1 the main stages of the co-design flow can be seen together with the general stages of traditional co-design flows based on processors and ICs.

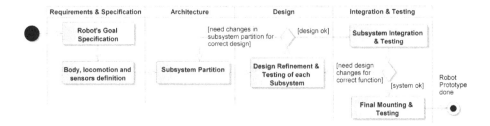

Fig. 1. Hardware-software co-design flow for the ExaBot. The vertical swimlines show the equivalent stages from traditional general co-design flow.

3   Robot Subsystems

3.1   Base Chasis and Locomotion

In order to fulfil the goals presented in previous sections a pre-built mechanical kit, the Traxster Kit [8], was selected as a base for the development of the ExaBot. This kit consists on a light, strong and small sized chassis, so it is transportable and can accommodate multiple sensors and processing units. As locomotion system it has two caterpillars, each connected to direct current motors with built-in quadrature encoders (see section 3.2). This allows in-place turning with simple motor commands, in contrast to other models (e.g. Ackerman steering), providing good stability and traction during motion, and helping to reduce odometry errors. Caterpillars were preferred instead of wheels because they allow to overcome small obstacles. This is especially important for outdoor rough terrains.

3.2   Sensors

Since the ExaBot should be adaptable to different applications, we can distinguish a base set of sensors intended for outreach activities and also a set of high-level sensors intended for research activities. There are also some sensors included for proper robot control. The base low-level sensors included are:

Proprioceptive Sensors: Wheel quadrature encoders (built-in in the Traxster kit motor) are used to sense the movement of each motor. With this information, a PID controller is implemented which allows velocity-based command to be processed by the robot. Furthermore, odometric readings can be used as a base for localization information (such as position and angle), to be fused to other sensors. To further monitor the motors, a current consumption sensor for the motor driving circuits are included. Finally, to control the battery charge level another sensor is used that measures the voltage values.

Range-Finders: While lasers are the most precise and reliable range finder sensors, their cost exceed by several times the intended final cost of the robot and therefore were not considered for the educational configuration. Thus, the range finders currently used on the ExaBot are infrared (IR) sensors an sonar sensors. These type of sensors are cheap and can return a single distance reading, which simplifies programming reactive behaviors. By mixing IR and sonar range-finders, both short-range punctual and mid-range wide sensors can be included in the same robot, allowing for different use cases. In particular, the ExaBot was given a ring of 8 Sharp GP2D120 IR (4-30 cm) range finders and a Devanatech SRF05 sonar (1-400 cm).

Line-Following: To enable line-following behaviors, two infra-red light detectors are included in the bottom front of the ExaBot.

Bumpers: Contact switches are placed at the front in order to easily detect collisions and to program simple evasive reactive maneuvers.

Other Sensors: There are several ways to add new sensors into the ExaBot. For one, the pins of the microcontrollers are mapped efficiently to maximize useful free ports. Also, sensors that implement SPI communication can be added to the communications bus (see subsection 3.4). An example of this capability was the recent addition of an IMU package (3-axis gyroscope L3G4200D, 3-axis accelerometer ADXL345 and 3-axis magnetometer MC5883L). Other types of low-level sensors can be used to explore further education-related tasks.

Of course, all the sensors can be removed or moved around the robot. Moreover, for research activities, the robot can carry sensors commonly used for autonomous navigation methods, such as laser range-finders and cameras. The final configuration depends on the task at hand, as can be seen in section 4. In Fig. 5 several ExaBot configurations are presented.

3.3   Computational Power

The processing power can be divided in two levels: a low level control board (for sensor and motor control), and a high level processing unit for more complex tasks. For the low level processing units, we chose to use Microchip PICs, in particular the 18F family. These are cheap, widely used microcontrollers that provide all the necessary modules to control the selected sensors and actuators. The high level processing unit can be any small or medium-size embedded computer. So far we have used the ExaBot with an embedded TS7250 PC104, an embedded Mini-ITX board (AT5ION-T Deluxe), an Arduino board, a RaspberryPI computer, an Android smartphone and common laptops (see Section 4).

3.4   Control Board

In this section we briefly describe each subsystem of the control board, with particular interest in the reconfigurable schemes, and its interface to a high-level computer mounted on the robot.

Motor Control Subsystem. The motor control subsystem (Fig. 2(a)) is in charge of the DC motor speed control. The microcontroller (μC) outputs a PWM

(Pulse With Modulation) signal to the motor driver (H-Bridge) in order to maintain a desired speed, while reading the motor encoder values. The control loop consists of a Proportional Integrative Derivative (PID) controller. Appropriate PID constants were set using the Ziegler-Nichols method [9]. The duty and direction are set by the PID controller and are updated every period according to external commands received in higher levels. Experimental results show that the desired speed with this control is always achieved in less than 100 control loops, that is, less than 81 ms. As an extra safety measure, if the motor's current consumption exceeds a reference value, a fault circuit signal is enabled, which overrides all PWM output and hence stops the motors.

The chosen μC is a 24-pin PIC18F2431, since it is tailored for DC motor control being the only PIC from the family that has several PWM modules and a Quadrature Encoder Interface module for encoder sensing. For sensing current the ACS712 IC was used, together with an LM319 voltage comparator for fault signal generation from a reference voltage. The selected H-bridge driver is the L298.

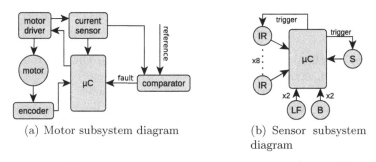

(a) Motor subsystem diagram   (b) Sensor subsystem diagram

Fig. 2. Motors and sensors subsystem diagrams

Sensor Control Subsystem. The sensor control subsystem (Fig. 2(b)) controls the installed low-level sensors: eight infrared range sensors, one sonar, two line-following sensors and two bumpers in the default configuration. For this subsystem, taking into account the need for many analog pins, a PIC18F4680 was selected.

The μC software implements the different control algorithms needed for each sensor type. In the case of IR sensors, since these produce an analog voltage (as a function of the sensed distance) which needs to be digitized and the μC has only one ADC module, a round-robin algorithm was implemented to read each sensor in turn. In the case of the sonar, distances are measured as a PWM signal (i.e., the length of the output signal driven high is proportional to the measured distance). To measure these pulses, one of the CCP modules of the μC was used. Finally, since bumpers and line-following sensors produce binary digital outputs, their output values are polled by simply using a timer. In order to meet the reconfigurability requirement, each sensor can be turned on or off. Since only sensors that are on are checked, this strategy also saves battery charge.

Communication Subsystem. To be able to reconfigure the main processing element, the ExaBot is capable of supporting different types of embedded or external computers. For a completely embedded solution, the ExaBot can be controlled by a specially designed connector, based on the SPI bus (Slave Peripheral Interface). This connector was designed to be used with an embedded PC104 computer, but any other embedded PC that supports SPI can be adapted. This configuration was used for many outreach applications, using the low-level sensors previously outlined. As another option, the ExaBot can be controlled through a serial interface (USART) by any type of external computer.

The communication subsystem is specially designed for these two communication options to be switched easily (Fig. 3). Hence, in both cases, the high-level application protocol is the same and only the low-level physical layers change. Moreover, the control board includes all the extra electronics for this change to be easily done (SPI signals multiplexing, and USART ICs). When the ExaBot is used with an external computer connected through the serial port (configuration A), the sensor μC is the master of the SPI communication. When an embedded computer such as the PC104 is used (configuration B), this computer is the master of the SPI bus and the three microcontrollers are the slaves.

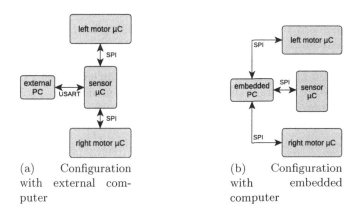

(a)   Configuration with external computer

(b)   Configuration with embedded computer

Fig. 3. Communication subsystem diagram

To enable transparent control regardless of the current configuration, the `libexabot` library was developed, which is capable of transparently handling the appropriate transport mechanism. The library presents the same interface either when running on the embedded PC through SPI or when running on an external computer and communicating via USART protocol. Furthermore, a similar version of the library is designed for handling remote control of the ExaBot via UDP (User Datagram Protocol) when the embedded computer is installed. In case an embedded computer is not desired, remote control can still be achieved by using a Bluetooth-to-serial dongle, for example to control the robot by a smartphone or a similar device. Since these libraries present a very simple C based interface, it can be used from any C/C++ application and even enables the development of high-level bindings for languages such as Python, Ruby, etc.

Finally, using the `libexabot` library a ROS (Robot Operating System) node was also developed which allows to control the ExaBot and to recieve odometry and data from the other sensors.

3.5   Final Product

Following the previously outlined design guides and requirements, a final Printed Circuit Board (PCB) was produced. Each subsystem was initially tested individually using prototyping boards along with specialized test firmware. A final dual-layer PCB was obtained, based on Surface Mount Design (SMD) components. In order to facilitate re-programming of the microcontrollers on the final PCB, the ICP (In-Circuit Programming) functionality was exported using RJ11 connectors in the board itself. Figure 4 shows the final control board of the ExaBot.

The control board can be powered either using external power (during testing) through a standard PC power supply, or using batteries. Since all subsystems were also analyzed in terms of power requirements, and in order to isolate the motor and control logic power lines, two battery lines are used. The ExaBot was so far tested with LiIon and LiPo batteries. Without an embedded PC, the robot can be powered for several hours using batteries of moderate capacity (around 2500mAh for control logic).

Fig. 4. Final ExaBot control board

In order to support the mounting of all sensors, the control board and batteries, the chassis was drilled and adapted as necessary. This also includes the development of many types of metal mounts which are capable of carrying a laser, low-level sensors and even a laptop on top of the robot.

The cost of the final ExaBot depends on which configuration is used, which in turn depends on which processing unit and sensors are installed. The cost of the main low-level control board including all the electronics and PCB printing, a base set of sensors, the chassis and batteries is approximately $250. However, the chassis (that includes the motors and encoders) amounts for more than half that cost.

4   Results and Applications

A short comparison of the ExaBot with similar robots available in the market can be found in Table 1. As already stated, the ExaBot is suited for research and education activities, is reconfigurable in both it's sensing and processing capabilities, and is much cheaper than similar commercial robots.

Table 1. Comparison between ExaBot and other similar mobile robots. R & E: Research and Education, NI: Not informed.

Robot	Functionality	Size	Cost ($)	Locomotion based on	Reconfig. sensors	Embedded PC	Max speed ($\frac{cm}{s}$)
Pioneer P3AT	Research	Large	6,000	wheels	yes	optional	70
Khepera III	R & E	Small	3,000	wheels	no	no	50
TurtleBot 2	R & E	Medium	2,000	wheels	no	yes	65
Lego EV3	Education	Medium	349	varies	yes	no	varies
Kilobot	Research	Small	100	vibration	no	no	NI
R-one	R & E	Small	NI	wheels	no	no	25
ExaBot	**R & E**	**Medium**	**250**	**caterpillars**	**yes**	**yes**	**50**

Different configurations of the ExaBot were used in several applications fulfilling all the goals the ExaBot was designed for. In the next subsections we briefly comment some of them.

4.1   Research

The main research activities with the ExaBot are related to autonomous visual navigation. In this context, the ExaBot was used as a platform for experiments in different works.

One work presents a real-time image-based monocular road following method [10]. To achieve real-time computation necessary for on-board execution in mobile robots, the image processing was implemented on a low-power embedded GPU. Hence, the ExaBot was configured to use a Mini-ITX board (AT5ION-T Deluxe)- that includes a 16 core NVIDIA GPU- as the main processing unit, and a FireWire camera (model 21F04, from *Imaging Source*) as the only exteroperceptive sensor (see Fig. 5, image **d**). The Mini-ITX board connects through RS232 to the control board as explained in section 3.4. The same configuration was used for experiments in a completed PhD Thesis [11], that proposes a hybrid method for navigation combining the aforementioned method to follow paths, with a landmark-based navigation method to traverse open areas.

Another work [12] presents the use of disparity and elevation maps for obstacle avoidance using stereo vision. In this work, a common notebook or netbook is used as the main processing unit and a low-cost Minoru 3D USB webcam as the only exteroperceptive sensor. The notebook connects through RS232 using a USB-to-serial dongle (see Fig. 5, image **c**).

An embedded method for monocular visual odometry was developed in a master thesis [13], using an Android-based smart-phone as the main processing unit and its embedded camera as the main sensor. For this work, yet another configuration of the ExaBot was used (see Fig.5, image **b**). Here, the Android cellphone connects either through Wi-Fi to the on-board PC104 or via Bluetooth directly to the control board.

Currently, further autonomous navigation experiments are being conducted with the aforementioned integration of a SICK TIM310 laser range finder, and a gyro-compensated magnetometer sensor.

4.2   Outreach

The ExaBot is also used in Educational Robotics courses, talks and exhibits. Educational Robotics proposes the use of robots as a teaching resource in K-12 education that allows inexperienced students to approach fields other than specifically robotics. A key problem in this context is to have an adequate easy-to-use interface between inexpert public and robots. For this, we developed a new behavior-based application for programming robots, specially the ExaBot [14]. Robotic-centered courses and other outreach activities were designed and carried out [15]. In the last years, three eight-week courses, five two-days courses, more than ten one-day workshops and talks were taught to different high school students using the developed programming interface and several ExaBots. For this work, the ExaBot was configured with the PC104 as the main processing unit and all the exteroperceptive sensors described in section 3.2: IR telemeters, sonar, bumpers and line-following sensors (see Fig. 5, image **a**).

4.3   Undergraduate Education

The ExaBot is also used in undergraduate and graduate courses at the Departamento de Computación, FCEN-UBA. In particular, it is used in the Robotics Vision course. This course covers topics regarding monocular and stereo vision applied to mobile robots. Several algorithms for autonomous robot navigation are implemented and tested by students using the ExaBot.

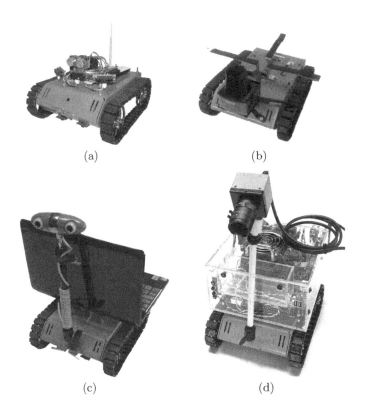

(a)   (b)

(c)   (d)

Fig. 5. Various configurations of the ExaBot: (a) with all low-level sensors and PC104 (b) with a laser range-finder (c) with a netbook and a 3D Minoru Camera, (d) with an embedded Mini-ITX board and a FireWire Camera

5   Conclusions and Future Work

In this paper, we present the design ideas, development and applications of the new mobile robot ExaBot. Our main goal was to obtain a multi-purpose low-cost robot- i.e., ten times cheaper than commercially available research robots- that could be used not only for research, but also for outreach and education.

The main requirement to achieve a low cost robot that can be used for such diverse fields is that the robot is highly reconfigurable. Hence, the ExaBot was designed with many sensors that can be optionally installed, and built-in sensor expansion ports. The high level processing unit and the communication protocol are also reconfigurable. In this manner, many different configurations of the ExaBot have been built and used; keeping the base cost of the robot at around $250. We have successfully used them for research activities, mainly in vision-based autonomous navigation; for undergraduate education; and for robotic-centered courses at K-12 education and other outreach activities.

As future works, we are planning to build the mechanical chassis ourselves to further lower costs. Moreover, further research experiments are planned using the recently incorporated sensors (laser scan and gyro-compensated digital compass), as well as new processing elements such as a BeagleBoard.

References

1. Adept Mobile Robotics: Pioneer 2-DX and 3-DX (2013)
2. K-team corporation: Khepera, Khepera II and Khepera III (2013)
3. Garage, W.: Turtlebot (2011)
4. McLurkin, J., Rykowski, J., John, M., Kaseman, Q., Lynch, A.: Using multi-robot systems for engineering education: Teaching and outreach with large numbers of an advanced, low-cost robot. IEEE Transactions on Education 56(1), 24–33 (2013)
5. Rubenstein, M., Ahler, C., Nagpal, R.: Kilobot: A low cost scalable robot system for collective behaviors. In: IEEE International Conference on Robotics and Automation (ICRA), pp. 3293–3298 (2012)
6. Benitti, F.B.V.: Exploring the educational potential of robotics in schools: A systematic review. Computers & Education 58(3), 978–988 (2012)
7. Gonzalez-Gomez, J., Valero-Gomez, A., Prieto-Moreno, A., Abderrahim, M.: A new open source 3d-printable mobile robotic platform for education. In: Advances in Autonomous Mini Robots, pp. 49–62. Springer, Heidelberg (2012)
8. RoboticsConnections: Traxster Kit (August 2012)
9. Bräunl, T.: Embedded Robotics - Mobile Robot Design and Applications with Embedded Systems, 2nd edn. Springer (2006)
10. De Cristóforis, P., Nitsche, M., Krajník, T., Mejail, M.: Real-time monocular image-based path detection. Journal of Real-Time Image Processing, 1–14 (2013)
11. De Cristóforis, P.: Vision-based mobile robot system for monocular navigation in indoor/outdoor environments. PhD thesis, Facultad de Ciencias Exactas y Naturales, Universidad de Buenos Aires (2013)
12. Piré, T., de Cristóforis, P., Nitsche, M., Berlles, J.J.: Stereo vision obstacle avoidance using disparity and elevation maps. In: IEEE RAS Summer School on Robot Vision and Applications (2012)
13. Gonzalez, S., González, E.: Smartphones como Unidad de Sensado y Procesamiento para la Localización de Robots Móviles Utilizando Odometría Visual Monocular. Master's thesis, Universidad de Buenos Aires (2013)
14. Caccavelli, J., Pedre, S., de Cristóforis, P., Katz, A., Bendersky, D.: A New Programming Interface for Educational Robotics. In: Obdržálek, D., Gottscheber, A. (eds.) EUROBOT 2011. CCIS, vol. 161, pp. 68–77. Springer, Heidelberg (2011)
15. De Cristóforis, P., Pedre, S., Nitsche, M., Fischer, T., Pessacg, F., Di Pietro, C.: A Behavior-Based Approach for Educational Robotics Activities. IEEE Transactions on Education 56(1), 61–66 (2013)

Adaptive Swarm Robot Region Coverage Using Gene Regulatory Networks

Hyondong Oh and Yaochu Jin

Department of Computing, University of Surrey
Guildford, Surrey, GU2 7XH, UK
{h.oh,yaochu.jin}@surrey.ac.uk

Abstract. This paper proposes a morphogenetic pattern formation approach for collective systems to cover a desired region for target entrapment. This has been achieved by combining a two-layer hierarchical gene regulatory network (H-GRN) with a region-based shape control strategy. The upper layer of the H-GRN is for pattern generation that provides a desired region for entrapping targets generated from local sensory inputs of detected targets. This pattern is represented by a set of arc segments, which allow us to form entrapping shape constraints with the minimum information that can be easily used by the lower layer of the H-GRN. The lower layer is for region-based shape control consisting of two steps: guiding all robots into the desired region designated by the upper layer, and maintaining a specified minimum distance between each robot and its neighbouring robots. Numerical simulations have been performed for scenarios containing either static and moving targets to validate the feasibility and benefits of the proposed approach.

Keywords: Pattern formation, Target entrapment, Swarm robots, Gene regulatory networks.

1 Introduction

Developing self-organising multi-agent systems has become a primary research area in recent decades due to their attractive properties such as robustness to faults and damages, adaptability to unknown environments and cost efficiency. In particular, considerable attention has been paid to multi-robot shape or region formation as a basic functionality for achieving the missions such as search and rescue, deployment of sensor network, and collective transport. Pattern formation algorithms for swarm robots can be largely divided into four categories: 1) behaviour-based control, where behavioural rules are employed onto individual robots with relative importance [1]; 2) leader-follower and virtual structures where the (virtual) leaders are identified and the followers follow the leaders with a set of formation constraints [2]; 3) potential field in which the robot moves through the gradient of a potential field defined by a sum of attractive and repulsive forces [3,4]; and 4) biologically-inspired approaches such as using morphogen gradient [5,6], a pheromone/hormone model [7] or a gene regulatory network (GRN) [8,9].

M. Mistry et al. (Eds.): TAROS 2014, LNAI 8717, pp. 197–208, 2014.

Despite the large body of research on multi-robot pattern formation, relatively little work has been performed on the control of a large number of robots for generating adaptive shapes interacting with an unknown and dynamic environment. Swarm Chemistry [10], a computational model of particle swarms following certain kinetic rules is a good example for a swarm pattern formation approach that can self-organise and self-repair. However, it requires velocity of nearby particles, which is difficult to obtain, and it is not straightforward to design the kinetic rules producing the desired patterns. Mamei *et al.* [5] used the concept of morphogen gradient diffusion to achieve pattern formation, but patterns are limited to polygonal shapes, and robots are assumed to be able to pass through each other without a physical size. A hierarchical two-layer GRN (H-GRN) is introduced for target entrapping, where the first layer is responsible for adaptive pattern generation, while the second is a control mechanism that drives the robots on to the generated pattern [11]. This H-GRN concept is demonstrated by extracting a certain number of points from an entrapping pattern and guiding robots to a nearest point. This one-to-one matching between robots and points could be a hard constraint in uncertain and dynamic environments. Cheah *et al.* [3] have introduced a region-based shape control based on potential field, where each robot in the group stays within a region as a group while maintaining a minimum distance from each other. It addressed the convergence analysis, and considered various shapes of the desired region. However, the target shape needs to be predefined with limited complexity, which may become inadequate for handling unknown environmental changes.

To address the above limitations of the existing work, this paper proposes a morphogenetic approach to a target entrapping problem by integrating a two layer H-GRN with the region-based shape control for swarm robots. The morphogenetic approach to collective systems exploits the genetic and cellular mechanisms that govern biological morphogenesis [8] which refers to the biological process in which cells divide, grow and differentiate, and finally resulting in the mature morphology under GRNs and morphogens. Here, GRNs are models of genes and the interaction of gene products that describe the gene expression dynamics [12]. Morphogen gradients, concentration gradients of substances present in the environment, play an important role to trigger the specific gene expression in morphogenesis. The upper layer of the proposed H-GRN is responsible for pattern generation which provides a morphogen gradient to the lower layer. The lower layer implements a region-based shape control strategy that guides all robots into the desired region from the upper layer, and maintains a specified minimum distance between each robot and its neighbouring robots as well. The minimum distance between robots is adapted to evenly distribute robots into the desired region. Using this region-based shape control for target entrapment instead of boundary coverage as in our previous work enables robots to entrap the targets more tightly and herd them into a desired shape or position by directly contacting and pushing the targets.

The rest of this paper is structured as follows. Section 2 presents a problem statement including the connection between multicellular organisms and multi-agent systems. Section 3 introduces a H-GRN model with a region-based shape control strategy, consisting of an upper layer for region generation and a lower layer for robot guidance into the region. Section 4 presents numerical simulation results from scenarios containing either pre-defined stationary or moving targets. Conclusions and future work are given in Section 5.

2   Problem Statement

This paper considers the problem of entrapping stationary or moving targets using a swarm of robots, consisting of two tasks: region generation and region-based shape control. Based on the target location, the entrapping region is generated, and the robots are deployed into the generated region. Similar to [11], we distinguish between *organising robots* that can detect at least one target and are responsible for the region generation, and *non-organising robots* that have not yet detected any target. The non-organising robots will follow their neighbouring robots until they detect a target or receive information on the target from organising robots. It is assumed that global (or at least local) 2D position is available for the robots, and when the robots are close to the targets, they can obtain the position of targets either by direct sensing or from other organising robots through local communication.

As we exploit the GRNs found in multicellular organisms to control swarm robots, a metaphor between cells and robots is required. To this end, each cell is considered as a single robot where protein concentrations of genes in the cell correspond to the position and internal states. Each protein has the following three roles: i) auto-regulation which regulate the gene expression level that produces the protein, thus controlling the robot's behaviour ii) reaction to a certain morphogen gradient from the environment (in this paper, from targets to be entrapped), and iii) diffusion into other cells to avoid collision of the robots. In the following sections, we will show in detail how to use these GRN functions for dealing with a target entrapping problem.

3   H-GRN Model with Region-Based Shape Control

This study uses a two-layer hierarchical GRN (H-GRN) model [11,13], however, with a different objective of region-based shape formation for reliable target entrapping, as illustrated in Fig. 1. The upper layer of the H-GRN is for pattern generation that provides a desired entrapping pattern based on local sensory inputs that report the position of the targets. In the figure, the protein concentration p represents the environmental input (i.e. target positions), which will serve as the input of the upper GRN and activate g_1, g_2 and g_3. In particular, the concentration of g_3 takes the role of morphogen to form the desired region or shapes around targets which will be transmitted to the lower layer. Note that the dynamics of the GRN in the upper layer is activated only in the organising robots that are able to detect targets, while the non-organising robots simply follow the movement of neighbouring organising robots. Once the target pattern

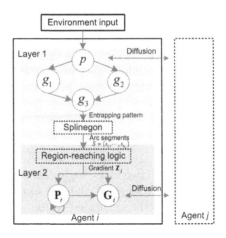

Fig. 1. Illustration of a two-layer H-GRN structure for target entrapping

is generated by the upper layer, it will function as the input to the lower layer to trigger its dynamics. The lower layer is for region-based shape control to guide all robots into the desired region while maintaining a specified minimum distance between the robots. These functions are realised by a single GRN where proteins $\mathbf{G}$ and $\mathbf{P}$ are used to represent the current positions and internal states of the robots. Thanks to diffusion of proteins p and $\mathbf{G}$, the neighbouring robots can share the target information and maintain a desired distance between robots to avoid robot collision.

3.1   Upper Layer: Region Generation

Target Entrapping Pattern. For generation of the target entrapping pattern, each organising robot will utilise the following gene regulatory dynamics to generate the desired concentrations:

$$\frac{dp_j}{dt} = -p_j + \nabla^2 p_j + \gamma_j, \quad p = \sum_{j=1}^{n_t} p_j, \tag{1}$$

$$\frac{dg_1}{dt} = -g_1 + \mathrm{sig}(p, \theta_1, k), \tag{2}$$

$$\frac{dg_2}{dt} = -g_2 + [1 - \mathrm{sig}(p, \theta_2, k)], \tag{3}$$

$$\frac{dg_3}{dt} = -g_3 + \mathrm{sig}(g_1 + g_2, \theta_3, k), \tag{4}$$

$$\mathrm{sig}(x, z, k) = \frac{1}{1 + e^{-k(x-z)}}, \tag{5}$$

where p_j represents the protein concentration produced by the j-th target input γ_j, and p the sum of concentrations of all detected n_t targets. The integrated

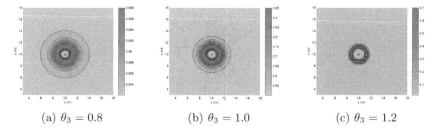

(a) $\theta_3 = 0.8$ (b) $\theta_3 = 1.0$ (c) $\theta_3 = 1.2$

Fig. 2. Examples of a target entrapping pattern from protein concentration g_3 generated by the upper layer of the H-GRN according to different thresholds θ_3 with fixed $\theta_1 = 0.25$ and $\theta_2 = 0.3$ where target is located at $(x, y) = (10, 10)$

protein p will activate the internal protein concentrations. k is a positive constant which determines the slope of sigmoid function, and g_1, g_2 and g_3 are protein concentrations, where g_3 defines target entrapping pattern. Note that g_3 is regulated by both g_1 and g_2 that are regulated by p only within a particular range of its concentration. In particular, protein g_3 can be regulated by protein p only when the concentration of p is between θ_1 and θ_2. This leads to an activating band of circle shapes depending on the thresholds of a sigmoid function θ_1, θ_2 and θ_3, as shown in Fig. 2.

Splinegone Representation. For the generated entrapping pattern to be used by the lower layer of the H-GRN, the contour of the target pattern needs to be extracted. This work uses a subset of a class of object termed Splinegons [14] to create a set of vertices that are connected by line segments of constant curvature, representing the curved nature of the target pattern as a set of arc segments. Firstly, a few representative points whose g_3 concentration is higher than a threshold value on the pattern are selected to describe the target pattern, and those points are connected by line segments of constant curvature. This allows us to form shape constraints to be conveniently used for a region-based shape control. The mathematical details for the constructing the whole Splinegon can be found in [14]. Figure 3 shows the procedure for Splinegon representation from a given target pattern using selected representative points and constant curvature generation. The generated region from the upper layer of the H-GRN can then be modelled as a set of circular arc segments $s_l \in S = \{s_1, \cdots, s_{n_s}\}$. For each arc segment, only the centre position and its curvature will be used for region-based shape control in the next section.

3.2 Lower Layer: Region-Based Shape Control

The lower layer of the H-GRN aims to guide all robots into the desired region generated by the upper layer and to maintain a specified minimum distance between robots. In the following, we first describe the GRN dynamics for driving the robots to the target region and then present the region-based shape control

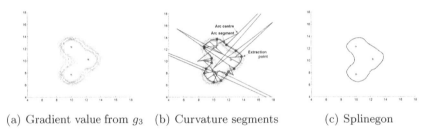

(a) Gradient value from g_3 (b) Curvature segments   (c) Splinegon

Fig. 3. Splinegon representation procedure from a target entrapping pattern generated by upper layer where targets are represented as a blue star

logic. At the end of this section, we discuss issues for implementing the region-based shape control within the GRN framework.

GRN Dynamics. The GRN in the lower layer is the same as the one used in our previous work [11], which was a modified GRN model used in [15]:

$$\frac{d\mathbf{G}_i}{dt} = -a\mathbf{z}_i + m\mathbf{P}_i \tag{6}$$

$$\frac{d\mathbf{P}_i}{dt} = -c\mathbf{P}_i + rf(\mathbf{z}_i) + b\mathbf{D}_i \tag{7}$$

where $i \in \{1, \cdots, N_o\}$ is the index of an organising robot, and protein types $\mathbf{G}_i$ and $\mathbf{P}_i$ correspond to a 2D position and internal state vector of robot i, respectively. a, m, c, r and b are constants to be optimised depending on the objectives of the task. $\mathbf{D}_i$ represents the concentration of protein $\mathbf{G}$ diffused out of the cell, indicating the density of robots and obstacles in the neighbourhood:

$$\mathbf{D}_i = \sum_{j=1}^{n_i} \mathbf{D}_i^j, \tag{8}$$

where n_i denotes the number of robots in the neighbourhood of robot i, and $\mathbf{D}_i^j$ represents the diffused protein concentration vector from robot j:

$$\mathbf{D}_i^j = \frac{\mathbf{G}_i - \mathbf{G}_j}{\|\mathbf{G}_i - \mathbf{G}_j\|}. \tag{9}$$

The diffusion process is activated only when the distance to the neighbour is less than a threshold r_n. In order to embed the desired shape into the regulatory dynamics, $f(\mathbf{z}_i)$ is defined as the following sigmoid function:

$$f(\mathbf{z}_i) = \frac{1 - e^{\alpha \mathbf{z}_i}}{1 + e^{\alpha \mathbf{z}_i}}, \tag{10}$$

where $\mathbf{z}_i$ represents a gradient value at the robot's current position, and $\alpha > 0$ determines the slope of the sigmoid function. This $\mathbf{z}_i$ regulates the concentration of both proteins $\mathbf{G}$ and $\mathbf{P}$ as a feedforward input in the GRN dynamics as in Eq. (7) so that robots could form a desired shape, which is defined as:

$$\mathbf{z}_i = \frac{dh_i}{d\mathbf{G}_i} \tag{11}$$

where h_i is the desired shape on which the robots need to be deployed defined by the upper layer. For instance, if the robots need to be deployed onto a circle of a radius R at a point $\mathbf{c} = [c_x, c_y]^T$, the shape function can be defined as:

$$h_i = [(G_{i,x} - c_x)^2 + (G_{i,y} - c_y)^2 - R^2]^2. \tag{12}$$

Note that if a circle radius R approaches to zero, then the robot is to be deployed onto a point $\mathbf{c}$ rather than a circle.

Region-Reaching Control Logic. The region-based shape control logic is developed in which the desired region is specified by an inequality function as follows:

$$F(X) \leq 0, \tag{13}$$

where $F(X)$ is a continuous scalar function with continuous first partial derivatives. In this paper, the target pattern is represented as a set of arc segments:

$$S_l(\mathbf{G}_i) = (G_{i,x} - c_{l,x})^2 + (G_{i,y} - c_{l,y})^2 - \left(\frac{1}{\kappa_l}\right)^2 = 0, \tag{14}$$

where $[c_{l,x}, c_{l,y}]^T$ and κ_l are the centre position and the curvature of the l-th arc segment, respectively. Thus, inequality functions for region shape control can be defined by $S_l(\mathbf{G}_i) \leq 0$ for keeping robots inside an arc segment or $-S_l(\mathbf{G}_i) \leq 0$ for pushing robots towards outside an arc. Inequality functions for the entire complex shape using the Splinegon representation can then be expressed as:

$$F(\mathbf{G}_i) = [\pm S_1(\mathbf{G}_i), \pm S_2(\mathbf{G}_i), \cdots, \pm S_{n_s}(\mathbf{G}_i)]^T \leq 0. \tag{15}$$

In this study, for an efficient region shape control, each robot finds its nearest arc segment $s_{l_i^*}$ at its current location rather than considering all arc segments consisting of the Splinegon:

$$s_{l_i^*} = arg \min_{s_l \in S} (\|\mathbf{G}_i - s_l^m\|), \tag{16}$$

where s_l^m is the middle point of l-th arc segment. Depending on the sign of the arc curvature and the position of targets, the shape control logic is divided into three cases with a gradient $\mathbf{z}_i$ and the inequality function of the nearest arc segment as follows.

Case 1: $\kappa_{l_i^*} \leq 0$

$$\mathbf{z}_i = \begin{cases} 0, & \text{if } S_{l_i^*}(\mathbf{G}_i) \leq 0 \\ \frac{dh_{1,i}}{d\mathbf{G}_i}, & \text{otherwise} \end{cases} \tag{17}$$

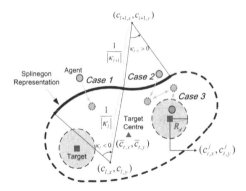

Fig. 4. Geometric relations for the region-based shape control logic

where $h_{1,i} = \left[(G_{i,x} - c_{l_i^*,x})^2 + (G_{i,y} - c_{l_i^*,y})^2\right]^2$. This gradient drives a robot towards the centre of its nearest arc segment, and once the robot enters a circular sector of a desired Splinegon, the gradient becomes zero, as illustrated in Fig. 4.

Case 2: $\kappa_{l_i^*} > 0$

$$
\mathbf{z}_i = \begin{cases} 0, & \text{if } -S_{l_i^*}(\mathbf{G}_i) \leq 0 \text{ and} \\ & \quad \|\mathbf{G}_i - \bar{\mathbf{c}}_t\| \leq \|\mathbf{G}_i - \mathbf{c}_{l_i^*}\| \\ \frac{dh_{2,i}}{d\mathbf{G}_i}, & \text{otherwise} \end{cases} \tag{18}
$$

where $h_{2,i} = \left[(G_{i,x} - \bar{c}_{t,x})^2 + (G_{i,y} - \bar{c}_{t,y})^2\right]^2$. Here, $\bar{\mathbf{c}}_t = [\bar{c}_{t,x}, \bar{c}_{t,y}]^T$ is the centre position of a target group. This gradient brings a robot towards the centre of a target group instead of an arc segment which is outside the Splinegon. Only if a robot is outside a circular sector $(-S_{l_i^*}(\mathbf{G}_i) \leq 0)$ as well as closer to the centre of target group than the one of an arc (i.e. $\|\mathbf{G}_i - \bar{\mathbf{c}}_t\| \leq \|\mathbf{G}_i - \mathbf{c}_{l_i^*}\|$), the gradient is zero.

Case 3: $\|\mathbf{G}_i - \mathbf{G}_t^j\| < R_d$ (for all $j \in \{1, \cdots, n_t\}$, regardless of the sign of $\kappa_{l_i^*}$)

$$
\mathbf{z}_i = \frac{dh_{3,i}}{d\mathbf{G}_i} \tag{19}
$$

where $h_{3,i} = [(1 + \epsilon)R_d^2 - (G_{i,x} - c_{t,x}^j)^2 - (G_{i,y} - c_{t,y}^j)^2]^2$. Here, ϵ is a positive constant. If a robot is inside a certain circular boundary of R_d around the target, the robot is pushed out of that area (which is larger than R_d, $(1+\epsilon)R_d$), in order to avoid any collision with the targets.

3.3 Implementation Issues

In order to implement the proposed algorithm in real robots in a changing environment, there are several issues to be addressed. First, the robots should be able to localise themselves with their own onboard sensors such as encoders or an inertial navigation system. Note that this localisation can be done in a local coordinate system via robot-robot communications by choosing a reference point as the origin. Second, as the total number of robots and targets is unknown to each robot, an appropriate neighbourhood size r_n for the diffusion process in Eq. (8) cannot be pre-defined in practice to ensure the convergence of the system or distribute the robots in the region as evenly as possible. This study adopts a similar method used in [16], which adjusts the initial guess of the distance with two physical constraints of the robot: the bumper range d_{min} and the sensor range d_{max}. If the current neighbour distance d is found to be small, then it is updated with a half-sum of d and d_{max} resulting in a distance increase, whereas if the distance is large, it is updated with d_{min}. Whether the current distance is close to optimal or not is determined by calculating the average number of neighbours after a sufficiently large time elapse for robots to stabilise inside the region for several iterations. Lastly, the effect of communication/sensing noise and delay should be carefully considered as this might lead to performance degradation or even instability of the shape control.

4 Numerical Simulations

Numerical simulations have been performed using scenarios containing either stationary or moving targets to validate the feasibility and benefit of the proposed algorithm. The number of robots used in the simulation is 100. Parameters for the upper layer of the H-GRN are set up as $\theta_1 = 0.25$, $\theta_2 = 0.3, \theta_3 = 1.2$ and $k = 20$, and for the lower layer as $a = 6.5$, $m = 4.2$, $c = 9.9$, $r = 4.3$ and $b = 3.5$. These values are obtained by an evolutionary optimisation run explained in [11] and fine-tuned for stabilisation of robots inside the desired region within a reasonable time. In addition, the maximum speed of the robot is bounded by 0.8 m/s considering the robots' physical capability.

4.1 Forming a Pre-defined Target Region

To entrap a stationary target, a pre-defined simple ring shape is first employed, assuming that the target is at the centre of a circle. Radius is 0.2 and 0.5 m for inner and outer circles, respectively, to define a region. A desired distance between the robots is 0.065 m, and the one between robots and obstacles is 0.3 m. Figure 5 shows snapshots of a situation where 100 robots are randomly distributed in the space, form the target region driven by the H-GRN, and then avoid two moving obstacles. By using a diffusion term $\mathbf{D}$ between robots and obstacles as in Eqs. (7)~(9), the GRN dynamics inherently adapts itself to the environmental changes, i.e., the moving obstacles. The movie clip for this can be downloaded at dl.dropboxusercontent.com/u/17047357/Ring_Obstacle.wmv.

In the following, we consider entrapment and tracking of a single moving target. If the trajectory of the target is known or can be estimated, it can be embedded in the GRN dynamics by modifying Eq. (6) as:

$$\frac{d\mathbf{G}_i}{dt} = -a\mathbf{z}_i + m\mathbf{P}_i + \mathbf{v}_t \qquad (20)$$

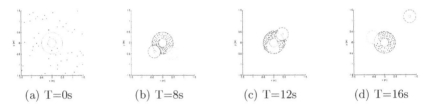

(a) T=0s   (b) T=8s   (c) T=12s   (d) T=16s

Fig. 5. 100 robots (blue points) forming a target shape denoted as a ring and avoiding moving obstacles, which are denoted by a red and a cyan small circle

where $\mathbf{v}_t$ is the velocity of the target to be followed. Figure 6 shows the numerical simulation result on tracking a moving target using a ring shape with or without information about the target movement. The trajectory of the moving target is denoted by a red line, which is specified by $\mathbf{c}_t = [c_x, c_y]^T = 0.1[t \; 3\sin(t/3)]^T$, where t represents time in seconds. It is assumed that each robot is equipped with a sensor that gives the range and direction to the target at 10 Hz. Sensory noise is set to zero-mean Gaussian noise with a standard deviation of $\sigma_r = 0.2$ m for the range and $\sigma_\phi = 2°$ for the direction. To obtain accurate estimates of the target motion, the decentralised extended information filter (DEIF) [17] is applied with a constant velocity target model. If the target velocity is unknown, it appears to be difficult for the swarm of robots to closely follow the moving target while forming the desired ring shape as shown in Fig. 6(a). By contrast, if the target position and velocity can be estimated by the DEIF filter, the robots successfully entraps the moving target, as shown in Fig. 6(b).

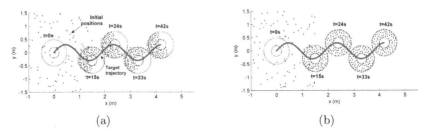

(a)   (b)

Fig. 6. Forming a ring-shape while tracking a target (a) The target position is known but the velocity is unknown. (b) The target position and the velocity are estimated.

4.2   Adaptive Formation of Complex Regions

To entrap multiple targets, the desired region generated by the upper layer of the H-GRN may become very complex and needs to be represented using Splinegon. Figure 7(a) shows a case of multiple stationary targets with an initial guess of a neighbourhood size of 0.55 m, which fails to stabilise the motion inside the region and maintain the given distance between robots. By adjusting the neighbourhood size with $d_{max} = 1.0$ m and $d_{min} = 0.1$ m through iterations of adaptation as shown in Fig. 7(c), the final neighbourhood size of 0.4234 m is obtained resulting in an even distribution of the robots, as shown in Fig. 7(b).

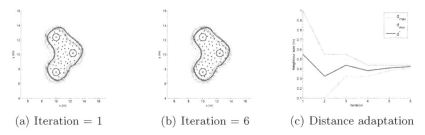

| (a) Iteration = 1 | (b) Iteration = 6 | (c) Distance adaptation |

Fig. 7. Entrapping of multiple static targets with neighbourhood size adaptation. By adjusting the neighbourhood size through iterations using the bumper range d_{min} and the maximum sensor range d_{max}, the final size of 0.4234 m is obtained at iteration 6.

Finally, Figure 8 shows the entrapping of multiple moving targets with a changing complex shape. As the targets move away from each other, the desired shape is dynamically changing from the upper layer of the H-GRN. After 20 seconds, the targets stop moving, and the robots are able to organise themselves into a region that surrounds the targets, as shown in Fig. 8(c). The movie clip can be downloaded at `dl.dropboxusercontent.com/u/17047357/Dynamic_Target.wmv`.

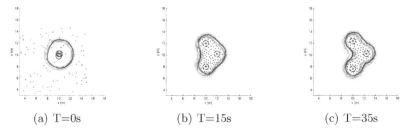

| (a) T=0s | (b) T=15s | (c) T=35s |

Fig. 8. Entrapping of multiple moving targets with a changing complex region. As the targets move away from each other, an entrapping shape is changing accordingly, and consequently the robots are organising themselves inside a shape.

5   Conclusions and Future Work

This paper presented a morphogenetic approach to region formation for entrapping targets using a swarm of robots. By adopting the H-GRN structure, the proposed algorithm has shown to be adaptable to environmental changes resulting from unknown target movements or obstacles. This capability of entrapping stationary or moving targets by forming a region can be applied to a variety of tasks such as contaminant/hazardous material boundary monitoring or isolation and transporting/herding targets into a pre-defined pattern or goal position. As future work, the idea of evolving the H-GRN based on simple network motifs for more flexible and robust pattern generation will be investigated, instead of using a predefined structure as in this paper. Moreover, relaxation of the assumption in the present model that relies on organising robots will be studied for a more realistic distributed self-organising system.

Acknowledgments. This work was funded by the European Commission 7th Framework Program, Project No. 601062, SWARM-ORGAN.

References

1. Balch, T., Arkin, R.: Behavior-based formation control for multirobot teams. IEEE Transactions on Robotics and Automation 14, 926–939 (1998)
2. Alur, R., Das, A., Esposito, J., Fierro, R., Hur, Y., Grudic, G., Kumar, V., Lee, I., Ostrowski, J.P., Pappas, G., Southall, J., Spletzer, J., Taylor, C.J.: A framework and architecture for multirobot coordination. In: Experimental Robotics VII. Lecture Notes in Control and Information Sciences, vol. 271, pp. 303–312 (2001)
3. Hsieh, M., Kumar, V., Chaimowicz, L.: Decentralized controllers for shape generation with robotic swarms. Robotica 26, 691–701 (2008)
4. Olfati-Saber, R.: Flocking for multi-agent dynamic systems: Algorithms and theory. IEEE Transactions on Automatic Control 51, 401–420 (2006)
5. Mamei, M., Vasirani, M., Zambonelli, F.: Experiments of morphogenesis in swarms of simple mobile robots. Applied Artificial Intelligence 18, 903–919 (2004)
6. Yeom, K., Park, J.H.: Artificial morphogenesis for arbitrary shape generation of swarms of multi agents. In: IEEE Fifth International Conference on Bio-Inspired Computing: Theories and Applications (BIC-TA), pp. 509–513 (2010)
7. Shen, W.M., Will, P., Galstyan, A.: Hormone-inspired self-organization and distributed control of robotic swarms. Autonomous Robots 17, 93–105 (2004)
8. Jin, Y., Meng, Y.: Morphogenetic robotics: An emerging new field in developmental robotics. IEEE Transactions on Systems, Man, and Cybernetics, Part C: Applications and Reviews 41, 145–160 (2011)
9. Taylor, T., Ottery, P., Hallam, J.: Pattern formation for multi-robot applications: Robust, self-repairing systems inspired by genetic regulatory networks and cellular self-organisation. Informatics Research Report (2006)
10. Sayama, H.: Robust morphogenesis of robotic swarms. IEEE Computational Intelligence Magazine 5, 43–49 (2010)
11. Jin, Y., Guo, H., Meng, Y.: A hierarchical gene regulatory network for adaptive multirobot pattern formation. IEEE Trans. on Systems, Man, and Cybernetics-Part B: Cybernetics 42, 805–816 (2012)
12. Jong, H.D.: Modeling and simulation of genetic regulatory systems: A literature review. Journal of Computational Biology 9, 67–103 (2002)
13. Jin, Y., Oh, H.: Morphogenetic multi-robot pattern formation using hierarchical gene regulatory networks. In: 1st Workshop on Fundamentals of Collective Adaptive Systems at ECAL (2013)
14. White, B., Tsourdos, A., Ashokaraj, I., Subchan, S., Zbkowski, R.: Contaminant cloud boundary monitoring using network of UAV sensors. IEEE Sensors Journal 8, 1681–1692 (2008)
15. Salazar-Ciudad, I., Jernvall, J., Newman, S.: Mechanisms of pattern formation in development and evolution. Development 130, 2027–2037 (2003)
16. Guo, H., Jin, Y., Meng, Y.: A morphogenetic framework for self-organized multi-robot pattern formation and boundary coverage. ACM Trans. on Autonomous and Adaptive Systems 7, 15:1–15:23 (2012)
17. Mutambara, A.: Decentralized Estimation and Control for Multisensor Systems. CRC Press LLC, Boca Raton (1998)

Communicating Unknown Objects to Robots through Pointing Gestures

Bjarne Großmann, Mikkel Rath Pedersen, Juris Klonovs, Dennis Herzog,
Lazaros Nalpantidis, and Volker Krüger

Robotics, Vision and Machine Intelligence Lab.,
Department of Mechanical and Manufacturing Engineering,
Aalborg University Copenhagen, Copenhagen, Denmark
{bjarne,mrp,juris,deh,lanalpa,vok}@m-tech.aau.dk

Abstract. Delegating tasks from a human to a robot needs an efficient
and easy-to-use communication pipeline between them - especially when
inexperienced users are involved. This work presents a robotic system
that is able to bridge this communication gap by exploiting 3D sensing
for gesture recognition and real-time object segmentation. We visually
extract an unknown object indicated by a human through a pointing
gesture and thereby communicating the object of interest to the robot
which can be used to perform a certain task. The robot uses RGB-D
sensors to observe the human and find the 3D point indicated by the
pointing gesture. This point is used to initialize a fixation-based, fast
object segmentation algorithm, inferring thus the outline of the whole
object. A series of experiments with different objects and pointing gestures show that both the recognition of the gesture, the extraction of the
pointing direction in 3D, and the object segmentation perform robustly.
The discussed system can provide the first step towards more complex
tasks, such as object recognition, grasping or learning by demonstration
with obvious value in both industrial and domestic settings.

Keywords: Human-Robot Interaction (HRI), Pointing Gestures, Object Extraction, Autonomous Mobile Robots.

1 Introduction

Autonomous robotic systems are increasingly integrated into daily life, not only
in industrial but gradually also in household environments. Therefore, more and
more people find themselves in situations where they have to interact with robots
– giving them instructions or requesting their assistance. However, the transition
from a user's abstract task description in his mind to a list of concrete actions
which can be executed by a robot is a big challenge; especially since the user is,
in most cases, not a robot expert.

In this work we present a system that is able to bridge this gap by exploiting
3D sensing for gesture recognition and real-time object segmentation. We have
employed the modular concept of *skills* [9] to merge pointing gesture recognition

M. Mistry et al. (Eds.): TAROS 2014, LNAI 8717, pp. 209–220, 2014.

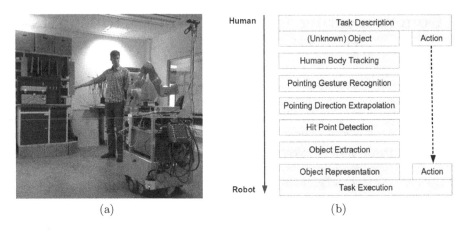

Fig. 1. Our scenario: communicating an unknown object to the robot by pointing (a) and steps for task delegation (b)

and object extraction on a robot, which in turn has allowed us to establish a communication pipeline from an inexperienced user to a robotic system for task execution, as in the scenario depicted in Fig. 1(a).

Skills, as discussed in our previous work [9], provide a modular and easily expandable way to describe tasks in an object-centric manner; the parameter of the skill is the object or location the action is applied on. Furthermore, the pointing gesture provides a simple and natural way for users to communicate the point of interest to the robotic system. A fixation-based visual segmentation step can then extract the actual object from that point of interest, by segmenting it from the environment, as shown in Fig. 2. Finally, we can use the derived object description to automatically fill in the missing parameter, i.e. object shape, orientation or location, for a successful skill parametrization and execution.

The whole sequence of steps that allow the delegation of a task from a user to our considered robotic system is shown in Fig. 1(b). These steps are further discussed in Sec. 3, while in Sec. 4 we experimentally evaluate the key components of this system in terms of robustness, accuracy and precision. The *contribution* of this work is twofold:

- First, we propose a system that can naturally bridge the communication gap between a human and a robot when it comes to the delegation of a task that involves previously unknown objects or locations.
- Then for achieving the first, the skill provides the segmentation process with the necessary affordance and goal (e.g. pick, place) information while the segmentation process provides in return the skill with the necessary affordance parameters.

It is our belief that such a system can constitute the first step towards enabling robots to perform even more complex tasks, such as object recognition, grasping or learning by demonstration, applicable both in industrial and domestic scenarios.

Fig. 2. Extraction of indicated object by pointing gesture

2   Background and Related Work

One of the most intuitive ways of communication is the use of gestures, which has been widely adopted for Human-Robot Interaction (HRI) [4]. Amongst the most basic but also essential gestures is pointing which is naturally used by humans to identify a certain object or location in the environment. The use of pointing gestures to interface with computer systems or robots is not a new concept though [8, 13]. Deriving directions out of pointing gestures has been widely researched and various techniques have been proposed to enhance the precision and accuracy by e.g. extrapolating the line between the user's eye and his fingertips [3] instead of his elbow and wrist, or even by training a sophisticated model of pointing directions using Gaussian Process Regression [1].

However, the accurate interpretation of the pointing gesture is only a part of communicating the object of interest to a robot and the deduced pointing direction is just a means to an end for identifying it. It involves 3D perception of the environment and extracting the corresponding object out of the scene. The robust transition from one single point to the whole object around that point is still an open research topic. Recently, the work by Quintero et al. [11] proposed an interface for selecting objects in a 3D real world situation by computing the hit points of the pointing direction on the target object. Even though they use the hit point for tracking an object with the CAMSHIFT algorithm, information about the object is only given implicitly. But when performing an action on an object, an explicit description of it is needed to execute the skill efficiently. Hence, the next natural step is to extract the target object out of the scene.

Dividing an image of a scene into meaningful partitions is the classical problem of image segmentation [2]. However, generalized automatic image segmentation without prior knowledge is an ill-posed problem, as the correctness of the segmentation strongly depends on the goal or intention of the user [6].

On the other hand, by putting such a goal-oriented segmentation process into a skill-based context, we can turn the skill into an object-centric perceptive skill which uses the computed hit point as the object (or location) to perform the action on. The interest of the user (the hit point) and the skill itself are thereby providing the necessary priors to the segmentation process such that only the region around the point has to be segmented. The benefit is mutual: the skill essentially provides the segmentation process with the necessary affordance and goal information, while the segmentation process provides in return the skill with the necessary affordance parameters.

An approach for extracting the outline of an object using a given point of interest has been recently proposed by Mishra and Aloimonos [5,6] who apply the GraphCut algorithm on a polar-space-transformed image using a given fixation point as the pole. This approach has been used and enhanced in our previous work [7] to create an algorithm for object extraction in near real-time and thereby making it a feasible approach to use in real world robotic scenarios. Instead of using only a few accurate, but complex and time-consuming stages during the algorithm, we implement a longer chain of less accurate but much faster modules. Thereby, the processing time of the whole system has been notably decreased and due to the additional stages, especially the GrabCut extension [12] applied in the end of the process, we are able to obtain results of the same or even higher quality.

3   System Description

We have developed and implemented our system on the mobile manipulator robot Little Helper. This robot is equipped with two RGB-D sensors, in particular the Microsoft Kinect and the Asus Xtion PRO Live cameras, for HRI purposes. Our scenario involves a user pointing towards an object, which is previously unknown to the robot (see Fig. 1(a)). The operation of our system can be coarsely divided into two phases. First, it continuously examines whether a pointing gesture is performed. When a pointing gesture is performed, it captures both a depth and RGB image that contain the point, where the pointing direction intersects with the closest physical object. Then, this point is used to initialize a fixation-based, fast object segmentation algorithm. The final result is the outline of the identified object of interest in that image.

3.1   Gesture Recognition and Fixation Point Generation

Human tracking is based on the continuous data stream from the Asus Xtion RGB-D sensor using the free cross-platform driver OpenNI and the NiTE skeletal tracking library [10]. The Asus Xtion camera is mounted on top of a pole on the robot as illustrated in Fig. 1(a) for a broader view of a scene, enabling upper human body visibility. In our pointing scenario, only two joints were considered for estimating the 3D pointing direction of the user's right forehand, namely his elbow and his wrist joints. A pointing gesture is only recognized when $a)$ a significant angle between a forehand vector and the vertical vector is detected, and $b)$ the forearm is held somewhat steady for $1.5s$. When a pointing gesture is registered, a pointing ray is defined as a 3D vector with origin at the wrist, and direction from the elbow to the wrist joints.

For finding the fixation point, we define a cylinder with a $1cm$ radius and the pointing vector being its center. We use this cylinder to search for intersection points of the pointing direction and the depth readings from the Xtion sensor. The point with minimal Euclidean distance from the intersection points to the location of the wrist is considered as a potential hit point with an object of interest. These hit points are iteratively retrieved and if no major deviations in 3D occur within one second, the 3D coordinates of a potential fixation point are calculated as a mean of the hit points from the last seconds frames.

Throughout this phase, visual feedback is provided to the user through a graphical user interface (GUI) to assist him. The main two windows of the GUI are shown in Fig. 3, where (a) is the tracking view, with real-time depth information from the Xtion camera and (b) is the object view with the depth measurements overlaid with the RGB image from the Kinect sensor. The tracking view shows the skeletal mapping of recognized main parts of the user's upper body. Once the pointing gesture has been detected, the intersection of the pointing direction and the depth readings from the Kinect is marked in the object view, shown in Fig. 3(b) as a red dot, indicating where the user is currently pointing. This information helps the user adjust the pointing direction and more accurately define fixation points on objects of actual interest.

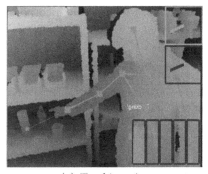

(a) Tracking view   (b) Object view

Fig. 3. Visual feedback provided through the GUI on the robot's front display

When a fixation point is extracted, the Xtion camera is switched off and the Kinect camera is enabled. This camera is mounted on the front part of the mobile robot platform, for a closer view of objects placed in front of the robot. Both RGB-D cameras, which have overlapping views, are not operating simultaneously in order to avoid any interference in the depth readings, as both uses structured infra-red light. Both cameras are extrinsically calibrated in order to accurately register their output with respect to the robot's coordinate system. The potential fixation point from the Asus Xtion camera is then projected on the 3D reconstructed scene as generated by the Kinect sensor. The closest 3D point from the point cloud of the Kinect sensor to the projected fixation point is considered as a final fixation point, which can be used as an input for the object extraction algorithm.

3.2   Object Extraction

In this work, we implement the fixation-based segmentation algorithm mainly based on our previous work [7]. Thereby, we solve the problem of inferring from a single point of interest given by the fixation point generation module to the whole object of interest. This is done by rephrasing the general segmentation problem as a binary labelling problem in log-polar space for a single object. The algorithm is shown in Algorithm 1.

The initialization of the algorithm requires two inputs: the combined RGB-D image of the scene and the mentioned hit point. The first step is the transformation of the input image to the log-polar space with the hit point as a pole. We then detect the edges on the RGB and depth image using an approximated first-order derivative of Gaussian kernel to create a probabilistic boundary map. Moreover, by computing the probabilistic boundary map in log-polar space, we imitate the blurred vision outside of the focus of a human visual system, as the transformation of a circular kernel to Cartesian coordinates results in an increasing kernel size proportional to the distance to the pole [7]. As this leads to higher boundary probabilities closer to the object, we implicitly incorporate a preference towards compact objects.

Algorithm 1. Basic Algorithm for Object Extraction

1: **function** OBJECTEXTRACTION($imageRGB, imageD, fixation$)
2:     $imagePolRGB \leftarrow toLogPolarSpace(imageRGB, fixation)$
3:     $imagePolD \leftarrow logPolarTransform(imageD, fixation)$
4:     $edgeRGB \leftarrow edgeDetect(imagePolRGB)$
5:     $edgeD \leftarrow edgeDetect(imagePolD)$
6:     $boundaryMap \leftarrow \alpha * edgeRGB + (1 - \alpha) * edgeD$
7:     $mask \leftarrow emptyImage$
8:     **for** 1 to 3 **do**
9:         $mask \leftarrow applyGraphCut(boundaryMap, mask)$
10:     **for** 1 to 5 **do**
11:         $mask \leftarrow applyGrabCut(boundaryMap, imagePolRGB, mask)$
12:     $maskCart \leftarrow toCartesianSpace(mask, fixation)$
13:     $object \leftarrow computeOutline(maskCart)$
14: **return** $object$

The probability boundary map is then used as a graph where each node represents a pixel and the weight of the links between those are initialized by the probabilities of the map. Then, a path through the probability boundary map is created by computing the minimum cut of the graph using the GraphCut algorithm. The path is the 'optimal' division of object and background pixels, i.e. the path itself defines the 'optimal' contour of the object in log-polar space in regard to the boundary map. The result of the graph cut algorithm is a binary mask with areas labelled as 'inside' or 'outside' the object.

Afterwards, an additional GrabCut algorithm is applied in order to compensate errors occurred during the edge detection and the graph cut. It uses the binary mask from the graph cut as an initialization and uses additional color information to refine the results. After the transformation back to Cartesian space, the contour of the object can easily be extracted from this mask.

4 Experimental Evaluation

We have performed several experiments with our developed system to assess the operation and the results of its key components. More precisely, we have evaluated the precision and the accuracy of the pointing recognition and object extraction modules under a variety of situations.

4.1 Pointing Recognition

For the pointing recognition, we have compared the detected pointing vector to the position of a known object. In this case, we have used a QR code which 3D pose can be accurately detected by a RGB-D camera. We have then conducted a total of 7 experiments by pointing at the center of the QR code from different angles and distances, and with different users, 50 times each. Both the detection of the QR code and the recognition of pointing gestures have been captured with the same camera in the same position, in order to eliminate errors due to the extrinsic calibration between multiple cameras. We used the combined experiments to measure the general accuracy and precision of the recognized pointing gesture, as shown in Fig. 4.

Fig. 4(a) shows the precision of the pointing gestures given by the relative angular deviation from the mean pointing direction. The histogram shows that the reproducibility for the pointing recognition system is quite high – the average error is less than 3∘ with a maximum error of 10°. Therefore, when pointing at an object at a distance of $1m$ from the wrist, the distance error of the hit point in regard to the object center is about $5cm$. In the general use-case, however, the distance between the user and the object is much smaller.

On the other hand, Fig. 4(b) shows the accuracy of the pointing gestures as the error with respect to the actually intended pointing direction (the center of the pointed object). It shows that the detected pointing directions are strongly biased – the mean error is shifted along the x-axis, suggesting a systematic angular error of 14° or about $24cm$ at a $1m$ distance. As the detection of the pointing direction is directly correlated to the used body tracking framework, it's accuracy and precision of the tracked elbow and wrist will be propagated to the pointing direction recognition. The quality of the skeleton tracking system depends on various parameters, such as the silhouette of the user, his clothing or his pose in relation to the camera. The erroneous pointing directions were not

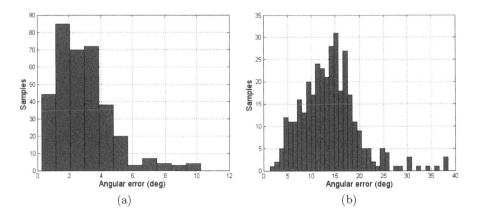

Fig. 4. Relative angular deviation from mean pointing direction (a) and angular deviation from ground truth direction of QR code (b)

centered at the same point relative to the ground truth for all of the 7 experiments, which suggests highly variant tracker performance. Additionally, there is the error introduced by the user himself, as the hand-eye-coordination plays an important role in pointing to the 'correct' spot. Hence, the user's pointing ability has a large effect on the mean angular deviation, as the comparison of different experiments with different users show.

However, as the accuracy of the pointing gestures indicates in Fig. 4(b), the 'blind' pointing gesture is not accurate enough to indicate a target object in the scene. The user need therefore feedback from the robotic system to control and correct his pointing gesture. In this work, we address this problem by directly showing the scene and the hit point in the GUI mounted on the robot. This allows us to easily adjust the pointing direction to the correct position.

4.2   Object Extraction

We have gathered two datasets with segmentation results of some common domestic and industrial objects, which were placed on a shelf in front of the mobile robot platform, in order to test the proposed approach with a variety of objects. We have compared the extracted outline of the different objects, using the hit points previously computed by the pointing recognition, to a manual segmentation of the objects provided by us. Using this comparison, we can compute the precision and recall metrics for each object individually, as seen in Fig. 5(a).

Additionally, we evaluate the performance of the object extraction algorithm by adapting the procedure used in the Pascal VOC competition and computing the ratio of overlapping area to the total area:

$$a = \frac{tp}{tp + fp + fn} \tag{1}$$

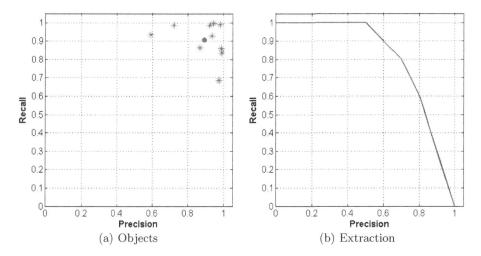

Fig. 5. The precision and recall values for each object, where the green dot represents the mean PR-value (a), and the PR-curve for the algorithm in terms of successful extraction (b)

where tp, fp and fn denoting the *true positives*, *false positives* and *false negatives* respectively. If a exceeds a certain threshold t is considered as a successful object extraction. By varying the threshold, we can derive a precision-recall-curve as shown in Fig. 5(b). With a threshold of $t = 0.5$ as used in Pascal VOC competition, we were able to extract all objects successfully.

Fig.6 shows an example of different object extractions demonstrating the difficulties and capabilities of the proposed algorithm:

a) A potato chips can with an irregular texture pattern which creates additional edges in the edge detection and makes it difficult to establish a proper color model for the Grab Cut algorithm. This leads to an oversegmentation of the object and thereby to a low recall value.
b) A transparent tape with a similar color as the board it is lying on. Due to missing boundary edges between tape and board, the algorithm tends to flood into similar-colored background objects and thereby resulting in a low precision value.
c) A small box with a distinct color and compact shape, but difficult depth map is extracted with a high precision and recall value.
d) A drill with a difficult shape and different colors which is successfully extracted by our algorithm. The extracted shape has almost the same quality as the manually marked outline.

Further qualitative results, some of which are shown in Fig. 7, support that the integrated system is able to provide fairly accurate outlines of most of the objects. This output information is expected to be sufficient for more complex tasks like object recognition, tracking or manipulation tasks, such as grasping.

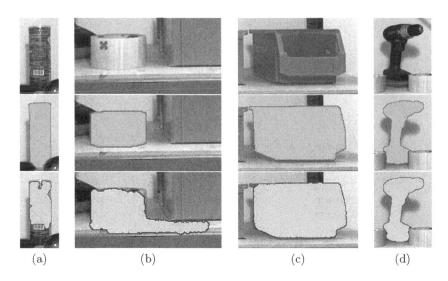

(a)   (b)   (c)   (d)

Fig. 6. The figure shows the extraction of (a) a chips can with the lowest recall, (b) a transparent tape with the lowest precision, (c) a box with the highest F-measure and (d) a drill with a complex shape. The top row shows the original image with the fixation point, the middle row shows the manually segmented ground truth and the bottom row shows the results of the extraction process

Fig. 7. Pointing and segmentation results of various domestic and industrial objects

5   Conclusion

We introduced a robust robotic system capable of identifying unknown objects indicated by humans using pointing gestures. By merging the concepts of object-based actions, pointing gestures and object extraction, we are bridging the communication gap between a non-expert user and a robotic system which is used to execute certain tasks.

In this work, we evaluated the accuracy and precision of the implemented pointing recognition which shows that pointing gestures are a viable way to communicate points of interest to a robotic system. However, due to noise in the human body tracking module itself and the pointing style that differs from person to person, the indicated point of interest is not accurate enough to work without a feedback from the robot. Therefore, we implemented a user interface which verifies the current pointing direction by showing the detected hit point in real-time. Furthermore, we evaluated the object extraction on its own with a given fixation point by the pointing recognition. The evaluation shows that the extraction algorithm provides accurate outlines for most objects in different scenarios which are expected to be feasible for estimating basic dimensions, such as width or height, of fixated objects and can be used to initialize more complex task, such as object recognition, grasping or learning by demonstration with obvious value in both industrial and domestic settings.

We intend to systematically and thoroughly evaluate the operation of the integrated system, in terms of overall robustness, accuracy and precision in realistic conditions, in the near future. Furthermore, we plan to include the extension of the presented system so as to be able to also recognize objects, rather than to just extract their contours. While our currently presented system can be used for delegation of simple tasks from humans to robots, we consider this further upgrade essential for making the system capable of dealing with much more complex tasks. Additionally, due to the modular, skill-based structure of the presented system, the pointing recognition and object extraction submodules could be easily replaced, e.g. by exchanging the pointing gesture recognition by an eye-tracking system applicable in a patient-at-home scenario where the user is potentially not capable of using pointing gestures. Using the presented algorithm as an intermediate step, it is also possible to integrate it into a fully skill-based system as a visual communication tool, thereby providing a smooth transition from the user's intentions of a task to it's final execution by a robot.

Acknowledgment. This work has been supported by the European Commission through the research project "Sustainable and Reliable Robotics for Part Handling in Manufacturing Automation (*STAMINA*)", FP7-ICT-2013-10-610917, "Robotics-enabled logistics and assistive services for the transformable factory of the future (*TAPAS*)", FP7-ICT-260026, and by the Danish Strategic Research Council through the project *Patient@home*.

References

1. Droeschel, D., Stuckler, J., Behnke, S.: Learning to interpret pointing gestures with a time-of-flight camera. In: ACM/IEEE International Conference on Human-Robot Interaction (HRI), pp. 481–488 (2011)
2. Ilea, D.E., Whelan, P.F.: Image segmentation based on the integration of colour-texture descriptors - A review. Pattern Recognition 44(10-11), 2479–2501 (2011)
3. Kehl, R., Van Gool, L.: Real-time pointing gesture recognition for an immersive environment. In: IEEE International Conference on Automatic Face and Gesture Recognition, pp. 577–582 (2004)
4. Krüger, V., Kragic, D., Ude, A., Geib, C.: The meaning of action: A review on action recognition and mapping. Advanced Robotics 21(13), 1473–1501 (2007)
5. Mishra, A., Aloimonos, Y.: Visual segmentation of simple objects for robots. In: Robotics: Science and Systems, Los Angeles, CA, USA (June 2011)
6. Mishra, A., Aloimonos, Y., Fah, C.L.: Active segmentation with fixation. In: IEEE International Conference on Computer Vision, pp. 468–475 (2009)
7. Nalpantidis, L., Großmann, B., Krüger, V.: Fast and accurate unknown object segmentation for robotic systems. In: Bebis, G., et al. (eds.) ISVC 2013, Part II. LNCS, vol. 8034, pp. 318–327. Springer, Heidelberg (2013)
8. Nickel, K., Stiefelhagen, R.: Visual recognition of pointing gestures for human-robot interaction. Image and Vision Computing 25(12), 1875–1884 (2007)
9. Pedersen, M.R., Nalpantidis, L., Bobick, A., Krüger, V.: On the integration of hardware-abstracted robot skills for use in industrial scenarios. In: 2nd International Workshop on Cognitive Robotics Systems: Replicating Human Actions and Activities (2013)
10. PrimeSense Inc.: Prime Sensor NITE Algorithms 1.5 (2011), http://www.primesense.com
11. Quintero, C.P., Fomena, R.T., Shademan, A., Wolleb, N., Dick, T., Jagersand, M.: SEPO: Selecting by pointing as an intuitive human-robot command interface. In: 2013 IEEE International Conference on Robotics and Automation (ICRA), pp. 1166–1171 (2013)
12. Rother, C., Kolmogorov, V., Blake, A.: "GrabCut": Interactive foreground extraction using iterated graph cuts. ACM Transactions on Graphics 23(3), 309–314 (2004)
13. Rusu, R.B., Holzbach, A., Bradski, G., Beetz, M.: Detecting and segmenting objects for mobile manipulation. In: Proceedings of IEEE Workshop on Search in 3D and Video (S3DV), held in conjunction with the 12th IEEE International Conference on Computer Vision (ICCV), Kyoto, Japan, September 27, pp. 47–54 (2009)

A Cost-Effective Automatic 3D Reconstruction Pipeline for Plants Using Multi-view Images

Lu Lou[1,3], Yonghuai Liu[1], Minglan Sheng[3], Jiwan Han[2], and John H. Doonan[2]

[1] Department of Computer Science, Aberystwyth University,
Aberystwyth, UK
[2] College of Information Science and Engineering, Chongqing Jiaotong University,
Chongqing, China
lul1@aber.ac.uk

Abstract. Plant phenotyping involves the measurement, ideally objectively, of characteristics or traits. Traditionally, this is either limited to tedious and sparse manual measurements, often acquired destructively, or coarse image-based 2D measurements. 3D sensing technologies (3D laser scanning, structured light and digital photography) are increasingly incorporated into mass produced consumer goods and have the potential to automate the process, providing a cost-effective alternative to current commercial phenotyping platforms. We evaluate the performance, cost and practicability for plant phenotyping and present a 3D reconstruction method from multi-view images acquired with a domestic quality camera. This method consists of the following steps: (i) image acquisition using a digital camera and turntable; (ii) extraction of local invariant features and matching from overlapping image pairs; (iii) estimation of camera parameters and pose based on Structure from Motion(SFM); and (iv) employment of a patch based multi-view stereo technique to implement a dense 3D point cloud. We conclude that the proposed 3D reconstruction is a promising generalized technique for the non-destructive phenotyping of various plants during their whole growth cycles.

Keywords: Phenotyping, multi-view images, structure from motion, multi-view stereo, 3D reconstruction.

1 Introduction

The phenotype of an organism emerges from the interaction of the genotype with its developmental history and its environment. This means that the range of phenotypes emerging even from a single genotype can be large [16]. The measurement of plant phenotypes, as they change in response to genetic mutation and environmental influences, can be a laborious and expensive process. Phenotyping, therefore, is emerging as the major bottleneck limiting the progress of genetic analysis and genomic prediction.

Recently, several methods have been developed to measure the 3D structure of entire plants non-destructively and then built accurate 3D models for analysis. These methods include laser scanning, digital camera photographing, and

M. Mistry et al. (Eds.): TAROS 2014, LNAI 8717, pp. 221–230, 2014.

structured light ranging. Stereovision was used to reconstruct the 3D surface of maize for measurement and analysis [11]. Kaminumaet et al. [13] applied a laser range finder to reconstruct 3D models that represented the leaves and petioles as polygonal meshes and then quantified morphological traits from these models. Quan proposed a method to interactively create a 3D model of the foliage by combining clustering, image segmentation and polygonal models [19]. Biskup designed a stereo vision system with two cameras to build 3D models of soybean foliage and analyzed the angle of inclination of the leaves and its movement over time [2]. 3D plant analysis based on a mesh processing technique was presented in [17], where the authors created a 3D model of cotton from high-resolution images using a commercial 3D digitization product named 3DSOM. Thiago developed an image-based 3D digitizing method for plant architecture analysis and phenotyping [23], and showed that state of the art SFM and multi-view stereovision are able to produce accurate 3D models, albeit with a few limitations.

The design of plant phenotyping systems means integrating and optimizing a measurement and phenotyping process with complementary processing and analyzing tools and makes it as efficient and controllable as possible. According to [4], the accuracy and precision of the treatment and measurement are fundamental concerns during any experimental procedure because accuracy is not only important when there is variation across individual genotypes during contrasting experiments, but is also critical when individual genotypes have multiple replicates that are evaluated across several batches.

This paper presents a cost-effective automatic 3D reconstruction method of plants from multi-view images, which consists of the following steps: (i) image acquisition using an off-the-shelf digital camera and turn-table; (ii) extraction and matching of local invariant features from overlapping image pairs; (iii) estimation of camera parameters and pose based on SFM; and (iv) employment of a patch based multi-view stereovision technique to implement a dense 3D point cloud.

2   Related Work

3D reconstruction based on multi-view images is also called Structure from Motion (SFM) that computes the camera poses and 3D points from a sequence of images. SFM uses natural features in images to estimate the relative translation and rotation between the camera poses of different images [9].

For the SFM technique, the camera pose is unknown and needs to be estimated from a set of given images. The first step of SFM framework is to employ local invariant feature detection and matching to produce a set of corresponding image points. Features such as interest (key) points are found in all images. One of the most popular features is SIFT, as introduced by [14]. SIFT features are found by searching for maxima in scale-space, constructed with Derivative of Gaussian filters and then a unique SIFT descriptor is constructed to represent each key point based on a histogram of gradient directions of points nearby. SIFT features are scale and rotation invariant and partially intensity and contrast change invariant. Alternative features can be used too, as long as they can be matched uniquely between images, such as Harris [22], SURF [3] and ASIFT [15]. In the second step, the interest points are matched between images using their descriptors. The matching pairs between images allows for estimating the relative transformation from one camera to another by satisfying projective geometry constraints [18][9]. While various features can be used for image matching, We have mainly tested SIFT and SURF for their repeatability and accuracy.

The best performing methods used for multi-view reconstruction of small objects are capable of challenging the accuracy of laser scanners. However, these approaches are generally not suitable for dealing with complex plant architecture. In particular, algorithms that partially rely on shape-from-silhouette techniques [10] [25], under the assumption that a single object is visible and can be segmented from the background, are not applicable to plant images due to unavailability of fully visible silhouettes - a result of occlusions. The patch based multi-view Stereo (PMVS) [8] algorithm proposed by Furukawa and Ponce is usually considered the state of the art amongst methods based on feature extraction and matching. It produces dense point clouds, and performs well for small objects as well as for urban scenes, and can recover significant geometry from such scenes. However, methods based on feature extraction and matching often show significant errors caused by severe occlusions, and areas of texture-less appearance.

3   The Proposed Method

3d sensing technologies provide the tremendous potential to accurately estimate morphological features and have been developed for non-destructive plant phenotyping based on laser scanners, structured light, multi-view stereo, etc.. But available 3D resolutions are currently focussed on specific a organ (e.g. leaves or roots) and usually tend to be qualitative rather than providing quantitative information and estimation of accuracy. Therefore, we have evaluated the existing methods and sought to build an efficient and automatic 3D reconstruction system that could cope with a diversity of plant form and size, while using equipment available to most biology labs. We exploited existing state of the art SFM and MVS methods and took account into both flexibility and practicability and propose a 3D reconstruction pipeline for plants using multi-view images. In the pipeline, we do not rely on the camera calibration to provide us with transformation, pose or distortion. Instead, we compute this information from the images themselves using computer vision techniques. We first detect feature points in each image, then match feature points between pairs of images, and finally run an incremental SFM procedure to recover the camera parameters.

3.1   Image Features Detection and Matching

The short baseline image sequences are captured by placing the plant on a turntable or slightly moving the camera around the plant as shown in Fig. 1(a). The first step is to find feature points in each image. We firstly use the SIFT keypoint detector and descriptor, because of its invariance to image transformations. A typical image contains several thousand SIFT keypoints. Next,for each pair of images, we match keypoint descriptors between the pair using the Approximate Nearest Neighbor (ANN) [5] as shown in Fig. 1(b). We had also tried the ASIFT (Affine-SIFT) detector and descriptor as alternative, The authors in [15] argued that ASIFT was better than such popular feature algorithms as SIFT, MSER and Harris-Affine on various datasets. Although many more features can indeed be detected and matched, it did not improve our experimental results.

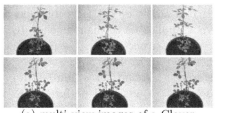

(a) multi-view images of a Clover   (b) Keypoints   detection   and matching

Fig. 1. Image acquisition (left), keypoints detection and matching (right)

3.2   Camera Pose Estimation Using SFM

Given a short baseline image/video sequence I with n frames taken by a turn-table or freely moving camera, we denote $I = \{I_t | t = 1, 2, ...n\}$, where I_t represents the color image captured at the frame t. The set of camera parameters for frame t in an image sequence is denoted as $C_t = \{K_t, R_t, T_t\}$, where K_t is the intrinsic matrix, R_t is the rotation matrix, and T_t is the translation vector. We robustly estimate a fundamental matrix for the image pair using Random Sample Consensus (RANSAC) [6]. During each RANSAC iteration, we compute a candidate fundamental matrix using the five-point-algorithm [7] and remove matches that are outliers to the recovered fundamental matrix. We get K_t and the focal length estimated from the EXIF tags of the .JPG image and finally solve the R_t and T_t. We employ the SFM method of Snavely et al. [21] by the open source software *Bundler*. But we improve *Bundler* in a few key aspects: (i) We sort the images as an ordered sequence according to incremental image name, and thereby reduce the computational cost from $O(n^2)$ to $O(n)$ in matching procedure ; (ii) We speed up the SIFT features detection based on GPU hardware, which is around 10 times faster than CPU; (iii) As alternative methods, we have also tried to use other feature detectors/descriptors to compare with SIFT in SFM, such as SURF [3], DAISY [24] and AKAZE [1].

3.3   Dense 3D Point Clouds Generation

After estimating the cameras' parameters, we employ a MVS algorithm (based on a open source software named PMVS [8]) to demonstrate that SFM and MVS can be used to produce accurate 3D point clouds of plants. PMVS is a multi-view stereo software that takes a set of images and camera parameters, then reconstructs 3D structure of an object or a scene visible in the images. The software takes the output from *Bundler* as input and produces a set of oriented points instead of a mesh model, where both the 3D coordinate and the surface normal are estimated at each oriented point. Compared with the other MVS software named CMPMVS [12], which takes the output of *Bundler* as input and produces a set of 3D meshes, PMVS excels in terms of computational cost and only uses one tenth/twentieth of the running time. However, we have also found that PMVS is inadequate for a complete, accurate and robust 3D reconstruction of typical plants, due to the thin, tiny or texture-less parts of plants.

4   Experimental Results and Discussion

Plant images were mainly captured using Canon digital cameras (Canon 600D) with $18 - 55mm$ focal lens, and Nikon digital cameras (D60 and D700) with $28 - 105mm$ focal lens and a network camera with 50mm focal lens. The images were stored in JPG format with $3184x2120$ or $3696x2448$ resolution. Videos were captured at $1920x1024$ with AVI format. In order to keep even illumination and uniform background, two diffuse lighting rigs and a black backdrop were used. The method was run on an Intel i7 laptop (Dell Precision M6700 with 16G RAM and Nvidia graphics card).

Fig. 2 shows sparse 3D point clouds (keypoints) produced by different feature algorithms in SFM. We found that (i) DAISY and AKAZE can produce many more 3D matching keypoints than SUFR/SIFT, but 3D keypoints produced by DAISY are mainly located in planar regions like leaves of plant or ruler (as reference), and 3D keypoints detected by AKAZE focused on the boundary or edges of plant or ruler ; (ii) DAISY and AKAZE can better estimate the camera's pose than SURF/SIFT in both the case of wide baseline where the rotation angle is bigger (in other words, the number of captured images is small) and the case of texture-less plants.

Table. 1 shows different output results when different features (DAISY, AKAZE, SURF and SIFTGPU) were used for keypoints detection and matching in 54 Brassica images, and dense 3D points produced by PMVS. In this experiment, we found that SIFT(SIFTGPU) has the best accuracy (minimal reprojection error) but the lowest inlier ratio, AKAZE has the best inlier ratio and better accuracy, and DAISY can produce the most keypoints and sparse 3D points but has the lowest accuracy.

To evaluate the precision of the proposed method, we used the commercial 3D modeling software (Artec Eva, 3DSOM and Kinect) to rebuild the same plant. We found that the Artec Eva handheld structured light scanner and Kinect did

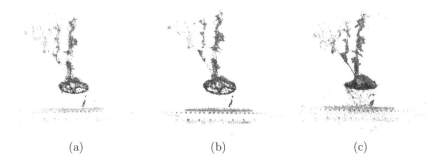

(a)   (b)   (c)

Fig. 2. Sparse 3D point clouds (keypoints) by different feature algorithms in SFM from 75 images of an Arabidopsis. (a) 16340 keypoints by SURF/SIFT. (b) 56784 keypoints by Akaze. (c) 33335 keypoints by DAISY.

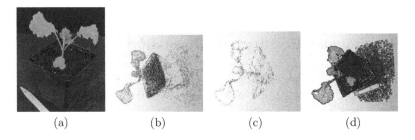

(a)   (b)   (c)   (d)

Fig. 3. (a) Raw image of an plant (one of 54 Brassica images). (b)(c) Sparse 3D point cloud by DAISY and AKAZE respectively. (d) Dense 3D reconstruction point cloud by PMVS.

not work well on plants, especially complex or tiny plants. 3DSOM is able to produce 3D mesh model for the simple plants, but its shortcomings include: (i) the subject plant must be placed on to the centre of a pattern board for calibration; (ii) It needs to be adjusted for some parameters and foreground/background segmentation must be undertaken manually; (iii) It requires more run time than PMVS and (iv) tiny parts of the subject are often omitted in the resulting 3D model. Comparison of results using different methods are shown in Fig. 4.

With the proposed method, no extra camera calibration or other strict environmental conditions are required, and data acquisition times are in the range 1-2 minutes per specimen, depending on plant complexity. We found that rotation steps of 3-6 degree increments are optimal and feature matching does not perform well with rotation steps of more than 9-10 degrees. The camera parameters, such as distance, pose, focus, lens, depth of field and time of exposure, etc., can be adapted widely to obtain high quality and clear images with an optimal resolution from a diverse set of plants. Only basic photography skills are required for effective capture and storage of data.

Table 1. Comparison of different features detection and matching

Item	DAISY	AKAZE	SURF	SIFTGPU
Number of features (mean)	7112	2839	1173	3610
Inlier ratio (mean)	23%	31%	20%	18%
Reprojection Error	0.371	0.256	0.331	0.242
Number of sparse 3D points	15479	3769	2426	3491
Number of dense 3D points	162510	161900	136274	153660

(a)   (b)

(c)   (d)

Fig. 4. 3D reconstruction of Arabidopsis and Maize. (a)(c) 3D Reconstruction by PMVS. (b)(d) 3D Reconstruction by 3DSOM and Kinect respectively.

Table 2. A typical 3D reconstruction running time on CPU and GPU

Item	number of images	time on CPU(minute)	time on GPU(minute)
1	180	91	42
2	120	50	16
3	60	32	6

Generally, the computational run time for 3D reconstruction was 5-30 minutes depending on the number of images. In the following case, the results showed the 3D model produced from 60 images was only slightly lower in quality than one produced from 180 images, the running time and the results of 3D construction were shown as Table. 2 and Fig. 5 respectively.

Fig. 5. 3D reconstruction results from different number of plant images. 71999 3D points were produced from 180 images (left); 68624 3D points were produced from 120 images (middle); 51675 3D points were produced from 60 images (right).

5   Conclusion

We describe an robust, economic and automatic 3D reconstruction pipeline based on multi-view images. Experimental results show that the proposed method is flexible, adaptable and inexpensive, and promising as an generalized groundwork for phenotyping various plant species. This advantage is significant for designing a cost-effective automatic 3D reconstruction framework and is equal to or better than methods presented to date in the literature, where laser scanners, structured light sensors and cameras were specifically established with some strict parameters, or only suitable for a single kind of plant. We have also found the DAISY and AKAZE can be used alternative of SIFT in the keypoints detection and matching.

Limitations of the current method include: (i) the camera pose estimation can fail if the plant leaves are smooth, texture-less or had self-similarity; (ii) computational cost is quite high and (iii) there are still some gaps, holes or noise in the final dense 3D point clouds.

Therefore, future work will focus on improving existing methods, combining the other techniques, such as employing multiple cameras to reduce further the occlusions, or using *apriori* knowledge (leaf symmetry, etc.) to improve the robustness of camera estimation as well as accuracy and completeness of final 3D reconstruction.

Acknowledgements. We acknowledge funding from the European Union FP7 Capacities Programme (Grant Agreement No. 284443: European Plant Phenotyping Network, an Integrating Activity Research Infrastructure project) and BBSRC NCG Grant Ref: BB/J004464/1.

References

1. Alcantarilla, P.F., Nuevo, J., Bartoli, A.: Fast explicit diffusion for accelerated features in nonlinear scale spaces. In: British Machine Vision Conf., BMVC (2013)
2. Biskup, B., Scharr, H., Rascher, U.S.: A stereo imaging system for measuring structural parameters of plant canopies. Plant, Cell & Environment 30 (2007)
3. Bay, H., Ess, A., Tuytelaars, T., Gool, L.V.: Speeded-up robust features (surf). Comput. Vis. Image Underst. 110, 346–359 (2008)
4. Cobb, J.N., DeClerck, G., Greenberg, A., Clark, R., McCouch, S.: Next-generation phenotyping: requirements and strategies for enhancing our understanding of genotype-phenotype relationships and its relevance to crop improvement. Theoretical and Applied Genetics 126, 867–887 (2013)
5. Arya, S., Mount, D.M., Netanyahu, N.S., Silverman, R., Andwu, A.Y.: An optimal algorithm for approximate nearest neighbor searching fixed dimensions. J. of the ACM 45(6), 891–923 (1998)
6. Fischler, M.A., Bolles, R.C.: Random sample consensus: A paradigm for model fitting with applications to image analysis and automated cartography. Commun. ACM 24(6), 381–395 (1981)
7. Stewenius, H., Engels, C., Nister, D.: Recent developments on direct relative orientation. ISPRS Journal of Photogrammetry and Remote Sensing 60, 284–294 (2006)
8. Furukawa, Y., Ponce, J.: Accurate, dense, and robust multiview stereopsis. IEEE Tran. PAMI 32, 1362–1376 (2010)
9. Hartley, R.I., Zisserman, A.: Multiple View Geometry in Computer Vision. Cambridge University Press (2004)
10. Hern, C., Esteban, N., Schmitt, F.: Silhouette and stereo fusion for 3d object modeling. Comput. Vis. Image Underst. 96, 367–392 (2004)
11. Ivanov, N., et al.: Computer stereo plotting for 3-d reconstruction of a maize canopy. Agricultural and Forest Meteorology 75, 85–102 (1995)
12. Jancosek, M., Pajdla, T.: Multi-view reconstruction preserving weakly-supported surfaces. In: Proc. CVPR, pp. 3121–3128 (2011)
13. Kaminuma, E., Heida, N., Tsumoto, Y., Yamamoto, N., Goto, N., Okamoto, N., Konagaya, A., Matsui, M., Toyoda, T.: Automatic quantification of morphological traits via three-dimensional measurement of arabidopsis. Plant 38, 358–365 (2004)

14. Lowe, D.G.: Distinctive image features from scale-invariant keypoints. International Journal of Computer Vision 60, 91–110 (2004)
15. Morel, J.M., Yu, G.: Asift: A new framework for fully affine invariant image comparison. SIAM J. Img. Sci. 2(2), 438–469 (2009)
16. Organisation, E.P.S.: White paper of plant phenotyping. Tech. rep. (2010)
17. Paproki, A., Sirault, X., Berry, S., Furbank, R., Fripp, J.: A novel mesh processing based technique for 3d plant analysis. BMC Plant Biology 12 (2012)
18. Prince, S.J.D.: Computer Vision: Models, Learning, and Inference. Cambridge University Press (2012)
19. Quan, L., Tan, P., Zeng, G., Yuan, L., Wang, J.D., Kang, S.B.: Image-based plant modeling. ACM Trans. Graphics 25, 599–604 (2006)
20. Seitz, S.M., Curless, B., Diebel, J., Scharstein, D., Szeliski, R.: A comparison and evaluation of multi-view stereo reconstruction algorithms. In: Proc. CVPR, vol. 1, pp. 519–528 (2006)
21. Snavely, N., Seitz, S.M., Szeliski, R.: Photo tourism: Exploring photo collections in 3d. In: Proc. SIGGRAPH, pp. 835–846 (2006)
22. Stephens, C.H.: Mike: A combined corner and edge detector. In: Proc. of Fourth Alvey Vision Conference, pp. 5–10 (1988)
23. Santos, T., Oliveira, A.: Image-based 3d digitizing for plant architecture analysis and phenotyping. In: Proc. Workshop on Industry Applications in SIB-GRAPI (2012)
24. Tola, E., Lepetit, V., Fua, P.: Daisy: An efficient dense descriptor applied to wide-baseline stereo. IEEE Tran. PAMI 32(5), 815–830 (2010)
25. Vogiatzis, G., Hern, C., Esteban, N., Torr, P.H.S., Cipolla, R.: Multiview stereo via volumetric graph-cuts and occlusion robust photo-consistency. IEEE Trans. Pattern Anal. Mach. Intell. 29, 2241–2246 (2007)

Improving the Generation of Rapidly Exploring Randomised Trees (RRTs) in Large Scale Virtual Environments Using Trails

Katrina Samperi[1] and Nick Hawes[2,*]

[1] Aston Lab for Intelligent Collectives Engineering (ALICE), Aston Institute for Systems Analytics, Aston University, Birmingham, UK
k.samperi@aston.ac.uk

[2] Intelligent Robotics Lab, School of Computer Science, University of Birmingham, Birmingham, UK
n.a.hawes@bham.ac.uk

Abstract. Rapidly exploring randomised trees (RRTs) are a useful tool generating maps for use by agents to navigate. A disadvantage to using RRTs is the length of time required to generate the map. In large scale environments, or those with narrow corridors, the time needed to create the map can be prohibitive. This paper explores a new method for improving the generation of RRTs in large scale environments. We look at using trails as a new source of information for the agent's map building process. Trails are a set of observations of how other agents, human or AI, have navigated an environment. We evaluate RRT performance in two types of virtual environment, the first generated to cover a variety of scenarios an agent may face when building maps, the second is a set of 'real' virtual environments based in Second Life. By including trails we can improve the RRT generation step in most environments, allowing the RRT to be used to successfully plan routes using fewer points and reducing the length of the overall route.

1 Introduction

Large scale virtual environments are becoming more complex over time. There is a challenge to provide an intelligent agent capable of autonomously building a map of large scale, populated and dynamic virtual worlds and then use these maps for navigating between places. Large persistent online worlds such as Second Life [14] are constantly changing and so providing an agent with a map in advance is not possible. The time taken to generate a map in this situation is important. If the agent spends too much time generating a map, then the server may consider the agent to be idle and so disconnect it from the world. The environment is also able to be changed at any point, with vast sweeping changes requiring little effort when compared to the real world. For this reason we are interested in agents that are able to generate a map of their environment in a very short period of time before using these to navigate a walking route between points.

* The research leading to these results has received funding from the European Community's Seventh Framework Programme [FP7/2007-2013] under grant agreement No. 600623, STRANDS.

M. Mistry et al. (Eds.): TAROS 2014, LNAI 8717, pp. 231–242, 2014.

Rapidly exploring randomised trees [13] and other roadmap representations are often used to try and generate a graph of interconnected points describing the configuration of free space in an environment. RRTs rely on the fast and random growth of a search tree that spans the environment from a given start point. While this random growth allows the tree to explore the entire environment given time, the length of time required to connect two points may not be ideal. RRTs also can stagnate when faced with highly cluttered environments or those with narrow corridors where selecting the correct set of points in order to discover a path between obstacles is important.

One factor that is often not considered when looking at map generation and use in virtual environments is that the environment can be populated. We can observe the movement of other avatars in a virtual environment more easily than in the real world, so how other people use the environment and the paths they take can be a useful source of information for the agent's map building process. These observations are known as 'trails'. Trails are a list of points an avatar has been observed at over time. They have previously been used in architecture [16], path finding for people in the real world [7,18] and to help agents navigate large real and virtual environments [1,21,25].

The contribution of this paper is a study into whether trails can be used as a useful source for RRT generation. This paper builds on our previous research on the effect of trails on probabilistic roadmaps [21]. In this study it was found that by using trails the agent could improve probabilistic roadmap generation by reducing the time required and the length of a planned route. Trails have not previously been used to generate RRTs and our hypothesis is that we can improve the generation of these roadmaps in a variety of environments. We investigate trails in both a group of hand generated environments, which cover a variety of situations an agent may face when building a map, and also in a set of online, Second Life based 'real' virtual environments.

The rest of this paper is structured as follows: We first look at related work in this area in Section 2. This covers the three RRT generation methods we will be using and trails in more detail. Following this, Section 3 looks at the preliminary experiments we performed in a set of generated environments. Section 4 details the experiment setup and results when we look at a set of Second Life based, 'real' virtual environments. Finally, Section 5 discusses our findings.

2   Related Work

Roadmaps are a type of map representation that reduces an environment down to a set of nodes, usually describing free space, and arcs, describing how to move between points. Probabilistic roadmaps [10] and rapidly exploring randomised trees [13] are two of the more popular types of roadmap representation. The difference between these maps is how they are generated and what they are typically used for. Probabilistic roadmaps are used for multi-query maps [10]. These are maps which will be used more than once to plan routes between points in the environment they were generated in whereas RRTs are generally single query maps, but are faster to generate [12].

The RRT algorithm generates a roadmap by simulating the growth of a tree [13]. The map starts from a single *root* node and grows outwards, adding branches to the existing structure as time goes on. A brief outline of the algorithm can be seen in Algorithm

1 [13]. Branches are added by selecting points over the entire environment, but these are not directly added to the map. The closest point in the tree is found and a branch is grown from the existing tree towards the selected point for a given distance. The selected point is then discarded. This gives a fully connected map, in which a path can always be planned between any two nodes in the tree.

Algorithm 1. The basic RRT generation algorithm

---

1. **Require:** Start point (s), End point (e), $list of points$, $list of branches$
2. Add (s) to list of points
3. **while** list of points does not contain a point close to (e) **do**
4.   Select a random point (p)
5.   Find the nearest neighbouring point (n) in the tree to (p)
6.   Calculate the point (p') a given distance from (n) towards (p).
7.   **if** n is free AND clearPath(n, p') **then**
8.   add (p') to the list of points
9.   add (n, p') to the list of branches
10.   **end if**
11. **end while**

We look at three varieties of RRT generation algorithm in this paper, the standard RRT [13], the greedy RRT [11] and the RRT Connect algorithm [11]. The standard RRT grows branches a fixed distance. This distance is set in advance. The greedy RRT differs on line six of the algorithm, by allowing branches to continue growing until they either reach the selected point, or an obstacle in the environment [11]. This has the advantage of only needing a small number of branches in clear or sparsely cluttered environments, but requires more collision checks along the length of the branch to check it is valid. Collision checking is one of the more costly operations in roadmap generation, so keeping the number of checks as low as possible speeds up map generation. The RRT connect algorithm also grows branches a set distance, but this algorithm grows two trees at once, one from the root node and one from the end point. The two trees grow towards the centre of the environment until their branches can be connected. This is considered to be one of the better RRT approaches as it tends to be faster to generate and find a valid path between points [11].

The disadvantage RRTs have is the generation time in large scale environments. RRTs have issues selecting the correct set of points in an environment which allow the tree to grow around obstacles in narrow corridors or highly cluttered environments [17].

The way points are selected affects the growth of the tree. The standard approach to point selection is to use random points [13]. This has the advantage of being fast and easy to implement, but random points do not always give an even spread over an environment. An alternative to this, while still keeping point selection random is to use a quasi-random number generator, such as the Halton sequence [4,6]. This method ensures that while the points are selected unpredictably, they cover the entire space of available numbers more evenly, usually giving a better roadmap [4,3]

It is our hypothesis that by biasing tree growth towards certain areas of the environment we will be able to faster generate an RRT that is able to find a short path between the root node and a given end point. To do this we can use *trails* to bias the point selection process.

Trails are the traces left as an agent (human, animal or AI) navigates between locations in an environment. Trails have been used by people for centuries to learn how to navigate new environments, find specific places and when hunting food. They are used to track the movement of people in history [2]. A digital trail is a set of locations an avatar has been observed at over time.

Trails have been used to observe and aid human navigation in virtual and real environments [19,8,9,15,23], as the starting point for a robot's map building process [25,1,21] and used to study motivation and behaviour of people in virtual spaces [24,1,22]. By observing the movement of avatars researchers have been able to study and identify potential busy locations in an environment[15] and where bottlenecks may occur [5].

3   Preliminary Experiments

We initially wanted to investigate whether trails would have a positive effect on RRT generation in any type of environment. To do this we generated a set of six environments, each encapsulating a different set of requirements from the world. Images of the generated environments can be seen in Figure 1. These are based on the examples used by Geraerts and Overmars in their paper [6] and cover some of the standard problems an agent may face with open areas and narrow passages in an environment.

As these environments are not populated by human controlled avatars we also needed to generate a set of trails for the agent to use. A trail in a populated virtual environment would be gathered by observing the movements of an avatar over time and recording their position. This gives a list of points and the order they were visited in. In the generated environments we simulated a trail by finding the shortest path between the bottom

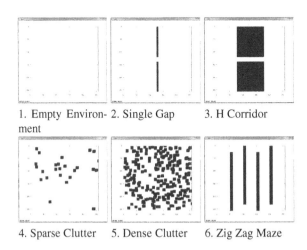

1. Empty Environment   2. Single Gap   3. H Corridor

4. Sparse Clutter   5. Dense Clutter   6. Zig Zag Maze

Fig. 1. The six test environments created. Areas in white are free space, areas in blue were blocked by objects.

left and top right corner in the environment and representing it as a set of points in a specific order. This is the same path that the agent is asked to plan in the experiment.

The agent was given the full set of objects that made up the world and the trails in advance. The task was to generate an RRT from a given start point in the environment. The RRT generation completed when either a maximum number of points had been selected and used to grown branches, or a branch end was added to the tree that was close enough to the given end point that would allow the two to be connected and a full route planned between the start and end points. The three RRT algorithms we looked at were the standard RRT algorithm, the greedy RRT algorithm and the RRT Connect algorithm. The maximum branch length was set at 10m for the standard and RRT Connect approaches. The greedy algorithm continued to grow branches until an obstacle was reached in the environment.

To examine whether trails had a positive effect on the tree growth we compared four different methods of point selection, random, Halton, trail bias and clustered trail points. This changes Algorithm 1 on line four. Rather than always selecting a point at random we select a point according to some particular point selection method. The standard approach is to use random point selection to generate the tree. Random points do not always give a good spread of points over the environment so we also wanted to look at a more even approach. Points selected according to the Halton sequence give an unpredictable selection of points over the entire environment.

We looked at two methods of using trails in point selection. The first biased point selection based on the presence of trails. A point was selected at random in the environment and the probability of this point being used to grow a branch in the tree was 10% plus the number of trail points within a 10m radius. There was always a 10% chance of keeping a point with no trail points nearby to ensure that the algorithm did not stagnate. An outline of the trail bias point selection algorithm can be seen in Algorithm 2.

Algorithm 2. Trail Bias Point Selection Algorithm

---

1. **Function:** Get Trail Bias Point (**Requires:** list of trail points)
2. **while** Not returned a point **do**
3. $p \leftarrow randompoint$
4. $neighbouringTrailPoints \leftarrow 0$
5.   **for all** trail point $t \in listoftrailpoints$ **do**
6.   **if** distance between t and p <10 **then**
7. $neighbouringTrailPoints$ ++
8.   **end if**
9.   **end for**
10. $probabilityOfKeepingPoint \leftarrow 10 + (10 * neighbouringTrailPoints)$
11.   **if** $probabilityOfKeepingPoint < randomNumber(0 \leq x \leq 100)$ **then**
12.   **return** p
13.   **end if**
14. **end while**

The second method clustered trail points together and used these as the points selected in the algorithm. There is a lot of noise present in trails in a real environment.

The movement of people is not always in perfect straight lines between points. For this reason we looked at previous work done on using trails for map building, and looked at clustering trails together. Each trail point was iterated over and if there was a neighbour within a short distance (<5m) the points were merged together. This is a form of the k-nearest-neighbour algorithm for clustering points [20]. The algorithm for clustering the trail points can be seen in Algorithm 3. The result of clustering is a skeletonised version of the trail map. These points were used directly as the points selected in line four of the RRT generation algorithm. Once the clustered trail points were exhausted, Halton points were used. Using only the clustered points would again lead to stagnation of the roadmap. Some randomness is required in all the point selection methods [21].

Algorithm 3. Trail point clustering algorithm

1. **Function:** Get Clustered Trail Points (**Requires:** list of trail points)
2. $listOfClusteredPoints \leftarrow 0$
3. **for all** trail point $t \in listoftrailpoints$ **do**
4. $listOfNeighbours \leftarrow$ get the list of all trail points within 5m radius of t
5.   **if** $listOfNeighbours.size() > threshold$ **then**
6. $center \leftarrow$ get the center point of the t and its neighbours
7. $listOfClusteredPoints.add(center)$
8.   **else**
9. $listOfClusteredPoints.add(t)$
10.   **end if**
11.   remove t from list of trail points
12. **end for**
13. **return** $listOfClusteredPoints$

The environments we looked at in the preliminary set of experiments were all 256m². This is the size that corresponds with a single Second Life region used in the 'real' virtual environments. The experiments were run 20 times for each combination of point selection method and RRT algorithm and the results looked at the average performance over these runs. The computer used to run the experiments was an Intel(R) Core(TM) i7-2600 CPU @ 3.40GHz running Windows 7 Enterprise edition.

3.1 Evaluation

If an RRT is given an endless period of time it will eventually find a route between any two points in the environment where one is available. We limited the time available to the agent for generating a map and planning the route to ten minutes. The environment the agent is inhabiting is online, and so speed when generating the map is one of the most important factors. Any map which had not found a route within ten minutes was considered unsuccessful. The maximum number of points the agent was able to attempt to include in the roadmap was set at 100,000. This was sufficiently large that the number of points should not be a limiting factor in the environment, but the algorithm would still have an exit point.

We wanted to compare the difference made by changing the point selection step in three RRT algorithms. To evaluate each combination of selection method and algorithm we looked at three factors: 1) The *success rate* of the algorithm at finding a route between the given start and end point. 2) The *number of points* in the roadmap, and 3) the *length* of the planned route. In an ideal map the success rate will be 100%, the number of points in the roadmap will be low and the length of the planned route short.

3.2   Results for Preliminary Experiments

We found that in general the best results were found using trail points in the RRT generation. The success rate for all the maps in these environments was 100%. A summary of the best results for each of the environments is below, in Table 1.

Table 1. Best results for each environment in the preliminary set. Bold text indicates the best point selection approach across all three RRT algorithms. Points is the point selection method used, length is the shortest planned route.

| | Normal | | Greedy | | RRT Connect | |
	Points	Length	Points	Length	Points	Length
Empty	Clustering	Clustering	Any	Any	**Clustering**	**Clustering**
Single Gap	Clustering	Clustering	Any	Any	**Clustering**	**Clustering**
H Corridor	Clustering	**Clustering**	Halton	Clustering	**Bias**	Clustering
Sparse Clutter	Clustering	Clustering	Random	Halton	**Clustering**	**Clustering**
Dense Clutter	Bias	Halton	Clustering	Clustering	**Halton**	**Bias**
Zig Zag Maze	Clustering	**Bias**	Halton	Random	**Halton**	Halton

Standard RRT Algorithm: Selecting points from the clustered trail set performed well in the Empty, Single Gap, H Corridor and Sparse Clutter environments. These maps used the least points and planned the shortest routes. In the dense clutter environment it was better to use biased trail point selection for the least number of points, or Halton points for planning the shortest distance. In the maze environment it was better to use clustered trail points for fewer points in the roadmap and biased trail points to plan the shortest paths.

Greedy RRT Algortithm: The greedy algorithm attempted to connect the start and end point in the roadmap directly as the first step. For this reason there is no difference made in the empty or single gap environment for this algorithm. A straight line path is possible and so this is the one found for all point selection methods. In the H corridor environment trails found the shortest path. Clustered trails performed the best in the dense clutter environment. However, in the maze environment it was better to use either Halton or random points to generate the map.

RRT Connect: In the empty, single gap and sparse clutter environments clustered trails out performed all other methods. In the H corridor environment it was better to use trail biased points to achieve the least number of points in the map. The shortest path still used clustered trail points. The dense clutter and zig zag maze have different results.

The shortest planned route used trail biased points in the dense clutter environment. However, in the maze environment using Halton points was better for both the least number of points and the shortest planned path.

3.3 Discussion

The dense clutter and maze environments are difficult for an RRT to plan routes in. In the dense clutter there are obstacles randomly placed around the environment forming many narrow passages to be traversed. In the maze environment the route has to change direction on itself constantly. Trail points only show where other avatars have been seen, so if they are relied on too heavily then the algorithms stagnate.

We looked at 18 different combinations of RRT algorithm and point selection method. In 13 of these cases it was better to use one of the two trail based point selection approaches to build the tree to ensure the least number of points are required. In 14 cases trail based point selection allowed for the shortest path to be generated. When comparing the different RRT generation algorithms we find that the RRT Connect algorithm is in the majority of cases the best. The maze environment was the only case where the shortest path was generated using the standard RRT algorithm. The least number of points still used the RRT connect algorithm. This validates that the RRT Connect algorithm is the best algorithm to use when generating trees in virtual environments.

4   The 'Real', Second Life Based, Virtual Environments

We then looked at four environments, each based on a single *region* in Second Life 256m^3 in size. The trails for these experiments were gathered by observing avatars in Second Life over a protracted period of time. The position of each avatar was recorded every half a second to build up a set of trails specific to each environment. We only considered walking agents in these experiments so objects and trails observed above ground level were filtered out of our evaluation.

The agent was given a full set of objects and trails that made up these environments in advance. These were gathered previously by having an agent sat in the world recording every object and avatar present. The number of trails and objects available in each environment therefore differs. Table 2 shows the difference between the number of objects and trails available in each environment, and Figure 2 shows the collision map for each environment. Again, the areas in blue on these maps are where our collision detector found obstacles. Areas in white are free space.

We looked at the same RRT algorithms and point selection methods as for the preliminary set of experiments, and the same evaluation criteria. The main difference between the preliminary and the Second Life based experiments is the number of obstacles in the environment and the number of trail points available. The preliminary environments had a very limited set of trails whereas trails observed from various real avatars may not always be as useful.

4.1 Results

The results in the Second Life based environments show that trail points can be useful for RRT generation, but not as much as in the preliminary environments. In the Second

Table 2. The total number of objects and trails in each environment:

Environment	Objects	Trail Information	
		Avatars	Total Trail Points
London Community	1476	23	921
Hyde Park	7096	4274	402036
Kensington	11770	61	2730
Knightsbridge	10201	183	10486
Mayfair	11765	1583	84475

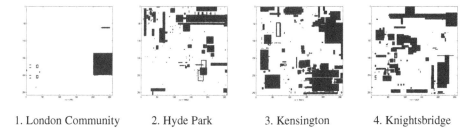

1. London Community   2. Hyde Park   3. Kensington   4. Knightsbridge

Fig. 2. The four Second Life based virtual environments. Areas in white are free space, areas in blue were blocked by objects.

Life based environments there are a lot more obstacles to be navigated around and the trails are not always as helpful. A summary of the results can be found in Table 3

Table 3. Best results for each environment in the Second Life virtual environments. Bold text indicates the best point selection approach across all three RRT algorithms. Points is the point selection method used, length is the shortest planned route.

	Standard		Greedy		RRT Connect	
	Points	Length	Points	Length	Points	Length
Hyde Park	Halton	Halton	**Halton**	**Clustering**	Random	Random
London Community	Random	Clustering	**Any**	**Any**	Bias	Halton
Kensington	Clustering	Bias	Random	Clustering	**Clustering**	**Clustering**
Knightsbridge	None	None	**Bias**	Bias	Bias	**Random**

The success rate for these maps was not as high as in the preliminary environments. The four environments are very different to one another. In the Hyde Park environment there are 57 trail points for each object in the environment. In Knightsbridge there is 1 and in Mayfair there are 7 trail points for each object. The difference in information for each of the environments changed the results. The success rates can be seen in Table 4.

It is much harder to extract general rules for each type of RRT generation algorithm as the environment properties have a much greater effect on the map generated. In the

Table 4. The success rates for each of point selection methods for each RRT generation algorithm

Hyde Park	Normal	Greedy	RRT Connect	Kensington	Normal	Greedy	RRT Connect
Random	100%	100%	100%	Random	100%	100%	100%
Halton	100%	100%	100%	Halton	100%	100%	100%
Bias	30%	100%	90%	Bias	100%	100%	100%
Clustering	60%	100%	100%	Clustering	100%	100%	100%
London Comm.				Knightsbridge			
Random	100%	100%	100%	Random	0%	100%	15%
Halton	100%	100%	100%	Halton	0%	100%	40%
Bias	100%	100%	100%	Bias	0%	100%	10%
Clustering	100%	100%	100%	Clustering	0%	100%	0%

preliminary environments the number of objects was always much lower than in the Second Life based environments.

Standard Algorithm: There is no general rule for which point selection method is best when using the standard algorithm. In each environment we get a different result for which of the point selection methods performed best in each situation. In the Knightsbridge environment none of the point selection methods were able to generate a map capable of planning the required route. The maps generated in the London Community and Kensington environments planned the shortest routes when using trails.

Greedy Algorithm: The greedy algorithm generally planned the shortest routes when using trails. It is also the only RRT generation algorithm that achieved a 100% success rate across all four environments. This is an unusual result as the RRT Connect algorithm is generally thought of as a better approach to RRT generation.

RRT Connect Algorithm: The RRT connect algorithm did not perform as well in the Second Life based environment when compared with the preliminary environments. In some cases, using trail points improved the performance of these maps using either the bias or clustering method. There is no clear preference for which method to use.

Unlike the preliminary environments, the Second List based environments are all vastly different. Because of this it was harder to extract a single rule of thumb for the best point selection method to use based on the algorithm used. In the simpler environment of London Community we could use any point selection method and a greedy approach to generating the RRT. This was able to plan a route directly from one side of the environment to another. As the environment becomes more complex we find the agent struggling to plan any routes at all, such as in Knightsbridge using the standard RRT algorithm. In this more complex environment we find that biasing trail points towards areas where trails are help the RRT generation.

5   Discussion

The preliminary experiments showed that in a variety of situations trails can be used to improve the generation of RRTs in virtual environments. It differs for each environment whether it is better to use the clustered trails or to bias randomly selected points to areas where trails have been observed.

By using trails we the RRTs generally require less points to complete a path between a given start and end point. The route planned is shorter than the route planned when using randomly or Halton based points. This is due to the human optimisation effect of trails, people tend to use shorter paths when navigating between points.

When we applied this to the Second Life, online populated environments we saw a much more mixed picture. These environments were much more complex than our preliminary set. In one environment, Knightsbridge, it was difficult for any RRT generation method other than the greedy RRT to generate a map capable of planning a route between the given start and end points. In general trails were still helpful for the RRT generation step as they tended to allow for planning shorter routes in the environment.

There is more work to be done in this area, we would like to use a combination of different point selection methods when generating RRTs to see if this has a better effect than just using one method. A combination of trail and Halton points has been shown to be better than using just one method when generating probabilistic roadmaps [21] and an investigation into whether this is also the case for RRTs will be interesting. Following on from our discovery that the greedy RRT algorithm was the only one to achieve a 100% success rate at being able to connect the start and end point in the Second Life environments, further investigation into whether this is an isolated incident, or if a greedy form of the RRT Connect algorithm is better in large scale virtual worlds.

References

1. Alempijevic, A., Fitch, R., Kirchner, N.: Bootstrapping navigation and path planning using human positional traces. In: 2013 IEEE International Conference on Robotics and Automation (ICRA), pp. 1242–1247. IEEE (2013)
2. Bradley, R.: The Prehistory of Britain and Ireland. Cambridge University Press (2007)
3. Branicky, M.S., LaValle, S.M., Olson, K., Yang, L.: Deterministic vs. probabilistic roadmaps. Submitted to the IEEE Transactions on Robotics and Automation (2002)
4. Branicky, M.S., LaValle, S.M., Olson, K., Yang, L.: Quasi-randomized path planning. In: IEEE International Conference on Robotics and Automation, ICRA 2001, vol. 2, pp. 1481–1487 (2001)
5. Chittaro, L., Ranon, R., Ieronutti, L.: Vu-flow: A visualization tool for analyzing navigation in virtual environments. IEEE Transactions on Visualization and Computer Graphics 12(6), 1475–1485 (2006)
6. Geraerts, R., Overmars, M.H.: A comparative study of probabilistic roadmap planners. In: Workshop on the Algorithmic Foundations of Robotics (2002)
7. Grammenos, D., Filou, M., Papadakos, P., Stephanidis, C.: Virtual prints: leaving trails in virtual environments. In: Proceedings of the Workshop on Virtual Environments 2002, p. 131. Eurographics Association (2002)
8. Grammenos, D., Mourouzis, A., Stephanidis, C.: Virtual prints: Augmenting virtual environments with interactive personal marks. International Journal of Human-Computer Studies 64(3), 221–239 (2006)
9. Jiang, J.-R., Huang, C.-C., Tsai, C.-H.: Avatar path clustering in networked virtual environments. In: ICPADS, pp. 845–850. IEEE (2010)

10. Kavraki, L.E., Svestka, P., Latombe, J.-C., Overmars, M.H.: Probabilistic roadmaps for path planning in high-dimensional configuration spaces. IEEE Transactions on Robotics and Automation 12(4), 566–580 (1996)

11. Kuffner Jr, J.J., LaValle, S.M.: Rrt-connect: An efficient approach to single-query path planning. In: Proceedings of the IEEE International Conference on Robotics and Automation, ICRA 2000, vol. 2, pp. 995–1001. IEEE (2000)

12. Kumar, S., Chakravorty, S.: Adaptive sampling for generalized probabilistic roadmaps. Journal of Control Theory and Applications 10(1), 1–10 (2012)

13. LaValle, S.M.: Rapidly-exploring random trees: A new tool for path planning, Technical report, Iowa State University (1998)

14. Linden Research Inc. Second life official site (2012), http://secondlife.com (last accessed January 8, 2014)

15. Miller, J.L., Crowcroft, J.: Avatar movement in world of warcraft battlegrounds. In: Proceedings of the 8th Annual Workshop on Network and Systems Support for Games, p. 1. IEEE Press (2009)

16. Myhill, C.: Commercial success by looking for desire lines. In: Masoodian, M., Jones, S., Rogers, B. (eds.) APCHI 2004. LNCS, vol. 3101, pp. 293–304. Springer, Heidelberg (2004)

17. Park, B., Choi, J., Chung, W.K.: Path reconstruction method for sampling based planners. In: 2010 IEEE Workshop on Advanced Robotics and its Social Impacts, ARSO, pp. 81–86. IEEE (2010)

18. Pouke, M.: Using gps data to control an agent in a realistic 3d environment. In: 2013 Seventh International Conference on Next Generation Mobile Apps, Services and Technologies (NGMAST), pp. 87–92. IEEE (2013)

19. Ruddle, R.: The effect of trails on first-time and subsequent navigation in a virtual environment. In: VR 2005: Proceedings of the 2005 IEEE Conference 2005 on Virtual Reality, pp. 115–122, 321. IEEE Computer Society (2005)

20. Russell, S.J., Norvig, P.: Artificial Intelligence: A Modern Approach. Pearson Education (2003)

21. Samperi, K., Hawes, N., Beale, R.: Improving map generation in large-scale environments for intelligent virtual agents. In: The AAMAS 2013 Workshop on Cognitive Agents for Virtual Environments. LNCS. Springer (May 2013)

22. Samperi, K., Beale, R., Hawes, N.: Please keep off the grass: individual norms in virtual worlds. In: Proceedings of the 26th Annual BCS Interaction Specialist Group Conference on People and Computers, BCS-HCI 2012, pp. 375–380. British Computer Society (September 2012)

23. Stoffel, E.-P., Schoder, K., Ohlbach, H.J.: Applying hierarchical graphs to pedestrian indoor navigation. In: GIS 2008: Proceedings of the 16th ACM SIGSPATIAL International Conference on Advances in Geographic Information Systems, pp. 1–4. ACM, New York (2008)

24. Yee, N., Bailenson, J.N., Urbanek, M., Chang, F., Merget, D.: The unbearable likeness of being digital: The persistence of nonverbal social norms in online virtual environments. Cyberpsychology & Behavior: The Impact of the Internet, Multimedia and Virtual Reality on Behavior and Society 10(1), 115–121 (2007)

25. Yuan, F., Twardon, L., Hanheide, M.: Dynamic path planning adopting human navigation strategies for a domestic mobile robot'. In: Intelligent Robots and Systems, pp. 3275–3281. IEEE (2010)

Intelligent Computation of Inverse Kinematics of a 5-dof Manipulator Using MLPNN

Panchanand Jha, Bibhuti Bhusan Biswal, and Om Prakash Sahu

Department of Industrial Design, National Institute of Technology, Rourkela, 769008, India
Jha_ip007@hotmail.com, bbbiswal@nitrkl.ac.in,
omprakashsahu@gmail.com

Abstract. This paper presents inverse kinematic solution of 5 degree of freedom robot manipulator. Inverse kinematics is computation of all joint angles and link geometries which could be used to reach the given position and orientation of the end effector. This computation is very difficult to attain exact solution for the position and orientation of end effector due to the nature of non- algebraic equation of inverse kinematics. Therefor it is required to use some soft computing technique for the solution of inverse kinematics of robot manipulator. This paper presents structured artificial neural network (ANN) model from soft computing domain. The ANN model used is a Multi Layered Perceptron Neural Network (MLPNN). In this gradient descent type of learning rules are applied. An attempt has been made to find the best ANN configuration for the problem. It was found that between multi-layered perceptron neural network giving better result and calculated mean square error, as the performance index.

Keywords: Inverse Kinematics, D-H Notations, MLPNN.

1 Introduction

The Robot manipulator is composed of a serial chain of rigid links connected to each other by revolute or prismatic joints. Each robot joint location is usually defined relative to the neighboring joint. The relation between successive joints is containing a 4x4 homogeneous transformation matrix that has orientation and position data of robots. Conversion of the position and orientation of robot manipulator end-effectors from Cartesian space to joint space is called as inverse kinematics problem. The corresponding joint values must be computed at high speed by the inverse kinematics transformation [1]. For a manipulator with n degree of freedom, at any instant of time the joint variable is denoted by $i = (t)$, $i = 1, 2, 3 ....n$ and position variables by $xj = x(t)$, $j = 1, 2, 3 ....m$. The relations between the end-effectors position $x(t)$ and joint angle (t) can be represented by forward kinematic equation

$$x(t) = f(\theta(t)) \tag{1}$$

Where, f is a nonlinear continuous and differentiable function.On the other hand, with the desired end effectors position, the problem of finding the values of the joint variables is inverse kinematics, which can be solved by,

M. Mistry et al. (Eds.): TAROS 2014, LNAI 8717, pp. 243–250, 2014.

$$\theta(t) = f^{'}(x(t)) \tag{2}$$

The simulation and computation of inverse kinematics using soft computing tech-nique is particularly useful where less computation times are needed, such as in real-time adaptive robot control [2], [3]. If the number of degrees of freedom increases, traditional methods will become more complex and quite difficult to solve inverse kinematics [4].Many research contributions have been made related to the neural network-based inverse kinematics solution of robot manipulators [5].

Although the use of ANN is not new in the field of multi-objective and NP-hard problem to arrive at a very reasonable optimized solution, the multi layered neural network (MLPNN) has been tried to solve inverse kinematics problem with 5-DOF manipulator [6]. MLP neural network is used to find inverse kinematics solution which yields multiple and precise solutions with an acceptable error and are suitable for real-time adaptive control of robotic manipulators [7].The study of previous work shows that the most of the researchers [8], have adopted me-thods like ANN, ANFIS etc. for simple problem. The features of MLPNN are suitable for solving multi parametric problem with desirable accuracy and hence this technique is suitable for the present problem having complexity and involv-ing multiple parameters. Therefore, the main aim of this work is focused on mi-nimizing the mean square error of the neural network-based solution of inverse kinematics problem. In other words, the angles obtained for each joint are used to compute the Cartesian coordinate for end effector. The training data of neural network have been selected very precisely. Especially, unlearned data in each neural network have been chosen, and used to obtain the training set of the last neural network.

2　　Mathematical Modeling of 5-dof Robot Manipulator

The Denavit-Hartenberg (D-H) notation and methodology are used in this section to derive the kinematics of robot manipulator. The coordinate frame assignment and the DH parameters are depicted in Fig. 1, and listed in Table 1 respectively, Xu et al., 2005.

Table 1. The D-H Parameters

C	θ_i (degree)	a_i (mm)	d_i (mm)	α_i(degree)
$O_0 - O_1$	θ_1	$d_1 = 150$	$a_1 = 60$	-90
$O_1 - O_2$	θ_2	0	$a_2 = 145$	0
$O_2 - O_3$	$-90 + \theta_3$	0	0	-90
$O_3 - O_4$	θ_4	$d_2 = 125$	0	90
$O_4 - O_5$	θ_5	0	0	-90

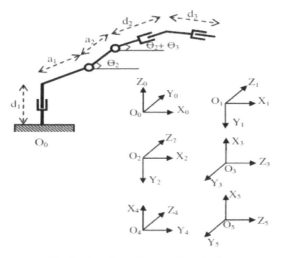

Fig. 1. Coordinate frames of manipulator

$$\theta_1 = a\tan 2(p_y - d_3 a_y, p_x - d_3 a_x)$$

$$(3)$$

$$\theta_2 = \text{atan2}(B_1, B_2) - a\cos\frac{(d_2 c_3 + a_2)}{\sqrt{B_1^2 + B_2^2}} + 2m\pi$$

$$(4)$$

Where,

$$B_1 = (d_2 c_3 + a_2) s_2 + (d_2 s_3) c_2$$

$$B_2 = (d_2 c_3 + a_2) c_2 - (d_2 s_3) s_2 \qquad \text{and} \qquad m = -1, 0 \text{ or } 1$$

$$\theta_3 = \pm a\cos\left(\frac{a^2 + r_z^2 - a_2^2 - d_2^2}{2a_2 d_2}\right)$$

$$r_z = -d_2 s_{23} - a_2 s_2 + d_1$$

$$(5)$$

$$\theta_4 = a\tan 2\left(-\frac{o_z}{c_{23}}, \frac{(o_z - c_1 s_{23} o_z / c_{23})}{s_1}\right)$$

$$(6)$$

$$\theta_5 = a\tan 2\{-(a_z c_{23} c_4 + s_{23} n_z), (n_z c_{23} c_4 - s_{23} a_z)\}$$

$$(7)$$

Where(X, Y, and Z) represents the position and $\{(n_x, n_y, n_z), (o_x, o_y, o_z), (a_x, a_y, a_z)\}$ the orientation of the end-effector.

It is obvious from the equations (3) through (7) that there exist multiple solutions to the inverse kinematics problem. By comparing the errors between these four generated positions and orientations and the given position and orientation, one set of joint angles, which produces the minimum error, is chosen as the correct solution [7].

3   MLP (Multilayered Perceptron) Neural Network Application for 5-dof Manipulator

It is well known that neural networks have the better ability than other techniques to solve various complex problems. Inverse kinematics is a transformation of a world coordinate frame (X, Y, and Z) to a link coordinate frame ($\theta_1, \theta_2, \theta_3, \theta_4$ and θ_5).

This transformation can be performed on input/output work that uses an unknown transfer function. MLP neural network's neuron is a simple work element, and has a local memory. A neuron takes a multi-dimensional input, and then delivers it to the other neurons according to their weights. This gives a scalar result at the output of a neuron. The transfer function of an MLP, acting on the local memory, uses a learning rule to produce a relationship between the input and output. For the activation input, a time function is needed [8].

A multi-layered perceptron with back-propagation algorithm is used for the present problem. The network is then trained with data for a number of end effector positions expressed in Cartesian co-ordinates and the corresponding joint angles. A block diagram of the proposed structure is shown in Fig. 2. The output signals are presented to a hidden layer neuron in the network via the input neurons. Each of the signals from the input neurons is multiplied by the value of the weights of the connection between the respective input neurons and the hidden neuron. The network uses a learning mode, in which an input is presented to the network along with the desired output and the weights are adjusted so that the network attempts to produce the desired output [8].

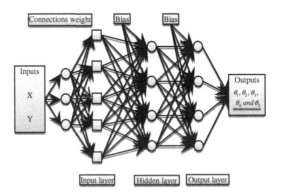

Fig. 2. Multi-layered perceptron neural network structure

A block diagram of the structure is shown in Fig. 2. Each of the signals from the input neurons is multiplied by the value of the weights of the connection, w_j, between the respective input neurons and the hidden neuron.

The aim of the training phase is to minimize this average sum squared error over all training patterns. The speed of convergence of the network depends on the training rate, η and the momentum factor, α. In this work, a two hidden layer neural network with three inputs, (X, Y, and Z), and four outputs, ($\theta_1, \theta_2, \theta_3, \theta_4$ and θ_5)was trained using the back-propagation algorithm described earlier, along a trajectory of the end-effector in the x-y plane.

4   Simulation Results and Performance Analysis

The proposed work is performed on the Matlab Neural Networks Toolbox. The training data sets were generated by using equation (3) through (7). A set of 1000 data were first generated as per the formula for the input parameter px, py and pz coordinates in mm. These data sets were the basis for the training, evaluation and testing the MLP model. Out of the sets of 1000 data, 900 were used as training data and 100 were used for testing for MLP.

The following parameters were taken: Learning rate 0.08; Momentum parameter 0.065; Number of epochs 10000; Number of hidden layers 2; Number of inputs 3 and Number of output 5. Back-propagation algorithm was used for training the network and for updating the desired weights. In this work epoch based training method was applied. The formulation of the MLPNN model is a generalized one and it can be used for the solution of forward and inverse kinematics problem of manipulator of any configuration. However, a specific configuration has been considered in the present work only to illustrate the applicability of the method and the quality of the solution vis-à-vis other alternatives methods.

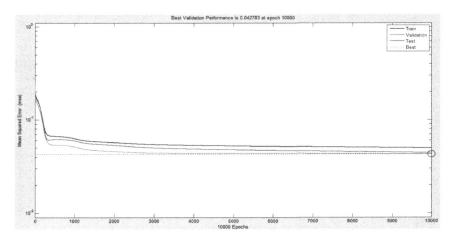

Fig. 3. Mean square error for θ_1

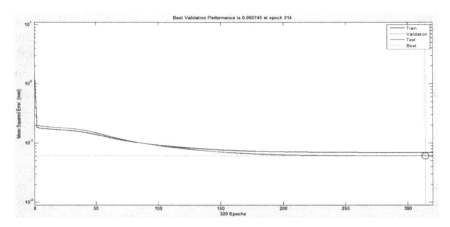

Fig. 4. Mean square error for θ_2

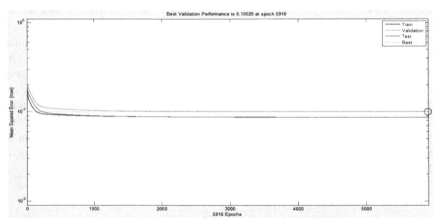

Fig. 5. Mean square error for θ_3

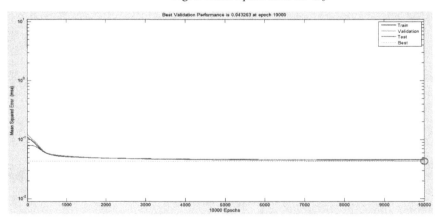

Fig. 6. Mean square error for θ_4

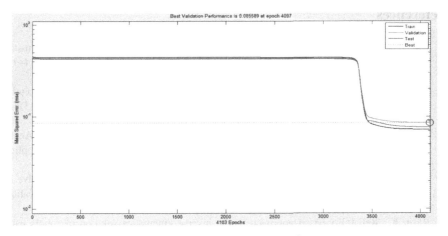

Fig. 7. Mean square error for θ_5

The mean square curve shown in Figure 3 through Figure 7 shows the building knowledge procedure for the new path which gives an indication for the success of the proposed algorithm. As shown in result, the used solution method gives the chance of selecting the output, which has the least error in the system. So, the solution can be obtained with less error as shown in Figures (3) through (7) for the best validation performance of the obtained data with the desired data. Generalization tests were carried out with new random target positions showing that the learned MLP generalize well over the whole space showing a deviation of 0.0037 of the error goal during the learning process. These errors are small and the MLP algorithm is, therefore, acceptable for obtaining the inverse kinematics solution of the robotic manipulator.

5 Conclusions

Mathematical models rely on assuming the structure of the model in advanced, thus resulting in sub optimal solution. Consequently many mathematical models fail to simulate the complex behavior of inverse kinematics problem. In contrast, ANN is based on the input/output data pairs to determine the structure and parameters of the model. Moreover, they can always be updated to obtain better results by presenting new training examples as new data become available. In the present problem the error value (mean square error) is 0.0037 which is very much acceptable when compare to the precision figures and repeatability error values of any typical manipulator. From the present study, it is observed that the MLP gives better results for inverse kinematics problem considering average percentage error as performance index. This artificial neural network based joint angles prediction model can be a useful tool for the production engineers to estimate the motion of the manipulator accurately.

References

1. De, X., Carlos, A.A.C., John, Q.G., Huosheng, H.: An Analysis of the Inverse Kinematics for a 5-DOF Manipulator. International Journal of Automation and Computing 2, 114–124 (2005)
2. Panchanand, J., Bibhuti, B.B.: A Neural Network Approach for Inverse Kinematic Solution of a SCARA Manipulator. International Journal of Robotics and Automation 3(1), 31–40 (2014)
3. Liu, W., Luo, Y., Yang, T., Shi, Z., Fu, M.: Numerical Study on Inverse Kinematic Analysis of 5R Serial Robot. 2010 International Forum on Information Technology and Applications (2010)
4. Chiddarwar, S.S., Babu, N.R.: Comparison of RBF and MLP neural networks to solve inverse kinematic problem for 6R serial robot by a fusion approach. Engineering Applications of Artificial Intelligence 23, 1083–1092 (2010)
5. Rasit, K.: Reliability-based approach to the inverse kinematics solution of robots using Elman's networks. Engineering Applications of Artificial Intelligence 18, 685–693 (2005)
6. Alavandar, S., Nigam, M.J.: Neuro-Fuzzy based Approach for Inverse Kinematics Solution of Industrial Robot Manipulators. Int. J. of Computers, Communications & Control 3, 224–234 (2008)
7. Hasan, T., Hamouda, A.M.S., Ismail, N., Al-Assadi, H.M.A.A.: Artificial neural network-based kinematics Jacobian solution for serial manipulator is passing through singular configurations. Advances in Engineering Software 41, 359–367 (2010)
8. Philip, D.W.: Neural Computing: Theory and Practice. Coriolis Group c/o Publishing Resources Inc. (1989)

Humanoid Robot Gait Generator:
Foot Steps Calculation for Trajectory Following

Horatio Garton, Guido Bugmann[*], Phil Culverhouse, Stephen Roberts,
Claire Simpson, and Aashish Santana

Centre for Robotics and Neural Systems, Plymouth University, Plymouth, England
gbugmann@plymouth.ac.uk

Abstract. During bipedal gait, a robot falls from one foot to the other. This motion can be approximated with that of an inverted pendulum with discrete movements of the contact point. We detail here how to use the linear inverted pendulum model (LIPM) for selecting successive contact points in such a way that the centre of mass (COM) of the robot flexibly follows a predefined set of waypoints on a straight or curved trajectory, allowing it to move forward, stop and revert its direction of motion in stable way. The use of a fixed step cycle duration reduces the mathematical complexity and the computational load, enabling real-time updating of gait parameters in a microcontroller.

1 Introduction

1.1 Bipedal Gait

Bipedal gait is composed of two cyclic movements. The frontal (left-right) oscillation moves the weight onto one foot (support foot) while the other (swing foot) travels in the air in order to touch the ground in a new, e.g. forward, position. The latter foot then becomes the support foot, allowing the other foot to swing too. The sagittal motion (forward-backward) is a forward or backward rotation around an axis that is approximately given by the position of the support foot. Observing humans shows numerous flexible ways of executing these two cycles. However, gait programming and control is still a difficult problem in humanoid robotics. Here, we are especially interested in gaits for small servo-driven humanoid robots. One approach used in the commercial Darwin robot [1] consists of a parameterized cyclic pattern generator, similarly to a gait developed at Plymouth University [2]. Such gaits are surprisingly effective and have set speed world records. They are also compact from the coding point of view and can be run on small microcontrollers. However, they lack flexibility and do not exploit the full potential of the robot. Another approach is exploiting the physics of the robot, modelled as a simple inverted pendulum, or its more constrained form, the Linear Inverted Pendulum Model (LIPM) [3,4]. Such models are also used

---

[*] Corresponding author.

M. Mistry et al. (Eds.): TAROS 2014, LNAI 8717, pp. 251–262, 2014.

to model human gait. In this approach, the commanded leg movements "follow" the natural gravity-driven swing of the robot on the standing leg, leading to a more stable gait, as can be shown by analyzing the Zero Moment Point (ZMP) of the robot (see section 1.4).

The problem when using inverted pendulum models is to select successive feet positions such that the natural swing leads to the robot performing actions selected by the user, like turning, starting, stopping, reversing, etc. In this paper we describe in details an original approach that provides a flexible solution at a low computational cost. One of the key features is the use of a fixed gait cycle duration. This simplifies the solution significantly, obviating the need for iterative methods such as in [4].

1.2   Plymouth Humanoids

Plymouth Humanoids are Plymouth University's robot football team. Using a team of five Drake robots, Plymouth competes in international robot competitions called RoboCup and FIRA. The Drake robots are designed, made and programmed by students and staff at Plymouth University. Figure 1 shows a drake robot. The gait for this robot must be dynamically controlled to produce forward, turning and sidestep motion at a given speed. These are complicated motions for a robot to accomplish; there is also the additional challenge of making the robot move at high speed.

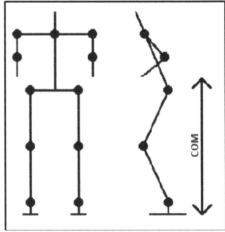

Fig. 1. Drake Robot   **Fig. 2.** Robot model in the walking position

At FIRA, the sprint event tests the robots ability to move forwards and backwards on a 3m track. The best robots are expected to complete the sprint in less than 25 seconds.

The drake robot is designed to conform to the RoboCup and FIRA specifications. The total height of the robot is 470mm, total weight is 3Kg. The height of the centre of mass (COM) in the walking position is 225mm. When the robot walks it has bent knees to allow the robot to extend its legs. The knee bends to shorten the total height of the robot by 25mm.

Due to the distribution of the mass of the upper body, it needs to be tilted forward to move the vertical projection of the centre of mass of the body above the hip joint (Figure 2). This reduces the torque on the hip pitch servos.

1.3   Linear Inverted Pendulum

The inverted pendulum model (IPM) of the robot consists of one large mass on the top of a massless pole. The walking motion of the robot is a combination of the frontal and sagittal motion cycles of the IPM. For an inverted pendulum of fixed length the equations of motions in the two planes are coupled and are difficult to resolve. A key finding by [3] is that, by introducing the constraint of a fixed height H of the mass, the two equations of motions become independent. This is the linear IPM (LIPM) (see Figure 3.) In practice, maintaining a fixed height is not a big constraint, and has the advantage of keeping the camera-carrying head more still. It is achieved by varying the length of the legs during the motion.

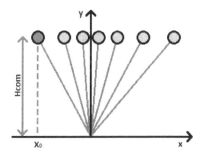

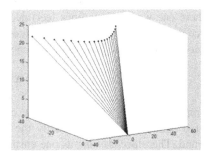

Fig. 3. Successive positions of the Linear Inverted Pendulum

Fig. 4. Linear inverted pendulum motion in two axes

The equation of motion is derived as follow. The mass undergoes a vertical gravitational force mg that exerts a torque $\tau = mgx$ on the pole, where x is the horizontal distance of the mass from the contact point. The torque exerts a horizontal force $F = \tau/H$ that accelerates the mass:

$$\ddot{x} = \frac{F}{m} = x\frac{g}{H} \tag{1.3.1}$$

The same equation is used in the frontal and sagittal planes. Its solution is:

$$x(t) = x_0 * \cosh\left(\sqrt{\frac{g}{H}} * t\right) + \frac{v_0}{\sqrt{\frac{g}{H}}} * \sinh\left(\sqrt{\frac{g}{H}} * t\right) \tag{1.3.2}$$

Where: x_0 = initial position of pendulum (m), measured from the foot position assumed to be $x=0$. v_0 = initial velocity of pendulum (ms-1). g = gravitational constant

(9.81ms-2). t = time (s). H = Height of COM (m). $\sqrt{\frac{g}{H}}$ is the time constant of the pendulum. It will be replaced by ω in this paper. The time derivative of (1.3.2) gives the velocity $v(t)$ of the mass:

$$v(t) = x_0 * \sqrt{\frac{g}{H}} * \sinh\left(\sqrt{\frac{g}{H}} * t\right) + v_0 * \cosh\left(\sqrt{\frac{g}{H}} * t\right) \qquad (1.3.3)$$

The 3-D motion of the COM can be generated by using equation (1.3.2) for the each of the sagittal and frontal motion (see e.g. figure 4).

These are the equations of motion of a point like mass at the end of a pole. The physical robot has an extended mass. For such multi-body pendulums, the time constant is different from the point like model. In practice, we found multiplying the point-like time constant (ω) by 0.7 produces the best gait, qualitatively assessed as having a good long-term stability and small-amplitude head oscillations.

The zero moment point (ZMP) [5] of the LIPM has an interesting property. The ZMP is the point p on the ground where the foot should be placed so that no torque is exerted on the ankle, i.e. the configuration is stable. For a robot that is standing still, p is exactly below the COM. For a robot that is, for instance, pushed back, the ZMP (where one would place ones feet) needs to be moved backward so that the forward force generated by gravity compensates exactly for the external push. For a pole with an accelerated mass, the ZMP p (in one dimension) is defined by [6]:

$$p = x - \frac{H}{g}\ddot{x} \qquad (1.3.4)$$

Comparing (1.3.4) with (1.3.1), one can see that p=0 for the free-falling linear inverted pendulum. That is exactly the point where the foot of the pole touches the ground. Therefore, as long as one ensures that the motion of the robot follows its natural pendulum motion, stability conditions are automatically satisfied.

In reality a physical robot is not a LIPM, but a multi-body system made of pendulums that exert internal forces on each other. The ZMP of such system moves away from the centre of the foot during gait, but with a displacement normally smaller than the dimensions of the foot, thus preserving the stability of a robot actuated on the basis of a simple LIPM [7,8].

2   Gait Generator

2.1   Waypoints Selection

There are many possible ways to design a gait based on equation 1.3.2 and 1.3.3. The approach selected here is to set target positions (waypoints) for the centre of mass to achieve after two steps, i.e. a full gait cycle. If the COM had only initial conditions and a goal position to move to, there would be an infinite number of possible solution trajectories, i.e. possible placements for the feet. To select one solution, firstly, the COM end velocity vector after the two steps must also be given, as described in section 2.2. Secondly, a time constraint is applied: One cycle of the gait, consisting of two footsteps, must be completed in a preset time T. This time is the same for all steps. This simplifies the inversion of the motion equations, because the sinh() and cosh() terms become constants. The robot walking speed is determined by T and by the distance between waypoints. We can achieve up to around 2 cycles/sec and distance of up to 15 cm per cycle. These values are close to the speed limits of the servo motors.

The robot is treated as a holonomic robot because it can move and rotate in two dimensions between two steps. At the start of each cycle of the gait, the robot is given a forward velocity a side velocity and a rotational velocity (forward speed FS, the side speed SS and turn speed TS). From these values a circular path is created and a new position and orientation for the COM can be calculated. Figure 5 shows how these commands are used to calculate the next COM position after one cycle time T. First, TS and SS define a motion direction V with unchanged robot orientation. TS then defines an arc or circle moving away from that direction and rotating the direction accordingly. The centre of the circle is placed on a line that is perpendicular to the velocity vector V (the vector sum of FS and SS). The length L of the arc from the start position to the next way point given by L = |V|*T. the radius r of the circle is defined by V and TS. TS is in rad s-1 (figure 5). It should be noted that the velocity V is the average velocity of the robot along its desired trajectory. This is different from the instantaneous velocity of the COM of the robot that has cyclic lateral and forward components vl and vf. The purpose of the gait generator in this paper is to control these components to make the COM pass through the desired waypoints.

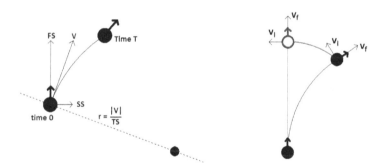

Fig. 5. COM positions before and after a cycle, including forward motion, rotation and side-stepping.

Fig. 6. Steps of the calculation of the COM velocity vector at the next way-point (image for forward and turn only).

2.2   Calculating the End-of-Cycle Velocity Vector

The velocity at the end of each cycle must be determined so that one set of left and right feet positions can be calculated. The end velocity V_{COM} is a vector in two dimensions.

The approach used here is to work out the V_{COM} for a straight line motion (TS=0), then to rotate the vector by how much the robot body would have turned at the end of the curve (figure 6).

The forward velocity at the end of the cycle is calculated on the basis of the average speed FS. The end-of-cycle lateral speed of the COM is based on the requested sidestep speed SS, as well as the desired separation of the feet. After the forward speed and side speed of the end of cycle COM position have been calculated, they are rotated by the angle the body of the robot will be facing.

2.2.1 Desired COM End-of-Cycle Lateral Velocity

The gait is designed to swing on the left foot first then on the right, this results in the end velocity being negative (moving to the left side of the robot, figure 7). The cycle time (T) is separated into two halves. In the second half the robot is swinging on the right foot so the starting position of the pendulum is -D/2 where D is the separation of the feet (assuming that the foot switch takes place when the robot exactly in the middle between the two feet). At this moment, the initial velocity v_0 points towards the right. So, at the end of the swing on the left foot, the position is defined by equation 1.3.2, which can be solved for the initial velocity v_0 at the start of the swing on the left foot.

$$-\frac{D}{2} = -\frac{D}{2} * \cosh\left(\omega * \frac{T}{2}\right) + \frac{v_0}{\omega} * \sinh\left(\omega * \frac{T}{2}\right) \qquad (2.2.1.1)$$

$$v_0 = \frac{-\omega * D * \left(1 - \cosh\left(\omega * \frac{T}{2}\right)\right)}{2 * \sinh\left(\omega * \frac{T}{2}\right)} \qquad (2.2.1.2)$$

Of course, during the gait cycle, v_0 is also the initial velocity of the swing on the right foot. When a side-step velocity is specified, it must be added to equation 2.2.1.2 defining the full end velocity in the lateral direction v_l.

$$v_l = \frac{\omega * D * \left(1 - \cosh\left(\omega * \frac{T}{2}\right)\right)}{2 * \sinh\left(\omega * \frac{T}{2}\right)} + SS \qquad (2.2.1.3)$$

2.2.2 Desired COM End-of Cycle Forward Velocity

The required forward velocity of the COM at the end of a cycle is determined only by the requested forward speed FS of the robot. This corresponds to a required travel distance in one cycle of $F = * T$. If the robot is moving forward at speed FS, then the total distance it will move forward in one cycle is. We shall call this distance F. In the second half of the cycle, when the robot swings on the right foot, the COM will move from position –F/4 to F/4 in time T/2 (the positions are relative to the contact foot, assuming an equal distance travelled during each half-cycle). We can put this into equation 1.3.2 and solve for v_0 again.

$$\frac{F}{4} = -\frac{F}{4} * \cosh\left(\omega * \frac{T}{2}\right) + \frac{v_0}{w} * \sinh\left(\omega * \frac{T}{2}\right) \qquad (2.2.2.1)$$

$$v_0 = \frac{\omega F \left(1 + \cosh\left(\omega \frac{T}{2}\right)\right)}{4 \sinh\left(\omega \frac{T}{2}\right)} \qquad (2.2.2.2)$$

If you used this v_0 in equation 1.3.3 and t = T/2 and x_0 = -F/4, the velocity at time T/2 would be the same as v_0. So the end velocity in the forward direction v_f is equal to v_0 in equation 2.2.2.2.

$$v_f = \frac{\omega * F * \left(1 + \cosh\left(\omega * \frac{T}{2}\right)\right)}{4 * \sinh\left(\omega * \frac{T}{2}\right)} \qquad (2.2.2.3)$$

Figure 8 shows the motion of the COM in the forward direction in relation to the right foot. Note that the velocity is nearly constant.

2.2.3 Applying the Rotation

The vector for the end of cycle velocity (V_{COM}) must be rotated by the same angle the COM position is facing at the end of the cycle α_2 (figure 6).

$$v_{COM} = \begin{bmatrix} v_l & v_f \end{bmatrix} * \begin{bmatrix} \cos(\alpha_2) & \sin(\alpha_2) \\ -\sin(\alpha_2) & \cos(\alpha_2) \end{bmatrix} \qquad (2.2.3.1)$$

α_2 is given by $\alpha_2 = \alpha_1 + T*TS$, where α_1 is the orientation of the robot at the start of the cycle.

2.3 Feet Placement Calculations

One cycle of the robot gait consists of a swing on the left foot then a swing on the right foot. Figure 7 shows the trajectory of the COM. The COM has an end position x_e and end velocity v_e. In each of the planes of motion the robot has an initial position x_0 and initial velocity v_0. (In what follows, the variable x is used for both planes, as the equations are exactly the same). The left foot and right foot placements must be selected to accomplish the cycle in time T and to reach the desired COM end state.

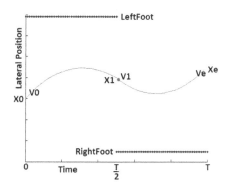

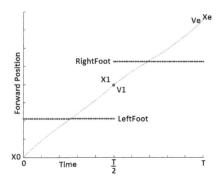

Fig. 7. Lateral foot positions and COM in time (T=0.5s, sidestep=20mm, lateral distance between feet = 90mm

Fig. 8. Forward foot positions and COM in time – sagittal position during one step. The velocity is nearly constant. Stride length = 70mm.

At time T/2 the COM position is at x_l with velocity v_l. This is the initial conditions for the swing on the right foot. The following equations can be created from Equation 1.3.2 and 1.3.3 for the mid cycle position and the mid-cycle velocity:

$$x_1 = (x_0 - L) \cosh\left(\omega \tfrac{T}{2}\right) + \tfrac{v_0}{w} \sinh\left(\omega \tfrac{T}{2}\right) + L \tag{2.3.1}$$

$$v_1 = w(x_0 - L) \sinh\left(\omega \tfrac{T}{2}\right) + v_0 \cosh\left(\omega \tfrac{T}{2}\right) \tag{2.3.2}$$

And for the end cycle position and velocity:

$$x_e = (x_1 - R) \cosh\left(\omega \tfrac{T}{2}\right) + \tfrac{v_1}{\omega} \sinh\left(\omega \tfrac{T}{2}\right) + R \tag{2.3.3}$$

$$v_e = \omega(x_1 - R) \sinh\left(\omega \tfrac{T}{2}\right) + v_1 \cosh\left(\omega \tfrac{T}{2}\right) \tag{2.3.4}$$

Where: L = Left foot position and R = Right foot position. Note that the dynamic part of the expressions operate with the pivot point (centre of ankle joint) at position x=0. This explains the subtractions and additions of L or R in the position expressions 2.3.1 and 2.3.3.

x_l and v_l can be substituted into equation 2.3.3 and 2.3.4. This gives two equations with two unknowns (L and R). The values of the sinh and cosh functions are constant, so this is just a linear simultaneous equation. This is the reason why a fixed frequency gait is useful. The equations can be rearranged to have L and R on the left hand side, then put into matrix form:

$$A \begin{bmatrix} L \\ R \end{bmatrix} = B \tag{2.3.5}$$

Where:

$$A = \begin{bmatrix} A_{11} & A_{12} \\ A_{21} & A_{22} \end{bmatrix} \tag{2.2.6}$$

$$A_{11} = -\cosh^2\left(\omega \tfrac{T}{2}\right) + \cosh\left(\omega \tfrac{T}{2}\right) - \sinh^2\left(\omega \tfrac{T}{2}\right) \tag{2.3.7}$$

$$A_{12} = 1 - \cosh\left(\omega \tfrac{T}{2}\right) \tag{2.3.8}$$

$$A_{21} = -2\,\omega \cosh\left(\omega \tfrac{T}{2}\right) \sinh\left(\omega \tfrac{T}{2}\right) + \omega \sinh\left(\omega \tfrac{T}{2}\right) \tag{2.3.9}$$

$$A_{22} = -\omega \sinh\left(\omega \tfrac{T}{2}\right) \tag{2.3.10}$$

and:

$$B = \begin{bmatrix} B_1 \\ B_2 \end{bmatrix} \tag{2.3.11}$$

$$B_1 = x_e - x_0 \left(\cosh^2\left(\omega \tfrac{T}{2}\right) + \sinh^2\left(\omega \tfrac{T}{2}\right)\right) - \tfrac{2v_0}{\omega} \cosh\left(\omega \tfrac{T}{2}\right) \sinh\left(\omega \tfrac{T}{2}\right) \tag{2.3.12}$$

$$B_2 = v_e - 2\,x_0\,\omega \cosh\left(\omega \tfrac{T}{2}\right) \sinh\left(\omega \tfrac{T}{2}\right) - v_0 \left(\cosh^2\left(\omega \tfrac{T}{2}\right) + \sinh^2\left(\omega \tfrac{T}{2}\right)\right) \tag{2.3.13}$$

Both sides of equation 2.3.5 can be multiplied by A^{-1} to obtain the solution:

$$\begin{bmatrix} L \\ R \end{bmatrix} = A^{-1}B \qquad (2.3.14)$$

The same equation is used to compute Lx and Rx, and then Ly and Ry. The detailed resulting expressions are not shown here.

During the motion cycle, the COM is made to rotate with a constant yaw angular velocity (by applying a rotation to the contact foot). If α_1 is the yaw angle of the COM at the start of the cycle and α_2 is the angle at the end, t is the time during the cycle and T is the total time of the cycle then the COM angle $\alpha(t)$ is given by:

$$\alpha(t) = (\alpha_2 - \alpha_1) * \frac{t}{T} + \alpha_1 \qquad (2.3.15)$$

The angles of the swinging feet must also be calculated for when they reach the ground. The angles of the left foot, α_l, and the right foot, α_r, are given by equations 2.3.16 and 2.3.17.

$$\alpha_l = \frac{1}{4}\alpha_1 + \frac{3}{4}\alpha_2 \qquad (2.3.16)$$

$$\alpha_r = \frac{3}{4}\alpha_1 + \frac{1}{4}\alpha_2 \qquad (2.3.17)$$

This formulation is somewhat arbitrary and using ½ and ½ instead of ¼ and ¾ actually gives a better good looking motion.

As the robot needs to dynamically change speed and direction while walking, the feet positions cannot be calculated too far in advance. On Drake, at the beginning of each cycle, the feet positions are calculated for the next cycle. The feet positions of the current cycle are known because they were calculated in the previous cycle. The previous feet positions are also stored.

2.4   Feet Lifting and Ground Speed Matching

The vertical lift of the foot from the ground is a fourth order polynomial (at4+bt3+ ct2+dt+e). The constants a to e are selected to make a curve that will have minimum impact on the ground. Figure 9 shows the vertical feet positions. The peak of the foot can be any value. On the drake robots the peak of the foot height is set to 30 mm.

Horizontally, each foot must move in the air from its previous position to its next position in time T/2. This is done smoothly using a third order polynomial equation (fig. 10, "foot position" curve).

In the world reference frame, the swinging foot moves forward. However, as the hip of the robot also moves forward, in the robot reference frame, the foot actually has to move backward before making contact with the ground (fig. 10. Curve "foot position from hip).

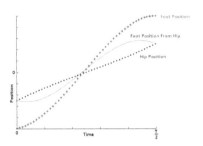

Fig. 9. Vertical feet positions

Fig. 10. Horizontal foot Movement during a half-cycle (start to end position)

2.5   Start Step

The start step consists of accelerating the COM to the right, then accelerating it back to the central position to give it the initial velocity required for a stable gait cycle. The acceleration curve is an elliptic curve which merges with the free pendulum curve at time Tm. From that time on, the robot is able to lift its left foot and start swinging on its right foot. In practice, the foot is made to lift at time 3T/4. An elliptic curve was chosen because of the flexibility to setting its slope (speed) at the merging point.

3   Results

Figure 11 shows four cycles of the robot COM motion and corresponding feet positions for a left curve. The green lines show the desired velocity at the end of each cycle. The red line shows the direction. Figure 12 illustrates an s-shaped trajectory through a superposition of video frames taken 1 second apart.

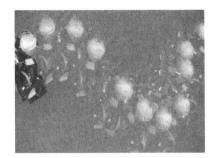

Fig. 11. Sequence of left and right foot positions corresponding to COM displacement curve shown in blue. The green lines show the direction of the COM velocity at the end of each cycle.

Fig. 12. S-shape trajectory executed by the robot. Superposition of snapshots taken 1 second apart.

A qualitative observation of the robot motion suggests some mismatch between the pure LIPM and the physical robot, e.g. feet impacts seem to be compensating for a slight oscillation speed error in sideways motion.

Our robot does not have ankle torque meters and the quality of the stability cannot be measured in terms of ZMP. The robot recovers from brief sideways and forward pushes, but this is a purely qualitative assessment of its stability.

In straight line, the robot has achieved speeds in excess of 0.3ms/s.

It is noteworthy that all the calculations described in this paper are completed on-board the robot on an ATxmega256a3 microcontroller once per cycle, in addition to calculations done every 20ms of the COM position calculated using the LIPM formulas, and the inverse kinematic of the feet trajectories.

4   Concluding Comments

This paper described how to calculate the positions of feet to make a robot follow a predefined trajectory flexibly combining translation and rotation. The constraints are a constant height of the robot and a constant time between steps. The calculations are based on a simple LIPM model and produce working solutions that are used on a real robot. The gait actions include forward and backward, straight and curved gaits, and start and stop steps.

Future work will evaluate the fitness of the simple LIPM model for our robot in demanding conditions, such as high-speed motion on irregular surfaces, possibly leading to the development of an appropriately improved model to increase the operational envelope.

References:

1. Ha, I., Tamura, Y., Asama, Y.: Gait pattern generation andstabilization for humanoid robot based on coupled oscillators. In: IEEE IROS, pp. 3207–3212 (2011)
2. Gibbons, P., Mason, M., Vicente, A., Bugmann, G., Culverhouse, P.: Optimization of Dynamic Gait for Bipedal Robots. In: Proc. 4th Workshop of Humanoid Soccer Robots held in Paris at IEEE Humanoids 2009 (2009)
3. Kajita, S., Kanehiro, F., Kaneko, K., Yokoi, K., Hirukawa, H.: The 3D Linear Inverted Pendulum Model: A simple modeling for a biped walking pattern generation. In: Proceedings of the 2001 IROS, Maui, Hawaii, pp. 239–246 (2001a)
4. Graf, C., Röfer, T.: A closed-loop 3D-LIPM gait for the RoboCup Standard Platform League humanoid. In: Zhou, C., Pagello, E., Behnke, S., Menegatti, E., Röfer, T., Stone, P. (eds.) Proceedings of the Fourth Workshop on Humanoid SoccerRobots in Conjunction with the 2010 IEEE-RAS International Conference on Humanoid Robots (2010)
5. Vukobratovic, M., Borovac, B., Surla, D., Stokic, D.: Biped Locomotion: Dynamics, Stability and Application. Springer (1990)
6. Choi, Y., You, B.J., Oh, S.R.: On the stability of indirect ZMP controller for biped robot systems. In: Proceedings of International Conference on Intelligent Robots and Systems, Sendal, Japan, vol. 2, pp. 1966–1971 (June 2004)

H. Garton et al.

7. Suleiman, W., Kanehiro, F., Miura, K., Yoshida, E.: Enhancing Zero Moment Point-Based Control Model: System Identification Approach. Advanced Robotics 25, 427–446 (2011)
8. Galdeano, D., Chemori, A., Krut, S., Fraisse, P.: Optimal Pattern Generator For Dynamic Walking in Humanoid Robotics. Transactions on Systems, Signals & Devices (to appear, 2014)

A Method for Matching Desired Non-Feature Points to Size Martian Rocks Based upon SIFT

Gui Chen, Dave Barnes, and Pan LiLan

Department of Computer Science, Aberystwyth University, SY23 3DB, UK
{chg12,dpb,lip8}@aber.ac.uk

In Mars rover missions ExoMars is the forthcoming ESA/Roscosmos 2018 mission, and one of the future goals is to cache rock samples for subsequent return to Earth as part of the Mars Sample Return (MSR) mission. Rocks on the martian surface are one of the most interesting science targets for geologists and planetary scientists. A geologist can conclude in which regions have similar rocks formed and been deposited at that time. Different physical parameters and locations can create distinct rocks, which are essential for the understanding of martian chemical elements and the geologic environment. Rocks since formation can tell us what has happened such as the influence of climate, erosion and transportation. However, rocks are one of the chief obstacles to endanger a rover's traverse if they are not detected ahead of time accurately. Therefore, the automated detection of the size of a rock is a valuable ability for an autonomous planetary rover.

The paper presents a novel approach to measure the size of a rock in a pair images. Here, we utilize the distance from top-most to bottom-most points and the distance between left-most and right-most points on the rock contour to indicate the bounding size of a rock. A rock detection method is employed to obtain the closed contour of the rock in the pair images [1]. In view of occlusions, top-most, bottom-most and right-most points are calculated on the contour of the rock in the left image, and similarly the left-most point is derived in the right image. Next, the SIFT-RANSAC method is used to properly match the prominent feature points in the pair of images. Based upon those feature points found by SIFT-RANSAC, the rough matching of non-feature points (i.e. top-most, bottom-most, left-most and right-most points) can be implemented by the proposed algorithm in [2]. Moreover, the fundamental matrix for epipolars combined with non-feature points is computed by those feature points taken by SIFT-RANSAC. The projection points of the rough matching points on the epipolar are derived. A correlation method is applied for the accurate matching points. The search range is 12 pixels at the two sides of the projection points using 3×3 windows on the epipolar according to the distortion model of the camera calibration (See Fig. 1). Finally, we utilize the four non-feature points to calculate the size of the rock using the stereo triangulation method.

The experiments were implemented using the Aberystwyth University Pan-Cam Emulator (AUPE) with two wide angle multi-spectral cameras (WACs) in our Trans-National Planetary Analogue Terrain Laboratory (PATLab). The intrinsic and extrinsic parameters of the camera are obtained by camera calibration [3]. 10 rocks of different sizes were used for our experiments. For ground truth of the rock size were measured using a micrometer. A comparison was achieved

M. Mistry et al. (Eds.): TAROS 2014, LNAI 8717, pp. 263–264, 2014.

between our method and the standard disparity equation method. Table 1 shows the experimental results, the average error of our method is 3.62%, whereas the standard disparity method produced an average error of 61.22%. Consequently, there is better accuracy with our method. The very large errors using the standard disparity method were not expected, and may be due to calibration errors. However as our method used the same calibrated images, it demonstrates how resilient our method is to possible systematic errors.

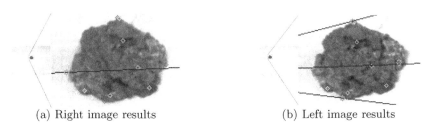

(a) Right image results   (b) Left image results

Fig. 1. Matching algorithm results. Pink points represent the matched points which are top-most, bottom-most, left-most and right-most points. Blue lines are the epipolar relative to the pink points. Yellow points denote the closest and secondary points obtained by SIFT with respect to pink points. Orange points express the rough matching points obtained using our proposed approach. Green points are the projection points on the epipolar. The accurate matching points are shown by the red points.

Table 1. Size parameters of the rocks

Rock Name	Distance of ToptoBottom (mm)			Distance of LefttoRight (mm)		
	Our Method	Disparity Method	Micrometer Value	Our Method	Disparity Method	Micrometer Value
Rock 1	105.47	170.06	103.46	161.52	249.02	158.96
Rock 2	67.86	109.57	66.16	96.86	146.47	96.15
Rock 3	85.88	126.01	84.43	98.35	157.61	95.68
CONGLOM ERATE	47.35	67.22	45.53	56.55	88.97	54.06
BRECCIA	49.71	72.03	46.73	58.33	95.08	57.75
SANDSTONE aolian	52.61	83.6	51.89	51.33	78.61	50.61
SANDSTONE torridonian	46.42	73.42	45.86	67.26	105.38	63.68
MUDSTONE	62.74	112.2	59.85	66.27	105.41	63.15
OXFORD CLAY	51.91	75.33	48.66	59.73	91.65	55.54
CRINOIDAL LIMESTONE	52.89	82.89	49.38	59.38	89.04	56.5

References

1. Gui, C., Barnes, D., Pan, L.: A SIFT-Based Method for Matching Desired Keypoints on Mars Rock Target. In: The International Symposium on Artificial Intelligence, Robotics and Automation in Space, i-SAIRAS (2012)
2. Gui, C., Barnes, D., Pan, L.: An Approach for matching Desired Non-Feature Points on Mars Rock Targets Based on SIFT. In: The Towards Autonomous Robotic System, TAROS Conference (2012)
3. Heikkila, J., Silven, O.: A Four-Step Camera Calibration Procedure with Implicit Image Correction. In: Proceedings of the 1997 IEEE Computer Society Conference on Computer Vision and Pattern Recognition (1997)

CogLaboration: Towards Fluent Human-Robot Object Handover Interactions

Ansgar Koene, Markus Rank, Roberta Roberts, Satoshi Endo, James Cooke,
Marilena Mademtzi, Chris Miall, and Alan M. Wing

University of Birmingham, Sensory Motor Neuroscience Lab, Center of Computational
Neuroscience and Cognitive Robotics, School of Psychology, B15 2TT, Birmingham, UK

Abstract. The CogLaboration project (http://coglaboration.eu/) was launched in 2011, as part of the EU FT7-ICT agenda for the development of service robotics and focuses on the fluent handover of objects between a robot and a human, which is considered to be a key requirement for providing successful and efficient robotic assistance to humans. The project addresses this challenge by integrating the study of human-human object handover interactions into the robot development process.

The Coglaboration project aims to both characterise and understand the mechanisms governing the exchange and collaborative handling of objects between pairs of humans. At the Sensory Motor Neuroscience laboratory (SyMoN) we have focused on examining the quantitative and qualitative characteristics of human-human object exchanges using tasks closely inspired by realistic conditions. This has included determining what is typical behaviour in such tasks, and what happens in unplanned and unanticipated situations. In our investigations we have used measures of upper limb motion (hand and arm trajectories) and the forces applied to objects, as well as using experimental manipulations to establish the sort of visual and tactile information used to correct arm and hand motion and to understand the way gestures are used to provide information or trigger object handover and shared handling procedures.

The same metrics that are derived from the studies of object exchange between humans are then used to evaluate the vision-driven robotic system, comprising a lightweight robotic arm (Kuka LWR) and a hand with tactile sensors, following a scenario-driven methodology.

Within the CogLaboration consortium, the University of Birmingham's SyMoN lab members are also involved in the evaluation of the performance of the robotic system prototype for Human-Robot interaction. This evaluation is based on a combination of motion tracking, touch and motion sensing data from the manipulated object and trial-by-trial ratings from the participants on the quality of the interactions.

Our investigation of human to human interactions consists of a series of studies, each of which provides detailed analysis of different aspects involved in handover and handling interactions.

M. Mistry et al. (eds.): TAROS 2014, LNAI 8717, pp. 265–266, 2014.

1. Motion tracking of object handover interactions, including: (i) impact of postural configuration on interaction dynamics; (ii) evaluation of changes in handover location on perceived interaction quality; (iii) unconscious adjustment to partner behavior and its implications for anticipating future behavior.
2. Variability and interpersonal adjustment in joint spatial pointing.
3. Comparison of hand placement and shape when grasping objects for handover vs. grasping for placement.
4. Grip force and object release time changes when tactile information is attenuated.
5. Changes in grip force profiles when handing over objects with the dominant or non-dominant hand.
6. Analysis of anticipatory and reactive grip force modulation during object handover based on grip force, load force and shear force data from handing over a force sensor instrumented object.
7. Error-based adaptation of grip force and movement profiles when handing objects to an unreliable receiving partner.
8. Analysis of relative information content of gestural features for non-verbal communication related to object handover interactions.
9. Behavioural adjustment and interpersonal interaction during two-person object handling with predictable compared with unpredictable load changes
10. Diffusion modeling of force perception during active/passive movement and resulting predictions for optimal time of force perception.
11. Identification of haptic and multimodal cues that people use to decide to release grasp, based on grip-force data from experiments using a robotic haptic interface for precisely controlled handovers of virtual objects.
12. Analysis of interpersonal timing interactions between multiple interacting participants.
13. Questionnaire based survey of requirements, concerns and desires related to object handover activities, including service robots, in daily living as perceived by elderly focus groups.
14. Evaluation of an object-embeddable touch and motion sensing device for monitoring how objects are handled during handover interactions.
15. Creation of a Wiki site for coordinating the sharing of data, models, methods and other information related to human and animal behavior (http://behaviorinformatics.bham.ac.uk).

The results of these human-human interaction studies are being used by the Cog-Laboration partners to design and refine the robotic system prototype for fluent human-robot object handover interactions that forms the central deliverable of the CogLaboration project.

Publications relating to this work can be found via the CogLaboration website (http://coglaboration.eu/) or the SyMoN lab site (http://symonlab.org/).

euRathlon Outdoor Robotics Challenge: Year 1 Report

Alan F.T. Winfield[1], M. Palau Franco[1], B. Brueggemann[2], A. Castro[3], V. Djapic[4],
G. Ferri[4], Y. Petillot[5], J. Roning[6], F. Schneider[2], D. Sosa[3], and A. Viguria[7]

[1] University of the West of England, Bristol, UK
[2] Fraunhofer FKIE, Bonn, Germany
[3] The Oceanic Platform of the Canary Islands (PLOCAN), Spain
[4] Centre for Maritime Research and Experimentation (CMRE), La Spezia, Italy
[5] Heriot-Watt University, Edinburgh, UK
[6] University of Oulu, Oulu, Finland
[7] Centre for Advanced Aerospace Technologies (CATEC), Seville, Spain

The aim of euRathlon[1] is to create real-world robotics challenges to test the intelligence and autonomy of outdoor robots in demanding mock disaster-response scenarios. Inspired by the Fukushima accident we are working toward a competition that requires autonomous flying, land and underwater robots acting together to survey the disaster, collect environmental data, and identify critical hazards. Leading up to this Grand Challenge in 2015 are directly related land and underwater robot competitions.

The first 2013 euRathlon land robotics competition, with a parallel workshop on field robotics, took place from 23-27 September 2013, in Berchtesgaden, Germany. The competition was organised around five scenarios (1) Reconnaissance and surveillance in urban structures, (2) Mobile manipulation for handling hazardous materials, (3) Search and rescue in a smoke-filled underground structure, (4) Autonomous navigation using satellite navigation, and (5) Reconnaissance and handling of explosive devices. A total of 14 teams from 7 countries took part, with each team entering between 1 and 4 scenarios. Of the 14 teams, 4 were from industry, and 10 from universities or research institutes. Each trial was witnessed and scored by judges, and benchmarking observers. The results have been published on the eurathlon2013 competition website[2]. For obvious safety reasons only technical staff, judges and the team actually competing had close access to robot trials, and so large screen CCTV displays provided live coverage of the trials as they took place. A parallel static display of robot systems and projects provided interest between trial runs.

Also in Berchtesgaden the 2013 workshop on field robotics provided an opportunity for attendees to deepen their understanding of outdoor autonomous robotics, to share ideas and experience, and to prepare themselves for future euRathlon competitions. Keynote speaker Shinju Kawatsuma, from the Japan Atomic Energy Authority, provided attendees with a fascinating and honest first-hand account of the use of robots in the Fukushima Daiichi NPP, following the March 2011 Tsunami.

---

[1] euRathlon (www.eurathlon.eu) is funded within the EU FP7, project 601205
[2] http://www.elrob.org/eurathlon-2013-results

M. Mistry et al. (Eds.): TAROS 2014, LNAI 8717, pp. 267–268, 2014.

Fig. 1. A montage of images from the euRathlon 2013 competition and workshop (photos: Aaron Boardley and Marta Palau Franco)

Benchmarking efforts focused on developing a set of benchmarks for the euRathlon 2013 land robotics competition as well as looking ahead to 2014 and 2015 competitions. We designed benchmarks that were used in euRathlon 2013 to assess robot performance. In parallel we have reviewed the state of the art in benchmarking in the robotics community. The National Institute for Standards and Technology (NIST), involved in designing benchmarks for the robotics community in the US for the last 3-5 years, provided a set of standard test rigs as part of their static display at euRathlon 2013, and a workshop talk.

In preparation for the euRathlon marine robotics competition in 2014 and the Grand Challenge in 2015 involving robots from the three domains: land, sea and air, the consortium has worked on the definition of the scenarios and the logistics. For the 2014 underwater robotics competition, the scenarios have been designed and published[3], and the venue chosen and date confirmed: the CMRE facility in La Spezia, Italy, 29 September – 3 October 2014. For the 2015 multi-domain Grand Challenge in October 2015, the candidate location is the island of Favignana, Sicily.

The 2013 euRathlon robotics competition and workshop was an outstanding success, establishing euRathlon in the landscape of emergency response competitions. (For YouTube daily video reports see[4].) While it would be too soon to expect wider impact, we believe we have laid the groundwork for years 2 and 3 of euRathlon, and its unique focus on smarter multi-domain multi-robot team working.

---

[3] http://www.eurathlon.eu/site/index.php/compete/scenarios2014/
[4] https://www.youtube.com/user/euRathlonVideos

FROctomap: An Efficient Spatio-Temporal Environment Representation

Tomáš Krajník, João Santos, Bianca Seemann, and Tom Duckett

Lincoln Centre for Autonomous Systems, University of Lincoln, UK
{tkrajnik,jsantos,tduckett}@lincoln.ac.uk, biaseemann@gmail.com

We present a novel software tool intended for mobile robot mapping in long-term scenarios. The method allows for efficient volumetric representation of dynamic three-dimensional environments over long periods of time. It is based on a combination of a well-established 3D mapping framework called Octomaps [1] and an idea to model environment dynamics by its frequency spectrum [2]. The proposed method allows not only for efficient representation, but also reliable prediction of the future states of dynamic three-dimensional environments. Our spatio-temporal mapping framework is available as an open-source C++ library and a ROS module [3] which allows its easy integration in robotics projects.

In order to act intelligently, an agent has to be able to reason about its surroundings and thus needs to model its environment. In mobile robotics, suitable world representation is an essential component that allows an intelligent robot to make decisions, plan future actions, estimate its location and cooperate with others. So far, the environment models used in mobile robotics have been tailored to represent static scenes and treat environment dynamics as unwanted noise. Thus, the research was aimed at efficient acquisition, representation and usage of static environment models.

One of the most popular environment representations is an occupancy grid, which allows for efficient probabilistic sensor fusion, motion planning, localization and exploration. The main drawback of occupancy grids is their low-memory efficiency because they represent large, empty areas of the environment by a large amount of empty cells. This is mitigated by the so-called Octomap [1] framework, that locally adapts the grid resolution to the level of detail required. The authors of Octomap show that the method can represent large-scale environments with a fine level of detail on standard computational hardware.

In order to act efficiently in changing environments, the robots need to model the environment dynamics as well. Since it is infeasible to store all the observations a robot makes over long periods of time, one has to represent the temporal domain in an efficient way. The work in [4] represents the environment by multiple temporal models, each representing a different timescale. The authors of [5] cluster the observations according to their similarity and create 'experiences' that capture the different appearances of the same spatial locations. The authors of [6,2] propose to use an occupancy grid and model the dynamics of individual cells independently. While [6] represents the occupancy of each cell with a hidden Markov model, [2] assumes that the variations observed are caused by hidden processes that are naturally periodic and identifies them by means of a Fast Fourier Transform. The authors of [2] show that their method (called

M. Mistry et al. (Eds.): TAROS 2014, LNAI 8717, pp. 269–270, 2014.
© Springer International Publishing Switzerland 2014

FREMen) achieves compression of the model's temporal domain by rates up to 10^6 while predicting the environment states with 90% accuracy.

We represent the occupancy of the cells stored in the Octomap by the spectral model proposed in FREMen. Thus, the efficiency of Octomap to model large spatial scales and the efficiency of FREMen to represent long periods of the time are combined in an efficient spatio-temporal environment model.

We consider the occupancy of each cell (voxel) stored in an Octomap as a binary function of time, i.e. the occupancy of ith cell is represented as $s_i(t)$. The approach in [2] considers the $s_i(t)$ as being influenced by a set of naturally periodic processes and proposes to represent $s_i(t)$ by a probabilistic function of time, which is a combination of harmonic functions which reflect the most prominent periodic processes that constitute the environment dynamics. Technically speaking, FREMen represents each function $s_i(t)$ by a set of the n highest values of its frequency spectrum $S_i(\omega)$ obtained by Fast Fourier transform and a Δ-encoded set $\mathcal{O}$ that represents time intervals where the Inverse Fourier Transform of $S_i(\omega)$ does not equal the original function $s_i(t)$, i.e.

$$s_i(t) = (\mathcal{IFT}(\mathcal{S}(\omega)) > 0.5) \oplus (t \notin \mathcal{O}), \qquad (1)$$

where $\oplus$ is a XOR operation. The FREMen model of each Octomap voxel is built iteratively; each time a state $s_i(t)$ is measured, Eq. (1) is calculated and if the observed occupancy does not equal (1), the time t is added to the outlier set $\mathcal{O}$. Whenever the set $\mathcal{O}$ grows too large, the spectrum S_i can be recalculated, which results in reduction of $\mathcal{O}$. Since Eq. (1) can be calculated for any value of t, (1) can be used to predict the state $s_i(t)$ as well.

Thus, our system takes a series of Octomaps observed over time and builds a temporal model of each observed voxel. After that, the system allows to calculate the state of the individual voxels and recover the Octomap for any given time.

We have applied the proposed approach to data collected in an office environment for a period of one week. The dataset consisted of 120960 Octomaps each representing 213192 spatial cells. The proposed system stored the spatio-temporal model in a 2MB file achieving a compression rate of 1:10000. Comparing the model's predictions to day-long measurements performed two weeks later showed that the model predicted the environment states with 98% accuracy.

Acknowledgement. The work is funded by the EU project 600623 'STRANDS'.

References

1. Hornung, A., et al.: OctoMap: An efficient probabilistic 3D mapping framework based on octrees. Autonomous Robots (2013)
2. Krajník, T., et al.: Spectral analysis for long-term mapping. In: ICRA (2014)
3. Krajník, T., et al.: (Froctomap), http://purl.org/robotics/froctomap
4. Biber, P., et al.: Dynamic maps for long-term operation of mobile service robots. In: Robotics: Science and Systems (2005)
5. Churchill, W., et al.: Experience-based navigation for long-term localiz. IJRR (2014)
6. Tipaldi, G.D., et al.: Lifelong localization in changing environments. IJRR (2013)

Combined Force and Position Controller Based on Inverse Dynamics: Application to Cooperative Robotics

Extended Abstract

Zhenyu Du[1], M. Necip Sahinkaya[2], and Pejman Iravani[1]

[1] Department of Mechanical Engineering, University of Bath, Bath, UK
p.iravani@bath.ac.uk
[2] Faculty of Science, Engineering and Computing,
Kingston University London, London, UK

1 Summary

This extended abstract presents a technique that enables the use of inverse dynamics for redundant systems, like cooperative robots. The advantage of this approach is simplified calculation of inverse dynamics that involves force demands. For the conventional method, one needs to properly separate the system according to the demanded forces. Once separated, the inverse dynamics need to be done on each subsystem. This is particularly difficult when the system is complicated, so more human input is needed. The method is applied in a feedforward controller. Experiments on two collaborating robots are carried out and results show improved performance.

2 Theory

The dynamic model of a system can be written using Lagrange's equations of motion:

$$\begin{bmatrix} \mathbf{M} & \mathbf{J}^T \\ \mathbf{J} & \mathbf{0} \end{bmatrix} \begin{bmatrix} \ddot{\mathbf{q}} \\ \lambda \end{bmatrix} = \begin{bmatrix} \mathbf{D} + \mathbf{Q} \\ -\dot{\mathbf{j}}\dot{\mathbf{q}} \end{bmatrix} + \begin{bmatrix} \mathbf{Eu} \\ \mathbf{0} \end{bmatrix} \tag{1}$$

where $\mathbf{q}$, λ, $\mathbf{Q}$, $\mathbf{J}$, and $\mathbf{u}$ are the vectors of generalized coordinates, Lagrangian multipliers, generalized inputs, constraint Jacobian matrix and control inputs respectively. The matrix $\mathbf{E}$ maps the control inputs on the generalized coordinates. The vector $\mathbf{D}$ contains the centrifugal, Coriolis and gravity forces and torques.

The conventional inverse dynamics only allows desired motions with the same degrees of freedom as that of the system. In order to overcome this, the definition of motion is extended to cover not only the acceleration of the generalized coordinates, but also the Lagrangian multipliers (or constraint forces) as follows:

$$\mathbf{B}\ddot{\mathbf{q}} + \mathbf{C}\lambda = \mathbf{y}(t) \tag{2}$$

where $\mathbf{B}$ is the matrix specifying the motion defining generalized coordinates, $\mathbf{C}$ is the matrix specifying force defining Lagrangian multipliers, and $\mathbf{y}$ are the desired acceleration/force functions describing the motion.

By substituting the rows corresponding to the control inputs in Eq. (1) with Eq. (2), the motion and the Lagrangian multipliers can be obtained in a similar manner as in

M. Mistry et al. (Eds.): TAROS 2014, LNAI 8717, pp. 271–272, 2014.

solving the forward dynamics. Then the control input can be calculated at each integration step using the rows of equations that were substituted in Eq. (1).

3   Experiments and Results

The system consists of two 2-link robot manipulators and a bar shaped load with a force sensor on a vertical plane, as shown in **Fig. 1**. Each robot has two degrees of freedom controlled by two joint motors. However, when carrying the load as shown in Fig. 1, the overall system has three degrees of freedom with four control inputs. The controller for each motor consists of three loops; position feedback, force feedback, and inverse dynamic feedforward input. For comparison, the experiment without the feedforward input is also carried out. A compression force demand of 0.5 N at the midpoint is set and two kinds of motions are executed. The first experiment includes a point to point motion with robot 1 holding position and robot 2 moving the load from $\theta=0$ to 0.5236 rad by a smooth acceleration profile. The motion lasts 0.4 seconds. A second experiment involves a sine wave on the y axis and a multi-frequency Schroeder Phased Harmonic Sequence (SPHS) signal on θ. The frequency and amplitude of the sine wave are set to 0.2 Hz and 0.06 m, respectively. The fundamental frequency of SPHS is 0.2 Hz with 10 harmonics. Results of two experiments are shown in **Table 1**. Both position and force tracking perform better with the new controller.

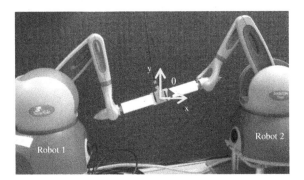

Fig. 1. System definitions

Table 1. RMS and the peak of the angular error (rad) and force error (N)

Motion	Controller	Angle RMSE	Angle peak	Force RMSE	Force peak
P2P	Pos + Force + FF	0.02359	0.0843	0.10917	0.1795
	Pos + Force	0.02890	0.0828	0.19408	0.3510
SPHS	Pos + Force + FF	0.03165	0.1636	0.08377	0.3566
	Pos + Force	0.03296	0.1849	0.11158	0.5557

4   Conclusion

Experimental results show improved performance with the proposed direct inverse dynamics based feedforward controller.

Learning Objects from RGB-D Sensors
for Cleaning Tasks Using a Team of Cooperative
Humanoid Robots

Marcelo Borghetti Soares, Pablo Barros, and Stefan Wermter

University of Hamburg, Department of Computer Science,
Vogt-Koelln-Strasse, 30, 22527, Hamburg, Germany
{borghetti,barros,wermter}@informatik.uni-hamburg.de
http://www.informatik.uni-hamburg.de/WTM

In this work, we address the problem of implementing cooperative search in humanoid robots (NAOs). The robots are taught to recognise a number of objects and then use their RGB-D sensors (attached to their heads) to search their environment for these objects. When an object is found they have to move to the target position and to perform a cleaning task in these objects (also, the location of recognized objects can help navigation). The challenge is threefold: 1) navigation/exploration, 2) real-time object recognition and 3) cooperation. This work will show preliminary results in object recognition and briefly discuss the approaches that will be employed for the entire system.

The scenario consists of a room in which some objects are spread (Figure 1 on the left). Initially, the robots (Figure 1 on the right) explore the environment until their RGB-D devices have detected objects. Then, the robots move to the objects, constantly trying to identify them. It is an assumption of our approach that these objects are located on tables or plane surfaces, so this strategy will help to detect potential objects before recognizing them. The robot has to extract the appropriate information from the point cloud, filter noise and correctly segment the objects. We are using RANSAC (RAndom SAmple Consensus) [1] to identify planes and VFH (View Point Feature Histogram) to collect a multidimensional descriptor (feature vector with 308 elements) that characterizes the object. The robots are, initially, completely unaware about the object's position. Thus, they cannot plan the best way to cooperatively distribute themselves in the environment, but they can model the influence of other robots as repulsive potential fields, in a purely reactive, though collaborative, way. Although a navigation approach based on RGB-D data is intended to be used, initially we will employ a ceiling camera (also used in [2]) that provides a global view of the environment.

Figure 2 shows some objects used in our experiments. We decided to create our own database to employ the same objects in a future stage of the project for online recognition. We have collected RGB-D data of 5 different types of objects (category) and 5 different objects of each type (instances) in 12 different views. The total number of images acquired was 300. We also applied rotation (roll, pitch and yaw) and added noise (in 10% of the points of each point cloud) to create a dataset of 4200 samples. From this total, 75% were used for training and 25% used for testing. A Multichannel Convolutional Neural Network from

M. Mistry et al. (Eds.): TAROS 2014, LNAI 8717, pp. 273–274, 2014.

Fig. 1. (Left) Final scenario to be used and (Right) NAO robot with RGB-D device

Fig. 2. Examples of RGB-D of 3 objects: box, book, cup

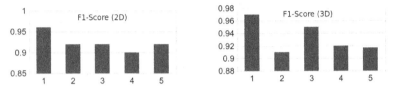

Fig. 3. *F1-score* results for Book(1), Box(2), Can(3), Cup(4) and Sponge(5)

our previous work [3], developed for gesture recognition, was used and trained with the *Backpropagation Algorithm*. The *F1-score* results for $2D$ images and $3D$ point clouds can be seen in Figure 3. These results show that our approach is promising, with *F1-scores* values greater than 0.9 for all objects. The next challenges that should be addressed are: 1) how to combine efficiently $2D$ and $3D$ information, 2) how to embed this recognition system in the robots to perform real time recognition and 3) treat noise and the presence of multiple objects. We expect that in the future the benefits of employing multiple robots for assistance living surpass the costs expended to acquire the technology.

References

1. Fischler, M.A., Bolles, R.C.: Random sample consensus: A paradigm for model fitting with applications to image analysis and automated cartography. Commun. ACM 24, 381–395 (1981)
2. Yan, W., Weber, C., Wermter, S.: A hybrid probabilistic neural model for person tracking based on a ceiling-mounted camera. JAISE 3, 237–252 (2011)
3. Barros, P., Magg, S., Weber, C., Wermter, S.: A Multichannel Convolutional Neural Network for Hand Posture Recognition. In: International Conference on Artificial Neural Networks. Springer, Heidelberg (September 2014)

Acoustic Based Search and Rescue on a UAV

Loai A. A. Sayed and Lyuba Alboul

SCentRo, Sheffield Hallam University, UK
loai.ahmed.s@gmail.com, L.Alboul@shu.ac.uk

The use of autonomous Unmanned Aerial Vehicles (UAVs) in search and rescue missions is increasing; they are used in surveying the environment, collecting evidence about missing or injured humans, or for damage assessment in disaster accidents. Sound source localisation is an important ability of the human auditory system and for a search and rescue team member it is reliable in the cases when they cannot see the target. Such ability, however, is degraded in robotic systems due to the noise generated by their actuators during their motion. Significant amount of previous research has been carried out in developing methods for sound source localisation and filtering noises generated by robots, but only a small portion of the research done has been focused on UAVs. Due to the huge noise generated by the rotors sound source localisation represents a real challenge.

1 Proposed System

We develop an auditory system for a quadcopter to localise missing or injured persons in a search and rescue mission through their screams [1], and an acoustic homing device that generates a powerful signal with frequency range 28KHz to 32KHz, as no environmental noises are expected in this frequency range.

Due to the noise generated by the rotors of the quadcopter the main research challenge was to filter the noises before localising the desired signal source. The whole system was designed and simulated in Matlab. All the possible noises were taken into account: the noises generated by the quadcopter rotors, the expected environmental noises in a search and rescue mission; the sound strength deterioration has been also considered. The method used is to a certain extent similar to the one proposed in [2]. Pre–recorded noise templates to remove the noise generated by the rotors have been used, but instead of applying a direct template subtraction, signals received by the four microphones mounted on the quadcopter are compared to a prerecorded noises library in order to determine the noise template with highest similarity. Then a frequency mask from the matching template is generated and all the frequencies covered by the mask from the received signals are removed. After filtering the only remaining signals are the screaming signal and the homing device signal. Then two parallel filtering processes are used, the first one removes all the frequencies in the signal except the screaming signal frequency band to localise its source and the second removes all the frequencies in the signal except the homing device frequency band.

For localising the signal source also two parallel processes are used, the purpose of the first process is to localise the screaming signal source and the second

M. Mistry et al. (Eds.): TAROS 2014, LNAI 8717, pp. 275–276, 2014.

process aims at the homing device signal. To estimate the time delay of arrival of the signals in both processes the signals from the four microphones are cross–correlated against each other in pairs [3], then assuming a constant sound velocity the sound source angle is calculated.

2   Results and Discussion

As the system is designed to localise both human screams and an acoustic homing device, the results consist of two main parts which are screaming sound source localisation and homing device localisation. For the human screams source localisation, the system was tested for multiple sound source angles with respect to the quadcopter's reference with a constant distance of one meter, between the sound source and the quadcopter's centre. The results showed maximum of 0.35 error percentage between the actual angle and the measured angle at the rotor's maximum speed. For the homing device localisation, the system was also tested for multiple locations for the sound source with variation of both the sound source angle and distance to the centre.The results showed maximum error percentage of 0.536 when the distance between the source and the quadcopter's centre was below 90m considering that the homing device was generating sinusoidal signals with amplitude gain equal to 100. When the distance exceeds 90m the system failed to provide the source angle with acceptable error percentage. The designed system has several limitations as the sound is considered to be propagating in the two-dimensional plane and the quadcopter is assumed to be stationary during the localisation process. The used filtering method limits ability of the system to detect and localise human screams when the source is more than 1.5 meters from the quadcopter centre and the screaming signal is of average strength.

3   Conclusion

The proposed system has been successful in removing the rotor's and the environmental noises, and in localising both the screaming and the homing device signal sources but with range limitations due to the aggressive filtering technique.

References

1. Begault, D.R.: Forensic analysis of the audibility of female screams. In: Audio Engineering Society Conference: 33rd International Conference: Audio Forensics-Theory and Practice, Audio Engineering Society (2008)
2. Ince, G., Nakadai, K., Rodemann, T., Hasegawa, Y., Tsujino, H., Imura, J.I.: A hybrid framework for ego noise cancellation of a robot. In: 2010 IEEE International Conference on Robotics and Automation (ICRA), pp. 3623–3628. IEEE (2010)
3. Murray, J.C., Erwin, H., Wermter, S.: Robotics sound-source localization and tracking using interaural time difference and cross-correlation. In: Proceedings of NeuroBotics Workshop, pp. 89–97 (2004)

Visual Commands for Tracking and Control

Loai A.A. Sayed and Lyuba Alboul

SCentRo, Sheffield Hallam University, UK
loai.ahmed.s@gmail.com, L.Alboul@shu.ac.uk

Tracking and following a mobile target (a robot or a person) is a very challenging and important task in autonomous mobile robotics applications, due to unstructured, dynamic and often unknown environment. We describe a method for following a master agent by a slave agent with aid of visual commands based on QR codes. Due to their data storage capacity QR codes are also used to control the slave robot speed and its position with respect to the master agent.

The QR code is mounted on the leader-agent, the slave agent is equipped with a web-camera capturing frames with $1280{\times}720$ resolution, which captures the image of QR code. The system uses the Zxing open-source library to read 2D barcodes developed in [3] and integrates it with MATLAB for further processing. After the QR code has been read an appropriate response to a command provided by the QR code is determined. The system is designed to respond to the set of commands: 'follow' to follow a master agent, 'move' to execute pre-set navigation commands and 'stop'. In the case of the 'follow' command the system executes tracking the QR code to determine the target coordinates and the distance to the target, which are used as an input for the controllers to calculate an appropriate navigation command for the slave robot to follow the master agent. Matlab's computer vision toolbox was used for feature detection and tracking [4]. The target reference features from a sample QR code with the built-in 'follow' command were extracted by means of Matlab's 'detectSURFFeatures' function. Then the same function was used for processing the captured frame and matching the obtained features with the reference features. The tracking process is performed only if there is a sufficient amount of matching features. It starts with estimating the target transformation matrix using the Matlab's function that employs the Random sample consensus(RANSAC) algorithm. After determining the transformation matrix of the target, the target size and centroid are used as feedback to the controllers. Two independent PID controllers were designed in Matlab. The first controller uses the target's size in the captured frames as feedback representing the distance to the target; its output signal is the robot's linear motion along its X-axis. The second Controller uses the target's coordinates as its feedback to align the robot with the target by setting the controller set-point to the frame's centre. Both PID controllers were tuned and their signal conditioned to accommodate the robot's rotors speed input.

The system was implemented on robots Pioneer-3AT that were operated on ROS. Communication between ROS and MATLAB was executed through the IPC-bridge [1] that allows Matlab acting as a ROS node with the ability to publish data to other ROS nodes. The IPC-Bridge supports the Geometry messages as standard ROS messages for navigation commands and is used to send the

M. Mistry et al. (Eds.): TAROS 2014, LNAI 8717, pp. 277–278, 2014.
© Springer International Publishing Switzerland 2014

control signal from Matlab to ROS to control the robot's motion. The Geometry message used is the Twist message that contains both the linear motion and the rotational motion control values.The IPC-Bridge ROS package were modified to publish messages directly to ROS ARIA that controlled all the robot's actuators.

Multiple experiments were carried out to test all the layers of the system. Firstly, the system was tested whether it was able to read the commands in the QR code at a maximum distance of 1.5 meters. Next the system was tested with two commands: 'move' and 'stop'. The robot responded to the first command by moving in the pre-set motion sequence, and the second command stops the robot and shuts down the system. As the system was designed for multi-agent operations, the 'follow' command was designed for a slave agent(robot) to follow a master agent. Initially, experiments were performed by moving the master agent along a straight line with multiple velocities; the slave agent was capable to match the master's velocity with an acceptable delay at all velocities bellow or equal to 0.25m/s. By increasing the master robot velocity to 0.3m/s the slave agent was able to follow the master agent but didn't always match its speed due to loss of target features in some of the processed images. With the velocity of the master robot increased further to 0.35m/s the slave robot failed to follow the master agent and stopped due to loss of features.

To test the slave agent ability to follow a master robot along a variable path, the master robot was remotely controlled, and its path data were plotted using ROS odometry data whilst the test area was mapped using ROS Gmapping package based on Open SLAM [2]; a laser scanner mounted on the master robot scanned the environment. The slave robot moved along the path of the master robot when it was moving along a polygonal path. However, as the slave robot doesn't follow the actual path of the master robot and the controllers are designed to maintain the distance and the target in the centre of the vision frame, when the master robot moved along curved paths there was a deviation of maximum 40 cm from the inside of the curved path. The system has some limitations: the slave robot may lose the features of the target or hits an obstacle, and stops, when the master robot turns in a close vicinity to an obstacle with a small turning radius or rotates around its Z-axes with rotation more than 80 degrees without linear motion. For path following, it is recommended to buffer the master robot positions and generate a path for the slave robot to navigate along, instead of direct vision output to the controllers.

References

1. Cohen, B.: Matlab and ROS (2011)
2. Grisetti, G., Stachniss, C., Burgard, W.: Improved techniques for grid mapping with rao-blackwellized particle filters. IEEE Transactions on Robotics 23(1), 34–46 (2007)
3. Mackintosh, A., Martin, A., Brown, B., Brunschen, C., Daniel, S.: Zxing, open source library to read 1d/2d barcodes (2009)
4. Tannenbaum, B.: Computer Vision with MATLAB, webinar demo files (January 2013)

A Novel Saliency Method Based on Restricted Boltzmann Machine (RBM) and Its Application to Planetary Exploration

Lilan Pan, Dave Barnes, and Chen Gui

Department of Computer Science,
Aberystwyth University, United Kingdom
{lip8,dpb,chg12}@aber.ac.uk

Perceiving surrounding interesting objects in an automatic way is a valuable capability for rover-based planetary exploration. Through analysing images captured by on-board cameras automatically, the regions of interest (ROI) can be obtained thereby implementing an autonomous capability to some degree. In order to detect the ROI from images, several bottom-up saliency-based methods have been applied (especially for Mars exploration) [1,2].

However, some traditional saliency methods cannot detect the ROI correctly. A possible explanation for this phenomenon may be that the traditional saliency methods have introduced some features independent or unrelated to the input image. Hence, we propose a saliency method which uses the features generated by the input image itself. This method begins by extracting random patches from a image. As we used single channel images, each patch had w-by-w pixels with only one channel. Here we set the w to 25. These patches were then represented as vectors with 625 elements. The vectors were regarded as the inputs of a network simplified from a deep belief network (DBN) [3]. The network was consisted of a 2-layer restricted Boltzmann machine (RBM). The first layer RBM converted the patch vectors to 100 features, and then these features were further abstracted to 10 features in the second layer. Like the DBN, the unsupervised training method was applied to train the weights and biases of each layer. Here we exhibit an example of first layer weights of a trained RBM in Fig. 1. The original image for extracting patches is shown in the top-left picture of Fig. 2.

It can be seen that the first layer RBM is like a series of convolution kernels. Some of the kernels are similar to Gabor filters to generate the features representing the structure of lamination. On the other hand, the second layer performs a further abstraction of the features from the first layer.

Ten feature maps were obtained by pooling all the patch vectors of the whole image through the trained DBN. Here, we named each feature map as F_i and then applied a max-min activating method to produce the final saliency map $SaliencyMap$:

$$SaliencyMap = \frac{\sum_{i=1}^{10}(1 - F_i) \times (\max(F_i) - \min(F_i))}{\sum_{i=1}^{10}(\max(F_i) - \min(F_i))}$$

An image which contains the content of layered sedimentary features was used to test the performance of our algorithm. The saliency results of the traditional

M. Mistry et al. (Eds.): TAROS 2014, LNAI 8717, pp. 279–281, 2014.

Fig. 1. The weights of the first layer RBM

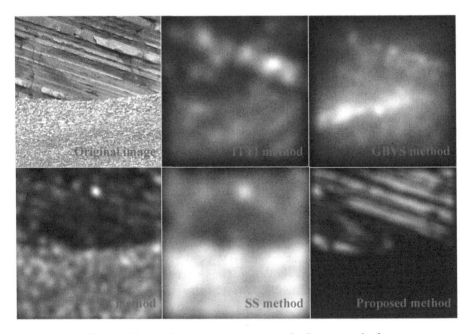

Fig. 2. The performance comparison of saliency methods

saliency methods including Itti's saliency (ITTI) [4], graph-based visual saliency (GBVS) [5], spectral residue (SR) [6] and image signature saliency (SS) [7] are presented as well. The performance comparison of all algorithms is illustrated in Fig. 2. By visual inspection, it can be seen that the saliency region of the proposed method is distinct and concentrates on the region containing layered sedimentary features which would be an interesting science target in actual Mars exploration.

References

1. Gao, Y., Spiteri, C., Pham, M., Al-Milli, S.: A survey on recent object detection techniques useful for monocular vision-based planetary terrain classification. J. Robotics and Autonomous Systems 62(2), 151–167 (2014)
2. Pan, L., Barnes, D.: An investigation into saliency-based Mars ROI detection. In: Pre. Symposium on Advanced Space Technologies in Robotics and Automation, Noordwijk, The Netherlands (2013)
3. Hinton, G., Osindero, S., Teh, Y.: A fast learning algorithm for deep belief nets. Neural Computation 18, 1527–1554 (2006)
4. Itti, L., Koch, C., Niebur, E.: A model of saliency-based visual attention for rapid scene analysis. IEEE Trans. Pattern Anal. Mach. Intell. (2011)
5. Harel, J., Koch, C., Perona, P.: Graph-based visual saliency. In: Neural Information Processing Systems (NIPS), pp. 545–552 (2006)
6. Hou, X., Zhang, L.: Saliency detection: A spectral residual approach. In: IEEE Conference on Computer Vision and Pattern Recognition (CVPR), pp. 1–8 (2007)
7. Hou, X., Harel, J., Koch, C.: Image signature: Highlighting sparse salient regions. IEEE Transactions on Pattern Analysis and Machine Intelligence (PAMI) 34(1), 194–201 (2012)

Author Index